Advances in Multiphase Flow and Heat Transfer

Vol. 4, 2012

Edited By

Lixin Cheng
School of Engineering
University of Portsmouth
United Kingdom

Dieter Mewes
Institute of Multiphase Process
Leibnia University of Hanover
Germany

CONTENTS

CHAPTERS

PREFACE

Multiphase flow and heat transfer have been found a wide range of applications in nearly all aspects of engineering and science fields such as mechanical engineering, chemical and petrochemical engineering, nuclear engineering, energy engineering, material engineering, ocean engineering, mineral engineering, electronics and micro-electronics engineering, information technology, space technology, micro- and nano-technologies, bio-medical and life science etc. With the rapid development of various relevant technologies, the research of multiphase flow and heat transfer is growing very fast nowadays than ever before. It is highly the time to provide a vehicle to present the state-of-the-art knowledge and research in this very active field.

To facilitate the exchange and dissemination of original research results and state-of-the-art reviews pertaining to multiphase flow and heat transfer efficiently, we have proposed the eBook series entitled ***Advances in Multiphase Flow and Heat Transfer*** to present state-of-the-art reviews/technical research work in all aspects of multiphase flow and heat transfer fields by inviting renowned scientists and researchers to contribute chapters in their respective research interests. The eBook series have now been launched and two volumes have been planned to be published per year since 2009. The e-books provide a forum specially for publishing these important topics and the relevant interdisciplinary research topics in fundamental and applied research of multiphase flow and heat transfer. The topics include multiphase transport phenomena including gas-liquid, liquid-solid, gas-solid and gas-liquid-solid flows, phase change processes such as flow boiling, pool boiling, and condensation etc, nuclear thermal hydraulics, fluidization, mass transfer, bubble and drop dynamics, particle flow interactions, cavitation phenomena, numerical methods, experimental techniques, multiphase flow equipment such as multiphase pumps, mixers and separators etc, combustion processes, environmental protection and pollution control, phase change materials and their applications, macro-scale and micro-scale transport phenomena, micro- and nano-fluidics, micro-gravity multiphase flow and heat transfer, energy engineering, renewable energy, electronic chips cooling, data-centre cooling, fuel cell, multiphase flow and heat transfer in biological and life engineering and science etc. The eBook series do not only present advances in conventional research topics but also in new and interdisciplinary research fields. Thus, frontiers of the interesting research topics in a wide range of engineering and science areas are timely presented to readers.

In volume 4, there are seven chapters on various topics relevant to multiphase flow and heat transfr. Chapter **1** deals with modeling of interfacial area transport in two-phase flows. The interfacial area concentration is an important parameter to characterize the interfacial transport of mass, momentum and energy. The dynamic modeling approach of interfacial area, namely, the interfacial area transport equation (IATE) is thus indispensable for an accurate prediction of two-phase flows using the two-fluid model. This chapter reviews the theoretical development of the IATE from two aspects: formulation of the transport equation and modeling of the closures. The first approach to arrive at the IATE is based on the statistical description of a large number of particles using the Boltzmann transport equation. This approach is straightforward to obtain the macroscopic equation of the interfacial area concentration. However, for flows with continuous interface such as annular flow, one has to resort to the second approach, the local instantaneous formulation to derive the macroscopic transport equation. The source and sink terms in the IATE are required to close the problem and they are divided into volume change term, phase change term and particle interaction term. Details on formulating IATE using both approaches and modeling of the closures are discussed.

Chapter **2** presents an experimental study on dynamic behavior of liquid droplet impacting on heated surfaces. The study of a single droplet impact and spread on a heated surface is motivated by its strong relevance to spray cooling technology. The generic view on the study of the spray cooling process exhibits the synthesis of fluid mechanics, heat transfer and surface thermodynamics. Due to the complex phenomena involved, no comprehensive theoretical models are available. The few works that appeared in the literature so far have been largely empirical; the applicability of several correlations proposed is limited. This eBook chapter reports an experimental study of fundamental aspects of the dynamic characteristics of a single droplet impacting on a heated surface. In this experiment, the entire dynamic process of a droplet from the moment of collision with the substrate surface including the rebound was visualized and analyzed using a high-speed CCD camera. The experimental study focused mainly on the spread of a liquid droplet under the influences of substrate temperature varying from 26°C to 240°C, the inclination angle of substrates at 0°, 30°, and 60°, the wettability of substrates with contact angles from 30° to 90°, the viscosity of liquids ranging from 0.00089 up to 0.9161 kg/m·s, and surfactants of different concentrations.

Chapter **3** deals with numerical simulation of multiphase flow and interface mass transfer on particle scale. More and more attentions have been paid to better understanding of the mechanisms of multiphase flow and interphase mass transfer on the mesoscale (particle scale), namely about solid particles, bubbles and drops. The behavior of particle swarms and effects of surfactant are subsequently studied to approximate real industrial processes. Besides, the sub-particle scale Marangoni effect, induced by interphase mass transfer, is also studied for seeking measures of mass transfer enhancement. The computational fluid dynamics and computational transport principles are being developed into reliable and efficient tools for improving the macroscopic performances of unit operations and process equipments.

Chapter **4** deals with the use of the ultrasonic technique for the study and online monitoring of multiphase Flows.Multiphase flows are very common in the petroleum, chemical, and nuclear industries, oftentimes involving harsh media, strict safety regulations, access difficulties, long distances, and aggressive surroundings; accordingly, there is a need to determine the dispersed phase holdup using noninvasive fast responding techniques. Moreover, knowledge of the flow structure is essential for the assessment of the transport processes involved. The ultrasonic technique fulfills these requirements and could have the capability to provide the information required. In this chapter, the current status of the ultrasonic technique in the context of multiphase flow metering (MFM) is thoroughly reviewed; the focus is on multiphase flows of interest in the oil industry. Specific aspects involving the application of the technique to gas-liquid and gas-liquid-solid flows as well as dispersions and slurries are addressed. It is expected that this chapter will provide the reader with a clear view of what is already known about the application of ultrasonics to multiphase flows, its potential, and the research needs so that the technique could be effectively brought to the field.

Chapter **5** presents a topic on direct contact condensation (DCC) of steam injected into water – new developments in condensation regime and steam plume length prediction. The condensation of steam injected into water is one of the least studied forms of condensation. Nevertheless, there is a range of devices which rely heavily on effective use of DCC, such as steam driven jet pumps, steam ejectors and safety valves in nuclear reactors. Therefore, correct prediction and modelling of DCC behaviour are crucial to obtain an optimised design of such devices. In this chapter the process of the DCC of steam injected into water is described in details, where different regions of the process will be presented. The process is compared with other modes of condensation and is described as one of the multi-phase flows. In addition, an analytical model for DCC is presented. Furthermore, the chapter focuses on the behaviour of injected steam, commonly termed as a regime. Different regimes of DCC are presented in details and parameters that determine the DCC regime are discussed. A three dimensional condensation regime diagram is presented, where the DCC regime is shown as a function of steam inflow rate, temperature of the water subcooling and steam injector diameter size. After that, the penetration distance of steam into water or a steam plume length is discussed and a two dimensional steam plume length diagram is presented. The diagram shows the length of the steam plume in relation to the steam Reynolds number and the condensation potential.

Chapter **6** presents a comprehensive review on drag reduction with surfactants and polymeric additives in multiphase flow. Addition of small amount of surfactants or polymeric additives may reduce the drag in multiphase flow systems. This chapter presents an overview on the available studies of drag reduction with surfactants and polymeric additives in multiphase flow including gas-liquid two-phase flow, liquid-liquid two-phase flow and gas-liquid-liquid three-phase flow. First, the research background of drag reduction technologies with additives is presented. Then, basic knowledge of drag reduction surfactants and polymeric additives is briefly described. Measured physical properties of surfactant and polymer solutions such as surface tension and viscosity are shown. Next, the drag reduction mechanisms of polymer and surfactant solutions are discussed. The, comprehensive review on the available studies on drag reduction with polymers and surfactants in multiphase flow is presented. The mechanisms are also reviewed. Based on the present review, future research needs are discussed and recommended.

Chapter **7** deals with an investigation on void fraction and flow patterns of two-phase flow in upward and downward vertical and horizontal pipes. A comparison of the performance of 54 void fraction correlations based on unbiased experimental data set of 3385 data points. A comprehensive literature search was undertaken for the available void fraction correlations and experimental void fraction data for upward and downward vertical and horizontal two-phase flows. The performance of the correlations in correctly predicting the diverse data set was evaluated. Comparisons between the correlations were made and appropriate recommendations were drawn. The analysis showed that most of the correlations developed are very restricted in terms of handling a wide variety of data sets. Based on this analysis void fraction correlations with the best predictive capability are highlighted.

As the founding editors of the eBook series, we are very happy to see that the e-books have been available to our readers since 2009 (two volumes per year). All the chapters have been reviewed by experts in the relevant areas. We are very much grateful to the authors who have contributed to the chapters and the reviewers who took their time and energy for their review. It is our great wishes if the eBook series are able to provide useful knowledge for our community and to facilitate the progress of the research in the field of multiphase flow and heat transfer.

We would like to express our gratitude to our families for their great support to our work on the eBook series.

Dr. Lixin Cheng
School of Engineering
University of Portsmouth
Anglesea Building, Anglesea Road
Portsmouth, PO1 3DJ
United Kingdom

Prof. Dieter Mewes
Institute of Multiphase Process
Leibniz University of Hanover
Callinstraße 36
D-30167 Hannover
Germany

List of Contributors

Antonio C. Bannwart

Universidade Estadual de Campinas (UNICAMP), Campinas, Brazil

Ricardo D. M. Carvalho

Universidade Federal de Itajubá (UNIFEI), Itajubá, Brazil

Jie Chen

Institute of Process Engineering, Chinese Academy of Sciences, China

Lixin Cheng

School of Engineering, University of Portsmouth, United Kingdom

Anka de With

Nuclear Research and Consultancy Group, Arnhem, The Netherlands

Afshin J. Ghajar

School of Mechanical and Aerospace Engineering, Oklahoma State University, USA

Takashi Hibiki

School of Nuclear Engineering, Purdue University, USA

Mamoru Ishii

School of Nuclear Engineering, Purdue University, USA

Liwen Jin

School of Mechanical and Aerospace Engineering, Nanyang Technological University, Singapore

Kai Choong Leon

School of Mechanical and Aerospace Engineering, Nanyang Technological University, Singapore

Lei Liu

State Key Laboratory of Multiphase Flow in Power Engineering, Xi'an Jiaotong University, China

Yang Liu

School of Nuclear Engineering, Purdue University, USA

Zai-Sha Mao

Institute of Process Engineering, Chinese Academy of Sciences, China

Dieter Mewes

Institute of Process Engineering, University of Hanover, Germany

Clement C. Tang

School of Mechanical and Aerospace Engineering, Oklahoma State University, USA

Chao Yang

Institute of Process Engineering, Chinese Academy of Sciences, China

Chun Yang

School of Mechanical and Aerospace Engineering, Nanyang Technological University, Singapore

Advances in Multiphase Flow and Heat Transfer

Vol. 4, 2012

CHAPTER 1

Modeling of Interfacial Area Transport in Two-Phase Flows

Yang Liu*, Takashi Hibiki and Mamoru Ishii

School of Nuclear Engineering, Purdue University, West Lafayette, Indiana, 47907, USA

Abstract: The interfacial area concentration is an important parameter to characterize the interfacial transport of mass, momentum and energy. The dynamic modeling approach of interfacial area, namely, the Interfacial Area Transport Equation (IATE) is thus indispensable for an accurate prediction of two-phase flows using the two-fluid model. This article reviews the theoretical development of the IATE from two aspects: formulation of the transport equation and modeling of the closures. The first approach to arrive at the IATE is based on the statistical description of a large number of particles using the Boltzmann transport equation. This approach is straightforward to obtain the macroscopic equation of the interfacial area concentration. However, for flows with continuous interface such as annular flow, one has to resort to the second approach, the local instantaneous formulation to derive the macroscopic transport equation. The source and sink terms in the IATE are required to close the problem and they are divided into volume change term, phase change term and particle interaction term. Details on formulating IATE using both approaches and modeling of the closures are discussed.

Keywords: Multiphase flow, two-fluid model, interfacial area concentration, interfacial area transport equation, bubble interaction mechanism.

1. INTRODUCTION

An accurate prediction of multiphase flow behaviors is very important for the design, operation and analysis of complicated industrial systems such as nuclear power plant, chemical process plant and various heat exchanger systems. Numerous models to describe multiphase flows have been developed and some are already widely used in the industry. Due to the complicated phenomena, multiphase flow models have to deal with a wide spectrum of scales ranging from the mesoscopic (<0.001m) behavior of small particles, to the overall performance of large industrial systems (~10m) [1]. It is generally believed that models which resolve finer scale, such as Direct Numerical Simulation (DNS) can yield better predictions since less empirical closures are involved. However, given the development of computational capability, it is not practical to perform DNS at the system level for complex multiphase flows in the foreseeable future [2]. For this reason, the two-fluid model which only concerns the averaged behavior of each phase is the most practical tool for the simulation of a large scale system.

As an Eulerian-Eulerian type model, the two-fluid model is obtained by averaging the single phase field equations of each phase to remove the high frequency spatial and temporal fluctuations. In this way, local instantaneous quantities of fluids, such as variations of pressure, velocity and temperature inside a bubble or a droplet, are not explicitly considered. The averaged quantities such as volume fraction of bubbles, averaged gas and liquid velocities, and averaged pressure appear as unknowns in the field equations in the two-fluid model. The distinct features of interfaces are smeared off during the averaging process. The interfacial transfer terms arise in the field equations to account for the interactions between two phases. Provided that the constitutive closures are accurate, the concept of the two-fluid model is consistent with the industrial concerns. Take nuclear power plant as an example, knowing the void fraction distribution in the reactor core, which is closely coupled with nuclear fission reaction is far more important than tracking the interface motion of individual bubbles.

The two-fluid model has been formulated based on various averaging techniques including time averaging, volume averaging, ensemble averaging, and combinations of these basic methods. For practical applications, the simplified three-dimensional two-fluid model can be written as [3].

***Address correspondence to Yang Liu:** School of Nuclear Engineering, Purdue University, West Lafayette, Indiana, 47907, USA; Tel: 765-496-3902; Fax: 765-494-9570; E-mail: liu130@purdue.edu

Lixin Cheng and Dieter Mewes (Eds)

Continuity equation:

$$\frac{\partial \alpha_k \rho_k}{\partial t} + \nabla \cdot (\alpha_k \rho_k \vec{v}_k) = \Gamma_k \ . \tag{1}$$

Momentum equation:

$$\begin{aligned} \frac{\partial \alpha_k \rho_k \vec{v}_k}{\partial t} + \nabla \cdot (\alpha_k \rho_k \vec{v}_k \vec{v}_k) &= -\alpha_k \nabla p_k + \nabla \cdot \alpha_k \left(\overline{\overline{\tau}}_k^{\mu} + \overline{\overline{\tau}}_k^{T} \right) \\ &\quad + \alpha_k \rho_k \vec{g} + \vec{v}_{ki} \Gamma_k + \vec{M}_{ik} - \nabla \alpha_k \cdot \overline{\overline{\tau}}_i \ . \end{aligned} \tag{2}$$

Enthalpy energy equation:

$$\begin{aligned} \frac{\partial \alpha_k \rho_k h_k}{\partial t} + \nabla \cdot (\alpha_k \rho_k h_k \vec{v}_k) &= -\nabla \cdot \alpha_k \left(\vec{q}_k^{\,k} + \vec{q}_k^{\,t} \right) + \alpha_k \frac{D_k}{Dt} p_k \\ &\quad + h_{ki} \Gamma_k + q''_{ki} / L_s + \Phi_k \ . \end{aligned} \tag{3}$$

where, Γ_k, $\vec{M}_{ik}$, $\overline{\overline{\tau}}_i$, q''_{ki} and Φ_k are the mass generation, generalized interfacial drag, interfacial shear stress, interfacial heat flux and dissipation, respectively. The term $1/L_s$ represents the surface area per unit volume. The subscript k denotes k-phase and i stands for the value at the interface. The interfacial jump conditions for mass, momentum and enthalpy energy are given as:

$$\begin{cases} \sum_k \Gamma_k = 0 \\ \sum_k \vec{M}_{ik} = 0 \\ \sum_k \left(\Gamma_k h_{ki} + q''_{ki} / L_s \right) = 0 \end{cases} \tag{4}$$

In the above field equations turbulent terms as well as interfacial transfer terms characterizing the mass, momentum and energy exchanges between two phases arise during the averaging process.

In two phase flows, the physical properties of two phases such as density, viscosity, thermal conductivity, *etc.*, are usually quite different. These differences in turn cause stronger variations of hydrodynamic and thermodynamic parameters between two phases, than those within each phase. For most two-phase flows under consideration, the interfacial terms are as important as (if not more than) the turbulent terms. For example, the turbulent shear stress of gas phase in a dispersed bubbly flow is almost negligible if compared to drag, lift, and other interfacial forces.

The interfacial transfer terms are strongly affected by the geometric characteristics of interfacial structures. To quantify this effect relevant to the interfacial transfer, a new parameter, the interfacial area concentration is defined as the surface area per unit volume [4]:

$$a_i \equiv \frac{A_i}{V} \ . \tag{5}$$

It can be shown that the interfacial mass and heat transfer rates are directly proportional to a_i. From the interfacial area concentration and void fraction, one can obtain the Sauter mean diameter $D_{Sm} \equiv 6\alpha/a_i$, which is equal to the particle diameter for spherical ones. As interfacial forces are closely related to bubble or droplet sizes, the interfacial momentum transfer terms can be also written as the product of a_i and certain driving potential [3]. Therefore, the interfacial area concentration is an important parameter in the two-fluid model since all the interfacial closures are directly related to it. Traditionally a_i was calculated based on flow regime dependent correlations. This approach has many shortcomings since flow regime is a qualitative description and it is often rather subjective. It cannot handle the dynamic nature of two-phase flow structures such as entrance effect, flow near the transition boundary, and flow structure with strong multi-dimensional distributions. When applied to system analysis code, the flow regime dependent correlations may induce numerical oscillations and instabilities due to the artificial bifurcation of these models.

In view of the above, the Interfacial Area Transport Equation (IATE) was developed by Kocamustafaogullari and Ishii [5] based on the concept of particle number density distribution. This approach starts from the Boltzmann transport equation which describes the statistical characteristics of a large number of dispersed particles. The transport equation of averaged interfacial area concentration can be obtained by first weighting the Boltzmann equation with the surface area of each particle, and then taking the integration over the entire particle volume range. The source and sink terms in the equation include two parts: volume change term and particle number change term. The latter one can be further divided into phase change term and particle interaction term due to particle coalescence and breakup. The IATE can model the evolution of interfacial structures across the flow regime transition boundaries through mechanistic modeling of various fluid particle interaction phenomena. Therefore, using IATE to predict the interfacial area concentration is not only more accurate and consistent with the two-fluid model, but also eliminates the artificial discontinuities stemming from the use of flow regime dependent correlations.

Another stream in deriving the interfacial area transport equation is based on the evolution of local instantaneous interfacial structures [6-9]. In this approach, the local instantaneous interfacial area concentration is expressed as a Dirac delta function. From the transport equation of phase indicator function, one can arrive at the macroscopic IATE after several steps of derivation. Comparing to the approach based on Boltzmann transport equation, this equation starts from the fundamental element of surface geometry. Therefore, it can be used to continuous interfaces existed in separated flows. A new interfacial area source term which cannot be handled by particle number density distribution also arises in the transport equation. On the other hand, as a macroscopic description of two phase flow, the particle number density distribution can yield better physical explanation of various terms in the transport equation. The development of the source and sink closures is also more straightforward to carry out.

The IATE based on either approach is purely a geometric equation of two-phase flow structures. The interfacial area development is usually strongly coupled with the hydrodynamic and thermodynamic behaviors of two-phase flows including phase change and bubble interactions. Therefore, the transport equation of interfacial area concentration itself cannot complete the model without proper consideration of these flow behaviors. Mechanistic models of source and sink terms related to particular two-phase phenomena must be developed to close IATE.

2. IATE BASED ON BOLTZMANN TRANSPORT EQUATION

The foundation of the IATE was initially established based on the Boltzmann transport equation of the particle number densities for dispersed two-phase systems [5]. Let $f(\vec{x},V,t)$ be the particle number density distribution function per unit mixture volume. This is assumed to be continuous and specifies the probable number density of fluid particles moving at a given time t, in a spatial range $\delta\vec{x}$ with its center-of-volume located at $\vec{x}$ with particle volumes between V and $V + \delta V$. The kinetic theory gives the following transport equation of particle number density distribution f:

$$\frac{\partial f}{\partial t}+\nabla\cdot\left(f\vec{v}\right)+\frac{\partial}{\partial V}\left(f\frac{dV}{dt}\right)=\sum_{j}S_{j}+S_{ph}\ . \tag{6}$$

In the right-hand side (RHS) of the equation, the S_j and S_{ph} are the particle source and sink rates per unit mixture volume due to j^{th} particle interactions (such as the disintegration or coalescence) and that due to phase change, respectively. By multiplying the above equation by the surface area of particles with volume

V, and then integrating over the entire particle volume range, one can obtain the macroscopic transport equation of a_i [5]:

$$\frac{\partial a_i}{\partial t} + \nabla \cdot (a_i \vec{v}_i) - \frac{2}{3}\left(\frac{a_i}{\alpha_g}\right)\left\{\frac{\partial \alpha_g}{\partial t} + \nabla \cdot (\alpha_g \vec{v}_g) - \eta_{ph}\right\} = \sum_j \phi_j + \phi_{ph} . \tag{7}$$

The definitions of interfacial area concentration and interfacial velocity in the above equation are given as:

$$a_i(\vec{x},t) \equiv \int_{V_{\min}}^{V_{\max}} f(\vec{x},V,t) A_i(V) dV , \tag{8}$$

and

$$\vec{v}_i(\vec{x},t) = \frac{\int_{V_{\min}}^{V_{\max}} f(\vec{x},V,t) A_i(V) \vec{v}_p(\vec{x},V,t) dV}{a_i(\vec{x},t)} . \tag{9}$$

The first two terms in the LHS of Eq. (7) represent the time rate of change of a_i and its convection. The other terms to balance the equation are essentially divided into two based on their mechanisms to alter interfacial area: volume change term as the third one in the LHS and number density change terms in the RHS of the equation.

In the volume change term, the volume source from nucleation or condensation η_{ph} is excluded from the total void fraction change. This is because η_{ph} comes with bubble number density change and it is considered in the term ϕ_{ph}. To simplify the volume change term, it has been assumed that the ratio of volume-equivalent diameter to surface-equivalent diameter is constant. This approximation is valid for spherical and cap bubbles, while for distorted or slug bubbles, additional shape factor should be considered. It is noted that the void fraction is propagated with the center-of-volume velocity $\vec{v}_g$. For dispersed bubbles, this velocity is equal to the center-of-area velocity $\vec{v}_i$ defined in Eq. (9). The ϕ_j's and ϕ_{ph} in the RHS of Eq. (7) represent the rate of change in the interfacial area concentration due to particle interactions and that due to phase change, respectively. They are defined by:

$$\phi_j \equiv \int_{V_{\min}}^{V_{\max}} S_j(\vec{x},V,t) A_i(V) dV, \tag{10}$$

and

$$\phi_{ph} \equiv \int_{V_{\min}}^{V_{\max}} S_{ph}(\vec{x},V,t) A_i(V) dV. \tag{11}$$

The macroscopic interfacial area concentration which takes all bubbles into account may not reflect the variation of the characteristics of individual bubbles in two-phase flows. These variations in shape and size cause substantial differences in the hydrodynamic behaviors of bubbles such as drag, lift and wall forces, as well as the bubble interaction mechanisms. For most two-phase flows, bubbles can be classified into five types: spherical, distorted, cap, slug and churn-turbulent bubbles. In view of their characteristics, they can be divided into two major groups, such that group 1 includes spherical and distorted bubbles, while group 2 includes cap, slug and churn-turbulent bubbles. For typical industrial systems such as nuclear power plant, the length scale of group 2 bubbles is usually comparable to (or even larger than) that of flow duct. Therefore geometric effect of the flow duct should be considered in developing the hydrodynamic and interfacial closures. This is not necessary for group 1 bubbles as the length scale is usually much smaller than that of flow duct. The boundary between the two groups is given by the maximum distorted bubble limit $D_{d,\max}$ [6]:

$$D_{d,\max} = 4\sqrt{\frac{\sigma}{g\Delta\rho}} \,, \tag{12}$$

where, σ, $\Delta\rho$ and g are surface tension, density difference and gravitational acceleration, respectively.

Given the above bubble classification and group boundary, the two group IATE can be obtained in the similar way as the one group IATE. The difference is that the integration should be carried out within the volume boundary of each group. In doing so, the volume change terms for two groups become:

$$\int_{V_{\min}}^{V_c} \left\{ A_i \frac{\partial}{\partial V}\left(f \frac{dV}{dt} \right) \right\} dV \simeq \left(\frac{\dot{V}}{V} \right)\left(-\frac{2}{3} a_{i1} + A_{ic} f_c V_c \right), \tag{13}$$

and

$$\int_{V_c}^{V_{\max}} \left\{ A_i \frac{\partial}{\partial V}\left(f \frac{dV}{dt} \right) \right\} dV \simeq \left(\frac{\dot{V}}{V} \right)\left(-\frac{2}{3} a_{i2} + A_{ic} f_c V_c \right). \tag{14}$$

In the above equations, the subscript "*c*" stands for the group boundary. Comparing to the one group IATE, an additional term, $\left(\dot{V}/V\right) A_{ic} f_c V_c$ which represents the inter-group transfer appears in the RHS of the above equations. The bubble number density distribution at the critical size must be known in order to determine this additional term. By assuming several representative distributions of group 1 and group 2 bubbles, Ishii and Kim [10] found that the inter-group transfer term could be approximated by:

$$\left(\frac{\dot{V}}{V} \right) A_{ic} f_c V_c \simeq \left(\frac{\dot{V}}{V} \right) a_{i1} C \left(\frac{D_{sc}}{D_{Sm1}} \right)^2 . \tag{15}$$

Here D_{sc} is the surface area equivalent diameter at the group boundary. The constant C is a coefficient related to the distribution of bubbles within each group and it should be bounded by:

$$0 \le C \le 2. \tag{16}$$

Taking Eq. (15) into consideration, the two group IATE can be written as follows [10]:

$$\frac{\partial a_{i1}}{\partial t} + \nabla \cdot \left(a_{i1} \vec{v}_{i1} \right) = \left(\frac{2}{3} - C\left(\frac{D_{sc}}{D_{Sm1}} \right)^2 \right) \frac{a_{i1}}{\alpha_1} \left[\frac{\partial \alpha_1}{\partial t} + \nabla \cdot \left(\alpha_1 \vec{v}_{g_1} \right) - \eta_{ph} \right] + \sum_j \phi_{j,1} + \phi_{ph} \,, \tag{17}$$

and

$$\frac{\partial a_{i2}}{\partial t} + \nabla \cdot \left(a_{i2} \vec{v}_{i2} \right) = \frac{2}{3} \frac{a_{i2}}{\alpha_2} \left[\frac{\partial \alpha_2}{\partial t} + \nabla \cdot \left(\alpha_2 \vec{v}_{g_2} \right) \right] + C\left(\frac{D_{sc}}{D_{Sm1}} \right)^2 \frac{a_{i1}}{\alpha_1} \left[\frac{\partial \alpha_1}{\partial t} + \nabla \cdot \left(\alpha_1 \vec{v}_{g_1} \right) - \eta_{ph} \right] + \sum_j \phi_{j,2} \,. \tag{18}$$

It can be seen that the same inter-group transfer term arises in both equations with opposite signs. This means that the inter-group transfer terms vanish if adding eqs. (17) and (18) together, which is consistent with one-group IATE. It is also noticed that the nucleation and condensation terms η_{ph} and ϕ_{ph} are excluded from the group 2 equation considering that small nucleation and condensation bubbles generally fall in group 1 category.

Eqs. (7), (17) and (18) establish the foundation of the one group and two group IATEs. The source and sink terms should be correctly modeled to account for various particle interaction and phase change mechanisms. This will be discussed later in this article.

The one group and two group IATEs developed from the Boltzmann transport equation can model the dynamic evolution of two-phase structures in dispersed flows such as bubbly, cap bubbly, slug, and churn-turbulent flows. However, they cannot be applied to separated flows such as annular flow. This is because there are continuous interfacial structures in separated flows. The evolution of these structures cannot be described by Boltzmann transport equation which relies on the statistical characteristic of large amount of particles. The interfacial area transport of separated flows must be formulated from a different basis than the particle number density distribution. The local instantaneous formulation of the evolution of interfacial structures can be employed for this purpose and it is discussed in the next section.

3. IATE FROM LOCAL INSTANTANEOUS FORMULATION

The interfacial area concentration a_i is defined as the surface area per unit volume [4] as shown in Eq. (5). This definition is intuitive from a macroscopic point of view. However, difficulties are encountered with the ordinary function representation if the local instantaneous value is of interest. This is because the volume measure of an interface is zero such that interfacial area given by Eq. (5) approaches infinity if the volume which contains finite surface approaches zero. This mathematical challenge can be overcome by the theory of distributions (or generalized functions) as demonstrated by various researchers [6, 11, 12]. The evolution of the local instantaneous interfacial area is independent of its carrier. In other words, it can be applied to gas dispersed flows, liquid dispersed flows, and flows with both phases continuous such as annular flow. Due to this advantage, the local instantaneous formulation has been proposed for the development of the interfacial area transport equation when the basic assumption to apply Boltzmann equation, *i.e.*, existence of large amount of bubbles or droplets does not hold for certain flows [9].

A smooth and sharp interface should be assumed in deriving IATE from the local instantaneous formulation. A smooth interface in three dimensional space can be represented by the level set of a differentiable function f [4, 13]:

$$f(\vec{x},t) = 0, \tag{19}$$

Function f is zero if evaluated at any point on the interface. The different phases in the flow field can be defined as follows:

$$f(\vec{x},t) \begin{cases} > 0, & \text{phase 1} \\ = 0, & \text{interface} \\ < 0, & \text{phase 2} \end{cases} \tag{20}$$

For multiple-particle systems, one may define f_j for the surface of each particle. A new function obtained from the product of all those f_j's has similar properties as described in the above equation. Therefore, the function f may represent both single and multiple interfaces.

Some interfacial parameters can be obtained from the level set representation [14]:

$$\vec{n} = -\frac{\nabla f}{|\nabla f|}, \tag{21}$$

and

$$|\vec{v}_i \bullet \vec{n}| = v_{ni} = \frac{|\partial f/\partial t|}{|\nabla f|}. \tag{22}$$

Here, $\vec{n}$ is the normalized gradient of *f*, and if $\vec{n}$ is evaluated at the interface, it represents the unit normal vector pointing from phase 1 to phase 2. v_{ni} is the magnitude of normal interfacial velocity. Given the definition of Eq. (21), its vectorial form becomes:

$$\vec{v}_i = -\frac{\partial f/\partial t}{|\nabla f|^2}\nabla f. \tag{23}$$

The phase indicator function χ, which is unity inside each phase while vanishes outside can be defined as:

$$\begin{cases} \chi_1(\vec{x},t) = H_f\left(f(\vec{x},t)\right) \\ \chi_2(\vec{x},t) = 1-\chi_1(\vec{x},t) = 1-H_f\left(f(\vec{x},t)\right) \end{cases} \tag{24}$$

where H_f is the Heaviside function. The local instantaneous interfacial area concentration can be written as the gradient of the phase indicator function:

$$\vec{a}_{ik}(\vec{x},t) = -\nabla\chi_k = |\nabla f|\delta\left(f(\vec{x},t)\right)\vec{n}_k, \tag{25}$$

where $\vec{n}_k$ is the unit normal vector pointing from *k*-phase towards the other phase. In this definition, an ordinary function cannot represent the derivative of a discontinuous function. Thus the Dirac delta function from distribution theory [15] is used instead. Kataoka *et al.* [12] originally defined the scalar value of the local instantaneous interfacial area concentration as: $a_i(\vec{x},t) = |\nabla\chi| = |\nabla f|\delta\left(f(\vec{x},t)\right)$. Although the practical interest is the averaged scalar value of interfacial area concentration, the vectorial form provides the direction of the interface. Another convenience coming from the definition in Eq. (25) is that the local interfacial area is defined for fluid fields on both sides of an interface. The magnitudes are the same while the directions are opposite at any point. The number of the interfacial area concentration fields becomes the same as the number of the fluid fields. For the two group classification of bubbles, the interfacial area of liquid phase a_{if} can be defined and the following equality should hold for a given volume:

$$a_{if} = a_{i1} + a_{i2}. \tag{26}$$

From the above definition, the local instantaneous interfacial area concentration is either zero or infinite at a local point. Its physical explanation can be made by taking the integral over a small volume in space. $\iiint_{\Delta x\Delta y\Delta z} \vec{a}_{ik}(\vec{x},t)\,dxdydz$ is a vector pointing in the direction of $\vec{n}_k$, with a magnitude equal to the surface area in the volume $\Delta x\Delta y\Delta z$. Therefore, the defined local instantaneous quantity is consistent with the macroscopic definition given in Eq. (5). With the local instantaneous definition, various averaged quantities can be obtained and two examples are shown in the following [12]:

Line average along z axis:

$$\tilde{a}_i^Z = \frac{1}{\Delta L}\sum_j \frac{1}{\cos\theta_z}. \tag{27}$$

Time average:

$$\tilde{a}_i^T = \frac{1}{\Delta T}\sum_j \frac{1}{|v_j|\cos\theta_j}. \tag{28}$$

Here, the '~' represents an averaging operation and details will be given in the following discussion. θ_z stands for the angle between the surface normal and the z direction. θ_j is the angle between the surface normal and the interfacial velocity direction. It can be seen that both of the above averaged quantities may have numerical difficulties when the angles become 90° such that zeros appear in the denominator. This is because the volume measure of any surface is zero as mentioned previously. In regard to this, the average must be taken over a finite volume to avoid an infinite interfacial area:

$$\iiint_{\Delta V} \left|\vec{a}_i(\vec{x},t)\right| dxdydz < \infty, \quad \Delta V \neq 0 . \tag{29}$$

The phase topological equation for any fluid field is given by [6, 8]:

$$\frac{\partial \chi_k}{\partial t} + \vec{v}_i \bullet \nabla \chi_k = 0 . \tag{30}$$

Taking the gradient of the above equation one obtains,

$$\frac{\partial \vec{a}_{ik}}{\partial t} + \nabla\left(\vec{v}_i \bullet \vec{a}_{ik}\right) = 0. \tag{31}$$

Taking the inner product of Eq. (31) with $\vec{n}_k$, the local instantaneous interfacial area transport equation can be written as [6, 8]:

$$\frac{\partial a_{ik}}{\partial t} + \nabla\bullet\left[\vec{a}_{ik} \bullet \vec{v}_i \vec{n}_k\right] + 2H_k \vec{a}_{ik} \bullet \vec{v}_i = 0 . \tag{32}$$

Here $2H_k$ is the total curvature of the surface: $\nabla\bullet\vec{n}_k = -2H_k$. Since the equation is given based on generalized functions, it cannot be directly applied in two-phase flow modeling due to the numerical difficulties. The interfacial area concentration at a local point is not the concern for most industrial applications. Instead, it is the averaged interfacial area concentration that is of practical interest for the application of the two-fluid model.

The averaging procedure is a low pass filtering process which eliminates unwanted high frequency signals from local instantaneous fluctuations of variables. Following Marle [11] and Labourasse *et al.* [16], a filter function $G(\vec{x},t)$ which is defined on domain $T\times\Omega:\left\{-\infty < t < \infty,\ \vec{x}\in R^3\right\}$ with compact support, is introduced for this purpose in multiphase flows. Various averaging can be realized by taking the convolution of the original parameter and filter functions, which can be given by the following general expression:

$$\begin{aligned} \tilde{\varphi}^A(\vec{x},t) &= \varphi(\vec{x},t) * G^A(\vec{x},t) \\ &= \int_{T\times\Omega} \varphi(x',y',z',t')\, G^A(x-x',y-y',z-z',t-t')\, dx'dy'dz'dt'. \end{aligned} \tag{33}$$

Again, '~' indicates an averaged value. Superscript 'A' denotes the averaging method with 'T', 'V' and 'S', representing time, volume and area averaging, respectively. An example of the filter function is given for time averaging as:

$$G^T(\vec{x},t) = \begin{cases} \delta(x)\delta(y)\delta(z)\dfrac{1}{\Delta T}, & |t| \le \dfrac{\Delta T}{2} \\ 0, & |t| > \dfrac{\Delta T}{2} \end{cases} \tag{34}$$

It can be shown that the convolution of any two-phase quantity $\varphi(\vec{x},t)$ and $G^T(x,y,z,t)$ is consistent with the original definition of time averaging given by Ishii [4]. In two-phase flow analysis, phase-weighted average is very important for certain parameters and it is defined as:

$$\tilde{\tilde{\varphi}}_k^A(x,y,z,t)=\frac{\widetilde{\varphi\chi_k}^A(x,y,z,t)}{\tilde{\chi}_k^A(x,y,z,t)}. \tag{35}$$

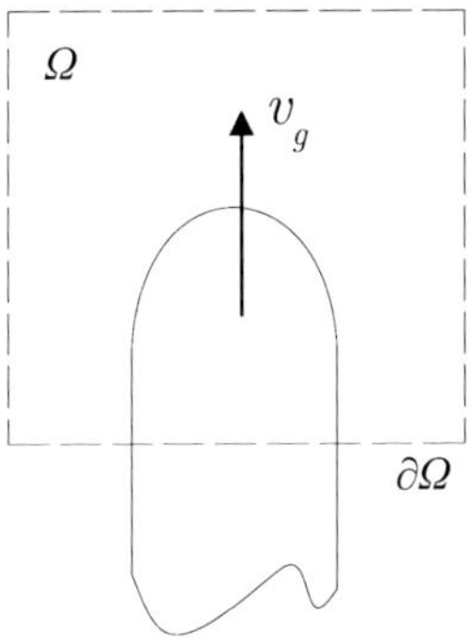

Figure 1: Schematic of a slug bubble entering the control volume.

The averaged interfacial area transport equation can be obtained by performing the convolution of the filter function with the local instantaneous form given in Eq. (32):

$$\frac{\partial\left(G^A*a_{ik}\right)}{\partial t}+\nabla\cdot\left[G^A*\left(\vec{a}_{ik}\cdot\vec{v}_i\vec{n}_k\right)\right]+2G^A*\left(H_k\vec{a}_{ik}\cdot\vec{v}_i\right)=0. \tag{36}$$

In the above derivation, nothing about the identity of the fluid has been assumed. This means Eq. (36) can be applied to any interfaces in addition to those carried by bubbles and droplets. The first term in the above equation is the time rate of change of the interfacial area concentration. The second term partially represents the convective term in the conventional sense. The third term may be regarded as an interfacial area source term, but this view is not completely correct as will be demonstrated. In Fig. **1**, a slug bubble is moving upward with its interface perpendicular to the control volume boundary $\partial\Omega$. The volume averaged interfacial area is considered to avoid any infinite interfacial area in domain Ω. With the volume averaging filter, Eq. (36) becomes:

$$\frac{\partial}{\partial t}\iiint_{\Omega}a_{ik}dxdydz+\oint\!\!\oint_{\partial\Omega}\vec{a}_{ik}\cdot\vec{v}_i\vec{n}_k\cdot d\vec{s}+2\iiint_{\Omega}H_k\vec{a}_{ik}\cdot\vec{v}dxdydz=0. \tag{37}$$

Here, Gauss's theorem is used to convert the volume integral to a surface integral. Similar volume averaged equation can be also derived from the surface transport theorem in a slightly different form [8]. The second term in Eq. (37) vanishes for the example shown here since the normal interfacial velocity at domain boundary is zero. By observation, the interfacial area inside the control volume Ω is increasing. This has to be taken into account by the third term of the equation which is due to the surface motion with its curvature.

It is noted that in the IATE derived from the Boltzmann transport equation [10], the increase of interfacial area in Ω appears in the convective term because the slug bubble is entering the control volume. Meanwhile, the source term vanishes considering that there is no bubble interaction, expansion or phase change. This means the second and the third terms in Eq. (37) cannot be regarded as the convective and source term in the conventional sense. To explain the difference, an example of a bubble entering the control volume can be employed. Morel [8] used this example to demonstrate the validity of Eq. (37). Here it is used to show the relation between the convective and source terms in IATE based on the Boltzmann equation and those based on the local instantaneous formulation. Later, the observation will be extended to any control volume using the surface divergence theorem.

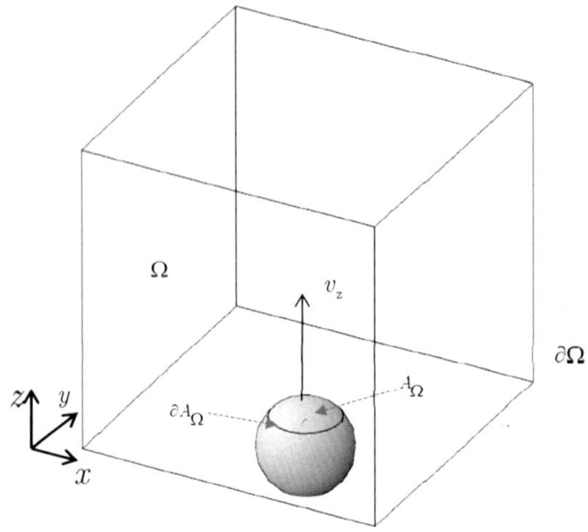

Figure 2: Schematic of a particle crossing the control volume boundary.

In Fig. **2**, a particle is shown moving across the control volume boundary $\partial\Omega$. A_Ω is the part of particle surface that falls in the control volume Ω. The boundary surface which meets the particle is perpendicular to the z axis. The particle is assumed to be solid and only moving in the z direction (motion in the other directions does not contribute to the interfacial area concentration change in Ω). From these conditions, one can obtain:

$$-2\iiint_{\Omega} \vec{a}_{ik} \bullet \vec{v}_z H_k dxdydz = \oint_{\partial A_\Omega} (\vec{v}_z \bullet \vec{n}_k) \frac{\sin\theta_z}{\cos\theta_z} dl \, , \tag{38}$$

where ∂A_Ω is the intersection curve between the particle surface and the domain boundary. θ_z stands for the angle between the surface normal and the z direction. For a particle that is moving away from the control volume, this equality holds, but a negative sign should be added in front of the line integral to represent the interfacial area decrease in the control volume. The second term in Eq. (37) can be written as:

$$\oiint_{\partial\Omega} \vec{a}_{ik} \bullet \vec{v}_i \vec{n}_k \bullet d\vec{s} = -\oint_{\partial A_\Omega} v_z \frac{\cos^2\theta_z}{\sin\theta_z} dl \, . \tag{39}$$

Combining equations (38) and (39) one gets:

$$\oiint_{\partial\Omega} \vec{a}_{ik} \bullet \vec{v}_i \vec{n}_k \bullet d\vec{s} + 2\iiint_{\Omega} \vec{a}_{ik} \bullet \vec{v}_i H_k dxdydz = -\oiint_{\partial\Omega} a_{ik} v_z ds \, . \tag{40}$$

The RHS term represents the interfacial area flux from the control volume boundary which is a function of the particle velocity. This is consistent with the convective term in the IATE based on Boltzmann equation. Therefore, for fluid particles that intercept the control volume boundary, the convective term in the conventional sense is taken into account partially by the second and partially by the third term in Eq. (37).

Using the surface divergence theorem [17], Eq. (37) can be extended even further. Assume the control volume Ω with any shape (not necessary a cube) intercepts a solid particle which has linear velocity $\vec{v}_0$ and angular velocity $\vec{\omega}$, one has:

$$\iiint_{\Omega} \nabla_S \left(\vec{v}_0 + \vec{r} \times \vec{\omega}\right) a_{ik} dxdydz = \oint_{\partial A_\Omega} \left(\vec{v}_0 + \vec{r} \times \vec{\omega}\right) \bullet \vec{\mu} dl \\ -2\iiint_{\Omega} H_k \left(\vec{v}_0 + \vec{r} \times \vec{\omega}\right) \bullet \vec{a}_{ik} dxdydz . \quad (41)$$

Here ∇_S is the surface divergence operator. $\vec{\mu}$ is the unit vector normal to the interception curve ∂A_Ω existed in the tangent plane of the particle surface. It points away from A_Ω. For solid particle, $\vec{v}_0$ and $\vec{\omega}$ are constant over the surface. It can be shown that the LHS in Eq. (41) vanishes and the equation can be rearranged as:

$$2\iiint_{\Omega} H_k \left(\vec{v}_0 + \vec{r} \times \vec{\omega}\right) \bullet \vec{a}_{ik} dxdydz = \oiint_{\partial\Omega} \left(\vec{v}_0 + \vec{r} \times \vec{\omega}\right) \bullet \left(\vec{N} - \vec{n}\left(\vec{n} \bullet \vec{N}\right)\right) a_i ds \\ = \oint_{\partial A_\Omega} \left(\vec{v}_0 + \vec{r} \times \vec{\omega}\right) \bullet \left(\vec{N} - \vec{n}\left(\vec{n} \bullet \vec{N}\right)\right) \frac{dl}{\sqrt{1 - \left(\vec{n} \bullet \vec{N}\right)^2}} . \quad (42)$$

Here $\vec{N}$ is the unit normal vector of the control volume surface. This equation converts the volume integral into a line integral which is a very useful property. For a particle which is completely enclosed in the control volume, it can be shown that the above integral is zero if every part of the interface is moving at the same linear and rotational speed. This is consistent with the macroscopic observation as the surface area remains constant for solid particles. However, most fluid particles or interfaces do not behave the same way as solid particles. Each part of the surface has its own fluctuating motion in relation to the average speed. Since the average speed does not increase the overall surface area, the interfacial area source must come from the additional motion of the local surface.

In light of the above consideration, define an averaged velocity $\vec{v}_{i,avg}$ satisfying the following equation:

$$\vec{v}_{i,avg} \bullet \oiint_{\partial\Omega} \vec{n}\left(\vec{n}_k \bullet \vec{N}\right) a_{ik} ds = \oiint_{\partial\Omega} \vec{v}_i \bullet \vec{n}\left(\vec{n}_k \bullet \vec{N}\right) a_{ik} ds . \quad (43)$$

Then Eq. (37) can be rearranged as:

$$\frac{\partial}{\partial t} \iiint_{\Omega} a_{ik} dxdydz + \oiint_{\partial\Omega} a_{ik} \vec{v}_{i,avg} \bullet \vec{N} ds + 2\iiint_{\Omega} H_k \vec{a}_{ik} \bullet \left(\vec{v} - \vec{v}_{i,avg}\right) dxdydz = 0 . \quad (44)$$

From Eq. (43), the interfacial transport velocity $\vec{v}_{i,avg}$ is not unique for any control volume at a given instant. It has multiple choices as long as the dot product in the LHS of Eq. (43) is equal to the integral in the RHS. Physically, the convective term vanishes for any surface perpendicular to the control volume boundary since the boundary cannot see the interfacial area flux no matter how fast the interface is moving. For a dispersed particle crossing the boundary, one can use the center of area velocity $\vec{v}_{ca}$ to approximate $\vec{v}_{i,avg}$ as $\vec{v}_{ca} \bullet \vec{n} \simeq \vec{v}_i \bullet \vec{n}$ at anywhere on the surface. Eq. (44) can be written as:

$$\frac{\partial}{\partial t} \iiint_{\Omega} a_{ik} dxdydz + \nabla \cdot \iiint_{\Omega} a_{ik} \vec{v}_{ca} dxdydz + 2\iiint_{\Omega} H_k \vec{a}_{ik} \bullet \vec{v}_{rc} dxdydz = 0 , \quad (45)$$

where the relative velocity is defined as: $\vec{v}_{rc} \equiv \vec{v} - \vec{v}_{ca}$. By definition of the following volume averaged quantities:

$$\begin{cases} \tilde{a}_i^V = G^V * a_{ik} \\ \tilde{\tilde{\vec{v}}}_i^V = \dfrac{G^V * \left(a_{ik} \vec{v}_{ca}\right)}{G^V * a_{ik}} \\ \tilde{\phi}_i^V = -2G^V * \left(\vec{a}_{ik} \bullet \vec{v}_{rc} H_k\right) \end{cases} \quad (46)$$

Eq. (45) can be put in a simpler form:

$$\frac{\partial \tilde{a}_i^V}{\partial t} + \nabla \bullet \left[\tilde{a}_i^V \tilde{\tilde{v}}_i^V \right] = \tilde{\phi}_i^V . \tag{47}$$

The above equation is the volume averaged IATE derived from the local instantaneous formulation of interface motion. The terms from left to right are the time rate of change term, convective term and source/sink term. It has a similar form as the IATE developed from the Boltzmann equation shown in Eq. (7). The only difference is that the source/sink term in Eq. (47) is expressed as the integral of microscopic surface motion with its curvature, while they are explicitly written in terms of expansion, particle interaction and phase change in Eq. (7).

The particle number density transport approach cannot model a continuous interface as it assumes the interface is carried by dispersed fluid particles. This is not the case for Eq. (47). Each term in the equation can be written as the volume integral of certain local interfacial parameters regardless of its category (dispersed or continuous). Thus it can be applied to both dispersed and continuous interfaces such as in annular flows. The convective velocity of the annular film can be defined in the same way as that in Eq. (43). The source/sink terms are simply the overall effect of the local surface motion with respect to the mean curvature. It is not only accurate from the microscopic point of view but can be also modeled by macroscopic parameters related to entrainment, deposition and wave characteristics.

The derivation of Eq. (47) only relies on the phase topological equation and a few mathematical manipulations. The macroscopic phenomena related to interfacial area development such as particle volume change, coalescence and breakup, phase change, *etc.*, should be predicted together with continuity, momentum and energy equations. However, given the mathematical form of $\tilde{\phi}_i^V$, the possible source/sink terms can be postulated from a geometrical point of view. This is summarized as follows:

$$\tilde{\phi}_i^V : \begin{cases} \text{with Particle Number Change} \begin{cases} \text{Bubble Interaction} \begin{cases} \text{Breakup} \\ \text{Coalescence} \end{cases} \\ \text{Phase Change} \begin{cases} \text{Nucleation} \\ \text{Condensation} \end{cases} \end{cases} \\ \text{without Particle Number Change} \begin{cases} \text{with Volume Change} \begin{cases} \text{Expansion} \\ \text{Contraction} \end{cases} \\ \text{without Volume Change} \{ \text{ Interface Distortion} \end{cases} \end{cases}$$

From the above classification, it can be seen that the term due to interface distortion is missing in Eq. (7). This term should exist since it is obvious that the surface area of a sphere is less than that of a cube with the same volume. Given that the particle number density function contains no information of particle shape, the IATE based on this method cannot model this particular term. For separated flows such as annular flow and horizontal wavy flow, this term may be very important since the interface distortion is the only source/sink for the continuous interface.

The local instantaneous interfacial area concentration defined previously may have difficulties under certain circumstances such as interface breakup, coalescence, entrainment and deposition. This is because singularities exist during these processes. However, the residence time of singularities is usually small such that it can be neglected while tracking the change of surface area. More importantly, the macroscopic change in surface area during particle interactions, rather than the motion of the interface, is the target for the interfacial area transport equations. The source/sink terms should be modeled by considering individual mechanisms in the context of statistical models of the ensemble.

4. MODELING OF INTERFACIAL SOURCE TERMS

In the last section, it is demonstrated that the interfacial area source comes from the local surface motion with respect to the mean velocity from a microscopic point of view. In order to obtain the macroscopic interfacial area source, one has to integrate the product of curvature and relative velocity over all the surfaces in the control volume. Provided that detailed three dimensional experimental data or DNS results are available, this formulation can be useful to extract the interfacial source terms from the local phase and velocity information. However, correlations depending on the macroscopic two-phase parameters must be given in order to close the interfacial area transport equation.

The simplest case to use Eq. (46) to develop the interfacial source and sink models may be the annular flow in a round tube. Assuming flow is axis-symmetric then liquid film interface depends solely on the film thickness. The interfacial area source and sink terms can be modeled by considering the characteristics of interfacial waves under various hydrodynamic and thermodynamic conditions. This one-dimensional problem may not be difficult to solve, however, it is not practical to carry out the integral over complicated surface structures for most other two-phase flows. Therefore, modeling of source terms should largely rely on the Boltzmann equation for dispersed flows. The source terms can be divided into volume change term, phase change term and particle interaction (coalescence or breakup) term as discussed previously. Comparing to the local instantaneous approach where all source terms are lumped into one, this classification facilitates the implementation of particular hydrodynamic and thermodynamic mechanisms. These mechanisms must be considered since interfacial area development is coupled with momentum and energy transport in two-phase flows.

The interfacial source due to volume change is expressed as a function of the void fraction and interfacial area concentration as shown in Eq. (7). These are unknowns in the field equations which do not require any closures. However, the void fraction change due to wall nucleation or condensation which involves bubble number change is considered as a separate source term thus being excluded from the volume change term. Proper wall nucleation model and condensation model should be given in order to close this term and the phase change terms as well.

4.1. Phase Change Term

The phase change effect is considered in two terms in the IATE. One is the volume change of existing bubbles due to either evaporation or condensation. In terms of interfacial structure, this process is no different from the bubble size change due to pressure and density changes. All these effects are also taken into account in the continuity equation. Therefore, this part of phase change is embedded in the volume change term. The second phase change effect is the surface area change due to generation of new bubbles or disappearance of the existing bubbles. The phenomenon either occurs at particular locations, such as wall nucleation, or it involves rapid change of interfacial structure, such as condensation induced bubble collapse. The bubbles involved in these processes are usually much smaller than average bubbles. Therefore, special treatment should be given to these processes considering that the scale and intensity of interfacial area changes are quite different from normal volume change effect.

The source term due to wall nucleation can be expressed as:

$$\phi_{nuc} = \frac{N_{nuc} f_d \xi}{A} \pi D_d^2 , \tag{48}$$

where N_{nuc}, f_d, ξ, A, and D_d are the active nucleation site density, bubble departure frequency, heated perimeter, cross sectional area, and bubble departure diameter, respectively. Additional models for N_{nuc}, f_d, and D_d should be developed in order to close the problem. The active nucleation site density N_{nuc} has been studied by many researchers. Hibiki and Ishii [18] developed a model as a function of the critical cavity size and the contact angle:

$$N_{nuc} = \bar{N}_{nuc}\left\{1-\exp\left(-\frac{\theta^2}{8\mu^2}\right)\right\}\left[\exp\left(f\left(\rho^+\right)\frac{\lambda}{R_c}\right)-1\right] . \qquad \textbf{(49)}$$

Here $\bar{N}_{nuc} = 4.72\times10^5$ sites/m^2, $\mu = 0.722$ radian, $\lambda = 2.50\text{x}10^{-6}$m, θ is the contact angle. The critical cavity radius R_c, is given by:

$$R_c = \frac{2\sigma\left\{1+\left(\rho_g/\rho_f\right)\right\}/\rho_f}{\exp\left\{i_{fg}\left(T_g-T_{sat}\right)/\left(\mathcal{R}T_gT_{sat}\right)\right\}-1} , \qquad \textbf{(50)}$$

where σ, T_g, T_{sat}, i_{fg}, and R are the surface tension, gas temperature, saturation temperature, latent heat and gas constant based on molecular weight, respectively. This model has been validated by a large number of active nucleation site density data taken in pool boiling and convective flow boiling systems. It shows that the active nucleation site density is a function of the critical cavity size and the contact angle, and the model can explain the dependence of the active nucleation site density on the wall superheat reported by various investigators.

The bubble departure frequency is usually modeled by considering the bubble waiting time and the bubble growth time. The bubble waiting time is the period when the liquid is heated until a new bubble appears on the active nucleation site. The bubble growth time accounts for the period from the inception of nucleation until bubble departs from the nucleation site. Recently, Situ *et al.* [19] proposed a correlation for the non-dimensional bubble departure frequency $N_{fd} \equiv f_d D_d^2/\kappa_f$ based on their own subcooled boiling data and other data available in the literature:

$$N_{fd} = 10.7 N_{qNB}^{0.634} . \qquad \textbf{(51)}$$

The non-dimensional heat flux N_{qNB} representing nucleate boiling heat transfer is defined as:

$$N_{qNB} = \frac{q''_{NB} D_d}{\kappa_f \rho_g i_{fg}} , \qquad \textbf{(52)}$$

where q''_{NB} is the heat flux and κ_f is the thermal diffusivity of liquid.

The bubble departure size can be estimated by considering the force balance on individual bubbles. For vertical flows, Situ *et al.* [20] assumed that the dominant forces were bubble growth force (added mass) and shear lift force at the moment of departure. The dimensionless bubble lift-off diameter ($D_{lo}^* \equiv \sqrt{C_l Re_b}$, C_l: lift coefficient) is then determined as a function of the Jacob number and the Prandtl number:

$$D_{lo}^* = 10.32 Ja^2 Pr_f^{-1} . \qquad \textbf{(53)}$$

The Jacob number Ja is defined as:

$$Ja \equiv \frac{\rho_f C_{pf}\left(T_w - T_{sat}\right)}{\rho_g i_{fg}} . \qquad \textbf{(54)}$$

For horizontal flows, Duhar *et al.* [21] argued that the bubble dynamics were dominated by buoyancy, drag and added mass forces upon departure. An analytic expression of the departure diameter is also obtained from the force balance approach. However, the model was only evaluated with their own n-pentane data with rather limited flow range.

Park *et al.* [22] studied the condensation effect on the interfacial area development. They divided the condensation process into a heat transfer controlled period and an inertia controlled period. In the heat transfer controlled period, the bubble shrinks at a moderate rate due to condensation heat transfer. This process does not involve any bubble number change, thus it is the counterpart to the evaporation induced interfacial area increase. Park *et al.* [22] proposed an interfacial sink term for this process:

$$\phi = -4(1-p_c)n_b\kappa_f Nu_c Ja \; . \qquad (55)$$

Here p_c is the probability of a bubble in the inertia controlled region; n_b is the bubble number density; Nu_c is the condensation Nusselt number. It should be mentioned that this term should not appear explicitly in the interfacial area transport equation. Rather, it is embedded in the void fraction change *via* the continuity equation shown in Eq. (1), where the interfacial mass transfer term Γ_k depends on the condensation heat transfer.

For inertia controlled period, the interfacial sink term can be written as [22]:

$$\phi_{cod} = -\pi D_{bc}^2 n_b \frac{1}{t_c} \; , \qquad (56)$$

where D_{bc} is the bubble diameter boundary between heat transfer region and inertia region. t_c is the bubble residence time which is calculated by:

$$t_c = \frac{D_{Sm}^2 - D_{bc}^2}{4} \frac{\rho_g i_{fg}}{Nu_c k_f \Delta T_{sub}} \; . \qquad (57)$$

This term should be treated separately from the void fraction change term in the IATE.

From the above discussion the total phase change term can be written as:

$$\phi_{ph} = \phi_{nuc} + \phi_{cod} \; . \qquad (58)$$

For typical boiling flows, the nucleation term usually appears near the heater surface as a boundary condition to the fluid field. The condensation term only exist in the bulk region in the subcooled boiling flow. The generation of new bubbles and disappearance of existing bubbles due to phase change usually take place while bubble size is small. Therefore, the phase change term ϕ_{ph} appears only in the group 1 IATE as shown in Eqs. (17) and (18).

4.2. Bubble Interaction Term

The bubble interaction, either coalescence or breakup largely exists in both adiabatic and diabatic two-phase flows. The flow regime transitions are usually caused by various bubble interactions. When small spherical and distorted bubbles are densely packed, they tender to coalesce and form large Taylor bubbles. This marks the transition from bubbly to slug flow regime. While the consecutive slug bubbles interact with each other, the transition from slug flow to churn-turbulent flow occurs [23]. Therefore, the bubble interactions play very important roles in the development of interfacial structures and they must be properly modeled to close the IATE.

There are two approaches to develop source and sink terms caused by various bubble interaction mechanisms: solving the algebraic equation of mean parameters (or assuming monosized particles); and solving the integral equation by considering particle size distributions. In the first approach, the source term defined in Eq. (10) is simplified as:

$$\phi_j \equiv S_j \Delta A_i \; . \qquad (59)$$

Here S_j is the particle number source and sink rate due to j^{th} interaction mechanism. ΔA_i is the surface area change during one particle interaction event. It is assumed that each coalescence and breakup event is a binary process which only involves two particles of the same size. Further assuming spherical shape for all bubbles, the interfacial area source term due to j^{th} bubble interaction mechanism can be approximated by:

$$\phi_j = \frac{1}{3\psi}\left(\frac{\alpha_g}{a_i}\right)^2 S_j \,. \tag{60}$$

Here ψ is a shape factor defined by:

$$\psi = \frac{1}{36\pi}\left(\frac{D_{Sm}}{D_e}\right)^3 , \tag{61}$$

where D_e is the volume equivalent diameter of bubbles. From Eq. (60), it can be noticed that each particle interaction event results in a change of surface area by $A_i/3$. A positive or negative sign should be assigned depending on whether the interaction mechanism is coalescence or breakup.

The mean parameter method simplifies the process to develop the source and sink closures. For group 1 bubbles, this assumption should be valid since both spherical and distorted bubbles have limited size variations. Their shapes are also close to spheres. Thus this approach has been widely used in the one group IATE development. However, the sizes of group 2 bubbles have much large variations comparing to group 1 bubbles. Their shapes also strongly depend on the flow channel geometry and bubble size. In a small diameter round tube, group 2 bubbles could be either cap or slug shape. In a narrow rectangular test section, the bubbles are two-dimensional confined cap shape. Therefore the monosized particle assumption is not proper for group 2 bubbles. In this case, the approach based on the integration over bubble distribution should be adopted in developing source and sink closures.

The bubble number density gain rate due to coalescence S_{GC} can be given for the bubbles with volume V as [5]:

$$S_{GC}(V) = \int_{V_{min}}^{V/2} \lambda(V-V',V')h(V-V',V')f(V-V')f(V')dV' , \tag{62}$$

where h(V-V',V') and λ(V-V',V') are the collision frequency and coalescence efficiency of bubbles of volume V-V' and V'. Similarly, the loss rate due to coalescence S_{LC} is given by:

$$S_{LC}(V) = -\int_{V_{min}}^{V_{max}-V} \lambda(V,V')h(V,V')f(V)f(V')dV' . \tag{63}$$

For the breakup events, the gain and loss rates of bubble number density are given as:

$$S_{GB}(V) = \int_{V}^{V_{max}} \beta(V',V)n(V')g(V')f(V')dV' . \tag{64}$$

and

$$S_{LB}(V) = -g(V)f(V) \,. \tag{65}$$

Here, g(V') is the breakup frequency of a particle having a volume of V'; β(V',V) is the distribution of daughter particles produced by breakup of a parent particle of V'.

Given the surface area A_i of the particle with volume V, the interfacial area source or sink term can be obtained by integrating the A_i weighted particle number density gain or loss rate:

$$\phi_j = \int_{V_{\min}}^{V_{\max}} S(V) A_i(V) dV \ . \tag{66}$$

The net change rate of particle number density is the sum of the gain rate and loss rate. For example, the bubble coalescence results in gain of large bubbles and loss of small bubbles. The net change of interfacial area is the combination of these two opposite effects. In practice, one must also consider the particular interaction mechanism and proper classification of two groups. Here the interfacial area source sink terms due to random collision in a confined channel is taken as an example to show the approach described above.

Sun *et al.* [24] derived the two-group IATE for a 20cm x 1cm narrow rectangular channel. The bubbles are assumed to be uniformly distributed within each group as shown in Fig. **3**. The group 1 bubbles can be assumed to be spherical in the test section. Confined cap shape with 50° wake angle is assumed for all group 2 bubbles as shown in Fig. **4** due to the unique test section geometry. Here G represents the 10mm gap of the test section. The boundary between group 1 and group 2 bubbles is determined when the radius R reaches the upper limit of distorted bubble regime $R_C = 2\sqrt{\sigma/g\Delta\rho}$. Therefore, the volume boundary between two groups is given by [24]:

$$V_C = 2.6G\frac{\sigma}{g\Delta\rho} \ . \tag{67}$$

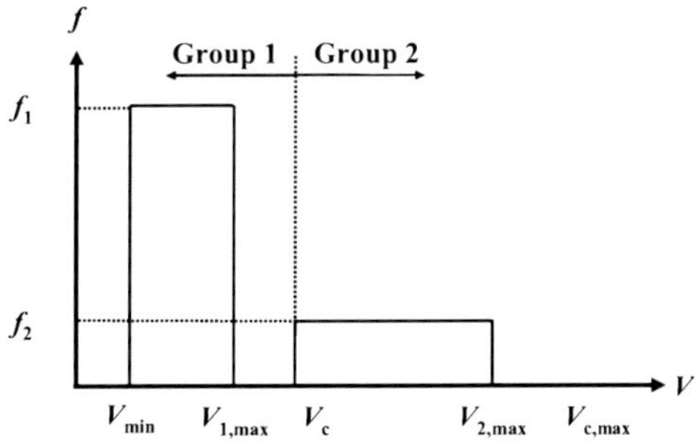

Figure 3: Simplified bubble number density distribution [24].

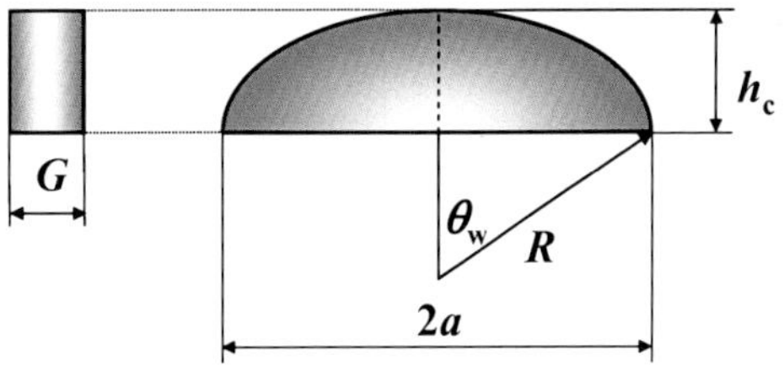

Figure 4: Confined cap bubble [24].

For random collision mechanism, four different cases were considered:

Case I, coalescence of two group 1 bubbles results in one group 1 bubble;

Case II, coalescence of two group 1 bubbles results in one group 2 bubble;

Case III, coalescence of one group 1 and one group 2 bubble results in one group 2 bubble;

Case IV, coalescence of two group 2 bubbles results in one group 2 bubble.

Take Case I as an example, the effective bubble collision frequency is given by [24]:

$$h(V-V',V') \simeq 1.84 m\varepsilon^{1/3}\left(\frac{\alpha_1}{\alpha_{1,\max}}\right)^{2/3}(V-V')^{7/9}. \quad \textbf{(68)}$$

Here V-V' is the volume of the larger bubble in the binary coalescence; ε is the turbulence intensity; $\alpha_{1,max}$ is the void fraction of Group 1 bubbles at maximum packing; m is the collision efficiency given by:

$$m = 1-\exp\left(-C_{RC1}\frac{\alpha_{1,\max}^{1/3}\alpha_1^{1/3}}{\alpha_{1,\max}^{1/3}-\alpha_1^{1/3}}\right), \quad \textbf{(69)}$$

where C_{RC1} is an adjustable coefficient determined by experimental data. If constant collision efficiency is assumed, the number density gain rate for bubbles of volume V can be obtained as:

$$\begin{aligned} S_{RC,G}^{(1)}(V) &= \int_{V_{\min}}^{V/2}\lambda(V-V',V')h(V-V',V')f(V-V')f(V')dV' \\ &\approx 0.75\lambda_{RC}^{(1)} m\varepsilon^{1/3} f_1^2\left(\frac{\alpha_1}{\alpha_{1,\max}}\right)^{2/3} V^{16/9}. \end{aligned} \quad \textbf{(70)}$$

Here the superscript $^{(1)}$ means that bubble interaction is within the range of group 1. For Case I, the volume of the resulted bubbles range from $2V_{min}$ to $2V_{m1}$. The interfacial area gain for this case can be carried out using Eq. (10) with proper integration range:

$$\begin{aligned} \phi_{RC,G}^{(1)} &= \int_{2V_{\min}}^{2V_{m1}} S_{RC,G}^{(1)}(V)A_{i1}(V)dV \\ &\approx 14.7 C_{RC}^{(1)} m\varepsilon^{1/3}\alpha_1^2\left(\frac{\alpha_1}{\alpha_{1,\max}}\right)^{2/3}\frac{1}{D_{Sm1}^{5/3}}. \end{aligned} \quad \textbf{(71)}$$

The interfacial loss rate due to coalescence can be divided into two parts: part 1 includes the smaller bubble in the binary coalescence process and part 2 includes the larger one. The loss rate of the part 1 bubbles can be estimated as:

$$\begin{aligned} \phi_{RC,L1}^{(1)} &= -\int_{V_{\min}}^{V_{m1}}\left(\int_{2V'}^{V_{m1}+V'}\lambda(V-V',V')h(V-V',V')f(V-V')f(V')dV\right)A_{i1}(V')dV' \\ &\approx -7.0 C_{RC}^{(1)} m\varepsilon^{1/3}\alpha_1^2\left(\frac{\alpha_1}{\alpha_{1,\max}}\right)^{2/3}\frac{1}{D_{Sm1}^{5/3}}. \end{aligned} \quad \textbf{(72)}$$

The loss rate of the part 2 bubbles is given by:

$$\begin{aligned} \phi_{RC,L2}^{(1)} &= -\int_{V_{\min}}^{V_{m1}}\left(\int_{V'+V_{\min}}^{2V'}\lambda(V-V',V')h(V-V',V')f(V-V')f(V')dV\right)A_{i1}(V')dV' \\ &\approx -11.0\, C_{RC}^{(1)} m\varepsilon^{1/3}\alpha_1^2\left(\frac{\alpha_1}{\alpha_{1,\max}}\right)^{2/3}\frac{1}{D_{Sm1}^{5/3}}. \end{aligned} \quad \textbf{(73)}$$

Hence, the net interfacial area sink for Case I can be determined from Eqs. (71), (72) and (73):

$$\begin{aligned} \phi_{RC}^{(1)} &= \phi_{RC,G}^{(1)} + \phi_{RC,L1}^{(1)} + \phi_{RC,L2}^{(1)} \\ &\approx -0.17\, C_{RC}^{(1)}\frac{\varepsilon^{1/3}\alpha_1 a_{i1}^{5/3}}{\alpha_{1,\max}^{2/3}}\left[1-\exp\left(-C_{RC1}\frac{\alpha_{1,\max}^{1/3}\alpha_1^{1/3}}{\alpha_{1,\max}^{1/3}-\alpha_1^{1/3}}\right)\right]. \end{aligned} \quad \textbf{(74)}$$

Similar derivation can be also applied to Case II to IV. When the coalescence involves two groups, the gain and loss rate are considered separately for each group. The surface area of the predefined shape for group 2 bubbles should be used in the integration. Finally, five interfacial source and sink terms can be obtained for the random collision mechanism:

Case I, $\phi_{RC,1}^{(1)}$;

Case II, $\phi_{RC,2}^{(11,2)}$;

Case III, $\phi_{RC,1}^{(12,2)}$ and $\phi_{RC,2}^{(12,2)}$;

Case IV, $\phi_{RC,2}^{(2)}$.

Here the subscript k = 1 or 2 represents the source or sink term in group k IATE. It is also noticed that only $\phi_{RC,2}^{(11,2)}$ is considered for Case II. This is because the loss rate due to coalescence of two group 1 bubbles is already accounted for in $\phi_{RC,1}^{(1)}$.

The integral approach takes into account more detailed processes of the bubble interaction phenomenon compared to the mean parameter approach. The uniform bubble distribution in each group, even though still a crude assumption, could partially reflect the fact that smaller bubbles may have a more significant contribution to the total interfacial area concentration. For most cap, slug, and churn-turbulent flows, experimental data has shown that interfacial area concentration of group 1 bubbles is much higher than that of group 2 even though void fraction may be less. By assuming certain shape of bubbles, this approach has been applied to round tubes and narrow rectangular channel [24, 25]. It should not be particularly challenging for other geometries such as rod bundle or annulus.

During the previous studies, a number of bubble coalescence and breakup mechanisms have been identified [5]: Random Collision (RC), Wake Entrainment (WE), Turbulent Impact (TI), Shearing-Off (SO), Surface Instability (SI), Rise Velocity (RV), Laminar Shear (LS), Velocity Gradient (VG), *etc.* Among them, the first five mechanisms, RC, WE, TI, SO and SI, were considered to be important for most two-phase flows and the previous modeling attempts on these mechanisms are summarized in Table **1**. The detailed mathematical models are summarized in reference [26]. Random collision mechanism describes the coalescence of neighboring bubbles driven by the turbulence in the continuous phase. Therefore, this mechanism is particularly pronounced in the flows with high turbulence intensity. It could occur within each group or across the group boundary as discussed previously. In air-water and steam-water two-phase upflows, the particle Reynolds number of a bubble is usually high enough to form a considerably large wake region behind it. Wake entrainment occurs when a trailing bubble falls in the wake region of a leading bubble and results in the subsequent coalescence of two bubbles. This mechanism can be applied to both groups and many models were developed as can be seen in Table **1**. In highly turbulent flows, bubbles could break into smaller ones when surface tension force cannot balance the force of turbulent eddies. This mechanism is called turbulent impact. It is an opposite effect of turbulence as compared to the turbulence induced random collision. The turbulent impact could occur only when turbulent energy exceeds certain limit which is usually specified by a critical Weber number in the models. Turbulent impact acts on both group 1 and group 2 bubbles. Therefore, inter-group transfer may be involved due to this mechanism. Another important breakup mechanism is shearing off of small group 1 bubbles from the skirt region of group 2 bubbles. It appears as a source term for group 1 and a sink term for group 2 in the two group IATE. This mechanism was found to be a major source of interface area concentration for certain conditions due to the generation of large amount of group 1 bubbles. For shearing off to occur, the Weber number defined on the diameter of group 2 bubbles has to reach certain limit such that the interfacial shear force can overtake the surface tension force. The last important mechanism is the breakup of group 2 bubbles due to surface instability. The resulted bubbles after breakup are still in the group 2 range. Therefore, this mechanism is an interfacial source for group 2 bubbles only.

Table 1: Summary of source/sink modeling of bubble interactions.

	Researcher	Mechanism	Approach	Experiment
One Group IATE	Wu *et al.* [27]	RC, WE, TI	Mean value	5.08cm pipe, air-water
	Hibiki and Ishii [28]	RC TI	Mean value	2.54cm, 5.08cm pipes, air-water
	Hibiki *et al.* [29]	RC, WE	Mean value	0.9cm pipe, air-water
	Kim *et al.* [30]	RC, WE, TI	Mean value	20cm x 1cm duct, air-water
	Yao and Morel [31]	RC, TI	Mean value	1.92cm pipe, heated R12
Two Group IATE	Hibiki and Ishii [32]	RC,WE, TI	Mean value	5.08cm pipe, air-water
	Fu and Ishii [25]	RC, WE, TI, SO	Distribution	4.83cm pipe, air-water
	Sun *et al.* [24]	RC, WE, TI, SO, SI	Distribution	20cm x 1cm duct, air-water

The significance of above identified bubble interaction mechanisms depends on flow conditions and geometry of specific test section as well. For instance, Hibiki and Ishii [28] only considered coalescence due to random collision and breakup due to turbulent impact for adiabatic bubbly flow in 2.54cm and 5.08cm round pipes. They could yield very good agreement with experimental data without considering the wake entrainment term in IATE. This is because the relatively high flow rates in their data make the contribution of wake entrainment almost negligible. Fu and Ishii [25] excluded the surface instability mechanism from two-group IATE developed on experimental data obtained in a 4.83cm round tube. They argued that the pipe diameter was much smaller than the maximum stable bubble limit. Group 2 bubbles were relatively stable such that surface instability induced breakup was not likely to occur. Therefore they only considered the breakup due to turbulent impact and shearing off mechanisms.

Most of the previous interfacial area source and sink terms were studied for gas dispersed flows up to churn-turbulent flow. The annular flows were hardly touched in terms of the development of IATE. Liu *et al.* [9] divided the flow field into four groups based on their characteristics: gas bubbles in liquid film; gas core; droplets in gas core; and liquid film. The three independent interfacial area groupings can be given accordingly: bubble interface in the liquid film (group 1); gas core/liquid film interface (group 2); and droplet interfaces in the gas core (group 3). Table **2** lists the possible intra-group and inter-group source/sink terms and they are briefly discussed here.

Group 1 in annular flow is actually an extension from group 1 bubbles in gas-dispersed flow, but it only exists in a confined region, *i.e.*, the liquid film. Some new inter-group mechanisms may appear: bubble bursting (sink) and liquid impingement (source). Overall the amount of bubbles inside the liquid film is small and is postulated to have a little effect on the film flow. For very thin films, bubble does not exist.

Table 2: Source and sink terms in annular flow.

	Inter-group source/sink	Intra-group source/sink
Group 1 Bubbles	Bubble distortion Bubble expansion/contraction Evaporation/condensation Bubble breakup and coalescence	Bubble bursting Liquid impingement
Group 2 Film-Core Interface	Change of wave characteristics, *e.g.*, wave length, roughness, frequency, *etc.*	Entrainment Deposition Bubble bursting
Group 3 Droplets	Droplet distortion Droplet breakup and coalescence Evaporation/condensation	Entrainment Deposition

Another dispersed group is group 3 which consists of the droplets in gas core region. Usually entrainment and deposition are the major source/sink terms. The droplet interaction may become very important in wispy-annular flow where a large amount of irregular drops exist in the core region. If the entrainment rate $\dot{\varepsilon}$ and deposition rate $\dot{d}$ in annular flow are known, the source and sink terms in one-dimensional IATE for droplets can be approximated by:

$$\phi = \frac{12(D-2\delta)}{D^2 \rho_f}\left[\frac{\dot{\varepsilon}}{r_{de}} - \frac{\dot{d}}{r_{dd}}\right]. \tag{75}$$

Here δ , r_{de} and r_{dd} are film thickness, radius of entrained and deposited droplets, respectively.

The Group 2 interface is a continuous interface between the liquid film and gas core. It provides important information in determining the interfacial shear stress and how the two continuous phases interact with each other. Since it is not carried by any particles, the source/sink terms should be modeled by the interfacial wave characteristics such as wave length, amplitude and frequency, *etc.* The interaction of this interface with bubbles and droplets may also change the shape locally for each event. The detailed models of these mechanisms are not seen in the literature. Initial attempts of determining one dimensional interfacial area concentration and interfacial velocity based on the local instantaneous formulation can be found in reference [9].

5. SUMMARY

As an important parameter characterizing the interfacial structure, an accurate prediction of the interfacial area concentration is indispensible in simulating two-phase flows using the two-fluid model. The interfacial area transport equation can model the interfacial area concentration dynamically and hence can overcome the shortcomings associated with the flow regime dependent correlations. In this article, it is shown that the IATE can be obtained from Boltzmann transport equation for dispersed flows. The local instantaneous description of surface evolution can be used to derive the IATE applicable to a broader range including continuous interfaces. The major difference between these two approaches lies in the interfacial source and sink terms. In the Boltzmann approach, the source and sink terms are classified as volume change term, phase change term and bubble interaction term. In the local instantaneous formulation, the interfacial source is expressed as an integral of the curvature weighted relative motion of local surface. The former approach is straightforward for the macroscopic description of the interfacial area development. This statistical description has also been widely used for the development of various source and sink terms. The local instantaneous formulation, however, must be employed for continuous surfaces such as annular flows. A new source term which is caused by the interface distortion is identified by this approach.

There are two ways in modeling the source and sink terms: solving the algebraic equation of mean parameters; and solving the integral equation by considering particle size distributions. The mean parameter approach usually assumes that particles are close to spherical shape and have small variations in size. This approach can be used to model interactions of group 1 bubbles and phase change terms. For group 2 bubbles, the bubble shape is no more spherical and it strongly depends on the channel geometry. The bubble distribution should be considered while developing the interfacial source and sink models. An example of the integral approach is shown for the random collision mechanism in the two group IATE in a confined channel.

The modeling of IATE and its closures have been carried out mostly for gas dispersed flows in the past studies. Not much attention was paid to separated flows due to the appearance of continuous interfacial structures, where both theoretical and experimental challenges exist. However, separated flows are very important for many industrial systems such as nuclear power plant and oil pipelines. As a dynamic approach, the IATE should be able to handle the flow regime transitions and predict the interfacial area over the entire spectrum of flow regimes. Therefore, these challenges must be emphasized and resolved with the assistance of the local instantaneous formulation.

NOMENCLATURE

A	Area
∂A	Boundary curve of surface A
a_i	Interfacial area concentration
C	Coefficient
D_d	Departure diameter of nucleation bubble
d	Diameter
$\dot{d}$	Deposition rate
f	Function
f_d	Bubble departure frequency
g	Gravitation acceleration
G	Filter function
h	Enthalpy
H_f	Heaviside function
H_k	Mean curvature
i	Latent heat
Ja	Jacob number
L	Length
M_{ik}	Generalized interfacial drag
m	Collision efficiency
$\vec{n}$	Unit normal vector to interface
$\vec{N}$	Unit normal vector to control volume boundary
Nu	Nusselt number
p	Pressure
p_c	Probability of inertia controlled region
q	Heat flux
Pr	Prandtl number
r	Radius
$\vec{r}$	Position vector
Re	Reynolds number
S	Number density source/sink
t	Time
T	Time domain
v	Velocity
V	Volume
We	Weber number
x	x coordinate
y	y coordinate
z	z coordinate

Greek Letters

α	Void fraction
Γ	Mass transfer due to phase change
δ	Dirac delta function
ε	Turbulence intensity
$\dot{\varepsilon}$	Entrainment rate
η	Void change due to nucleation and condensation
θ	Angle
κ	Thermal diffusivity
λ	Coalescence efficiency
$\vec{\mu}$	Unit normal vector to interception curve in tangent plane
ξ	Perimeter
ρ	Density
σ	Surface tension
τ	Shear stress
ϕ	Interfacial area source/sink term
φ	Function
χ_k	Phase indicator function
ψ	Shape factor
ω	Angular velocity
Ω	Control volume

Subscripts

c	Group boundary
ca	Center of area
cod	Condensation
f	Liquid
i	Interface
j	j^{th} interaction mechanism
g	Gas
G	Particle number gain
k	k^{th} group/phase
L	Particle number loss
max	Maximum
min	Minimum
nuc	Nucleation
ph	Phase change
rc	Relative to center
sat	Saturation

sub	Subcooled
Sm	Sauter mean diameter/radius

Superscripts

A	Average
S	Area-average
T	Time-average
V	Volume-average
Z	*z*-direction line average

Symbols

$\vec{f}$	Vector *f*
$\overline{\overline{f}}$	Tensor *f*
$\tilde{f}$	Average of *f*
$\tilde{\tilde{f}}$	Phase weighted average of *f*

REFERENCES

[1] M. A. Van Der Hoef, M. Van Sint Annaland, N. G. Deen, and J. A. M. Kuipers, "Numerical simulation of dense gas-solid fluidized beds: A Multiscale Modeling Strategy," *Ann. Rev. Fluid Mech.*, Vol. 40, pp. 47-70, 2008.

[2] R. T. Lahey Jr, "On the direct numerical simulation of two-phase flows," *Nucl. Eng. Des.*, Vol. 239, pp. 867-879, 2009.

[3] M. Ishii, and K. Mishima, "Two-fluid model and hydrodynamic constitutive relations," *Nucl. Eng. Des.*, Vol. 82, pp. 107-126, 1984.

[4] M. Ishii, *Thermo-Fluid Dynamic Theory of Two-Phase Flow*, Paris: Eyrolles, 1975.

[5] G. Kocamustafaogullari, and M. Ishii, "Foundation of the interfacial area transport equation and its closure relations," *Int. J. Heat Mass Transfer*, Vol. 38, pp. 481-493, 1995.

[6] D. A. Drew, "Evolution of geometric statistics," *SIAM J. Appl. Math.*, Vol. 50, pp. 649-666, 1990.

[7] C. Morel, N. Goreaud, and J.-M. Delhaye, "The local volumetric interfacial area transport equation: derivation and physical significance," *Int. J. Multiphase Flow*, Vol. 25, pp. 1099-1128, 1999.

[8] C. Morel, "On the surface equations in two-phase flows and reacting single-phase flows," *Int. J. Multiphase Flow*, Vol. 33, pp. 1045-1073, 2007.

[9] Y. Liu, M. Ishii, and J. R. J. Buchanan, "Foundation of the interfacial area transport equation in annular two-phase flows," in *Int. Conf. Multiphase Flow (ICMF 2010) Proc.*, Tampa, Florida. USA, 2010.

[10] M. Ishii, and S. Kim, "Development of one-group and two-group interfacial area transport equation," *Nucl. Sci.Eng.*, Vol. 146, pp. 257-273, 2004.

[11] C. M. Marle, "On macroscopic equations governing multiphase flow with diffusion and chemical reactions in porous media," *Int. J. Engng Sci.*, Vol. 20, pp. 643-662, 1982.

[12] I. Kataoka, M. Ishii, and A. Serizawa, "Local formulation and measurements of interfacial area concentration in two-phase flow," *Int. J. Multiphase Flow*, Vol. 12, pp. 505-29, 1986.

[13] J. A. Thorpe, *Elementary Topics in Differential Geometry*, New York: Springer-Verlag, 1985.

[14] R. Aris, *Vectors, Tensors, and the Basic Equations of Fluid Mechanics*, Englewood Cliffs, New Jersey: Prentice-Hall, Inc., 1962.

[15] G. Friedlander, and M. Joshi, *Introduction to the Theory of Distributions*, Cambridge: Cambridge University Press, 2003.

[16] E. Labourasse, D. Lacanette, A. Toutant, P. Lubin, S. Vincent, O. Lebaigue, J. P. Caltagirone, and P. Sagaut, "Towards large eddy simulation of isothermal two-phase flows: governing equations and a priori tests," *Int. J. Multiphase Flow*, Vol. 33, pp. 1-39, 2007.

[17] J. C. Slattery, L. Sagis, and E.-S. Oh, *Interfacial Transport Phenomena, Second Edition*, Springer, 2007.

[18] T. Hibiki, and M. Ishii, "Active nucleation site density in boiling systems," *Int. J. Heat Mass Transfer*, Vol. 46, pp. 2587-2601, 2003.

[19] R. Situ, M. Ishii, T. Hibiki, J. Y. Tu, G. H. Yeoh, and M. Mori, "Bubble departure frequency in forced convective subcooled boiling flow," *Int. J. Heat Mass Transfer*, Vol. 51, pp. 6268-6282, 2008.

[20] R. Situ, T. Hibiki, M. Ishii, and M. Mori, "Bubble lift-off size in forced convective subcooled boiling flow," *Int. J. Heat Mass Transfer*, Vol. 48, pp. 5536-5548, 2005.

[21] G. Duhar, G. Riboux, and C. Colin, "Vapour bubble growth and detachment at the wall of shear flow," *Heat Mass Transfer*, Vol. 45, pp. 847-855, 2009.

[22] H.-S. Park, T.-H. Lee, T. Hibiki, W.-P. Baek, and M. Ishii, "Modeling of the condensation sink term in an interfacial area transport equation," *Int. J. Heat Mass Transfer*, Vol. 50, pp. 5041-5053, 2007.

[23] K. Mishima, and M. Ishii, "Flow regime transition criteria for upward two-phase flow in vertical tubes," *Int. J. Heat Mass Transfer*, Vol. 27, pp. 723-37, 1984.

[24] X. Sun, S. Kim, M. Ishii, and S. G. Beus, "Modeling of bubble coalescence and disintegration in confined upward two-phase flow," *Nucl. Eng. Des.*, Vol. 230, pp. 3-26, 2004.

[25] X. Y. Fu, and M. Ishii, "Two-group interfacial area transport in vertical air-water flow - I. mechanistic model," *Nucl. Eng. Des.*, Vol. 219, pp. 143-168, 2002.

[26] M. Ishii, and T. Hibiki, *Thermo-Fluid Dynamics of Two-Phase Flow*, New York: Springer, 2005.

[27] Q. Wu, S. Kim, M. Ishii, and S. G. Beus, "One-group interfacial area transport in vertical bubbly flow," *Int. J. Heat Mass Transfer*, Vol. 41, pp. 1103-1112, 1998.

[28] T. Hibiki, and M. Ishii, "One-group interfacial area transport of bubbly flows in vertical round tubes," *Int. J. Heat Mass Transfer*, Vol. 43, pp. 2711-2726, 2000.

[29] T. Hibiki, T. Takamasa, and M. Ishii, "Interfacial area transport of bubbly flow in a small diameter pipe," *J. Nucl. Sci. Tech.*, Vol. 38, pp. 614-620, 2001.

[30] S. Kim, X. Sun, M. Ishii, S. G. Beus, and F. Lincoln, "Interfacial area transport and evaluation of source and sink terms for confined air-water bubbly flow," *Nucl. Eng. Des.*, Vol. 219, pp. 61-75, 2002.

[31] W. Yao, and C. Morel, "Volumetric interfacial area prediction in upward bubbly two-phase flow," *Int. J. Heat Mass Transfer*, Vol. 47, pp. 307-328, 2004.

[32] T. Hibiki, and M. Ishii, "Two-group interfacial area transport equations at bubbly-to-slug flow transition," *Nucl. Eng. Des.*, Vol. 202, pp. 39-76, 2000.

CHAPTER 2

Dynamic Behavior of Liquid Droplet Impacting on Heated Surfaces

Liwen Jin, Chun Yang* and Kai Choong Leong

School of Mechanical and Aerospace Engineering, Nanyang Technological University, 50 Nanyang Avenue, Singapore 639798, Republic of Singapore

Abstract: The study of a single droplet impact and spread on a heated surface is motivated by its strong relevance to spray cooling technology. The generic view on the study of the spray cooling process exhibits the synthesis of fluid mechanics, heat transfer and surface thermodynamics. Due to the complex phenomena involved, no comprehensive theoretical models are available. The few works that appeared in the literature so far have been largely empirical; the applicability of several correlations proposed is limited. This book chapter reports an experimental study of fundamental aspects of the dynamic characteristics of a single droplet impacting on a heated surface. In this experiment, the entire dynamic process of a droplet from the moment of collision with the substrate surface including the rebound was visualized and analyzed using a high-speed CCD camera. The experimental study focused mainly on the spread of a liquid droplet under the influences of substrate temperature varying from 26°C to 240°C, the inclination angle of substrates at 0°, 30°, and 60°, the wettability of substrates with contact angles from 30° to 90°, the viscosity of liquids ranging from 0.00089 up to 0.9161 kg/m·s, and surfactants of different concentrations.

Keywords: Droplet spreading, sessile droplet heat transfer, spray cooling.

1. INTRODUCTION

Scientific interest on the dynamic behavior of droplet-surface interactions stems from their relevance to numerous technological processes such as ink-jet printing, sputter coating, precision solder dispensing, pesticide spray and spray cooling. In droplet spray cooling technology, a stream of liquid droplets is used to cool heated surfaces through latent heat of evaporation, and such a technology is widely applied for controlling quench rate and material properties in heat treatable aluminum and steel alloys. Owing to the high heat transfer rates and low consumption of coolant (compared with other cooling techniques such as liquid jet cooling), droplet spray cooling is also a promising mode to steadily dissipate large amounts of heat and maintain uniform surface temperatures for electronic and laser devices. Unfortunately, the studies reported to date have mainly been empirical in nature, and a lack of predictive capability regarding spray cooling exists due to the complex phenomena involved [1-3].

It has been demonstrated that the study of the fundamental aspects of dynamic characteristics of a single droplet impact on heated surfaces not only can greatly simplify the problem but also can still capture the key features of the phenomenon towards gaining useful insights into the spray cooling mechanisms. In the literature, studies of single droplet impact hydrodynamics and heat transfer have been reported using theoretical analysis, numerical simulation, and experimental investigation approaches. Simplified theoretical analyses were conducted to obtain analytical expressions for certain important variables in the droplet impact and spread problem, *e.g.*, the maximum spread by Chandra and Avedisian [4], Scheler and Bousfield [5], and Pasandideh-Fard *et al.* [6], the prediction of rebound by Mao *et al.* [7], and the time of evaporation by Lee *et al.* [8]. Under certain assumptions such as constant thermophysical fluid properties, smooth substrate surface, no hysteresis, no solid-fluid interaction induced effects etc, the existing models can provide some approximate predictions of the processes of dynamic droplet impact. Complete numerical approaches were also employed to simulate the transient flow field during droplet spreading using CFD codes. Examples are the spreading process study by Trapaga and Szekely [9], the spreading with wetting

***Address correspondence to Chun Yang:** School of Mechanical and Aerospace Engineering, Nanyang Technological University, 50 Nanyang Avenue, Singapore 639798, Republic of Singapore; Tel: +65 6790-4883; E-mail: mcyang@ntu.edu.sg

Lixin Cheng and Dieter Mewes (Eds)

and surface tension effects by Hatta *et al.* [10], the spreading with heat transfer effect by Zhao *et al.* [11], the process of spreading and rebounding by Fujimoto and Hatta [12], the slashing process by Reiber and Frohn [13], and the complete process of spreading, recoil, and splashing for deposition of tin droplets with solidification by Bussmann *et al.* [14]. In addition, many experimental studies of droplet spreading hydrodynamics and heat transfer have been carried out visually using either single-shot flash photograph or high-speed camera techniques. Examples include a comprehensive study of the effect of substrate temperature Chandra and Avedisian [4], the droplet maximum spreading [5] and rebound [7], the resident time of evaporating droplets on a low temperature substrate [8], mapping of impact and heat transfer regimes as a function of heater surface temperature and the droplet Weber number [15], the effect of surface roughness [16, 17], microdroplet wetting dynamics [18, 19], to name just a few.

Theoretical advancement on the droplet impact and spreading was reviewed by deGeenes [20]. Significant unsolved questions still remain regarding the evolution of a droplet spreading, which is connected to physical chemistry (wettability), statistical physics (motion of the three-phase contact line), and long-ranged surface forces such as van der Waals interaction forces. To date, existing theories still cannot establish a correlation between microscopic and macroscopic contact angles and the motion of the three-phase contact line. No theory is available to adequately predict the effects of surface roughness and surfactant on the droplet impact and spreading processes. To this end, the objectives of this study are (1) to carry out thorough experiments to observe *in situ* the behavior of liquid droplets impact and spread on a solid surface under the influences of various parameters including the surface temperature, inclination angle of the impact surface, surface wettability (in terms of contact angle), liquid type, and surfactant, and (2) to generate experimental data that would enable validation and improvement of the spray cooling models.

2. EXPERIMENTAL INVESTIGATION

2.1. Experimental Setup

A schematic of the experimental setup and detailed diagram of the test unit are shown in Fig. **1**. The experimental setup was constructed to observe the deformation behavior of a single droplet impacting on a heated surface. The main components are 1) a microsyringe to form and release the droplet, 2) a flat surface on which the droplet fell, 3) a high speed camera to record the motion of the liquid droplet, 4) spotlights to provide illumination for the capture of images and 5) a temperature controller linked to a cartridge heater and a thermocouple to measure the temperature of the substrate.

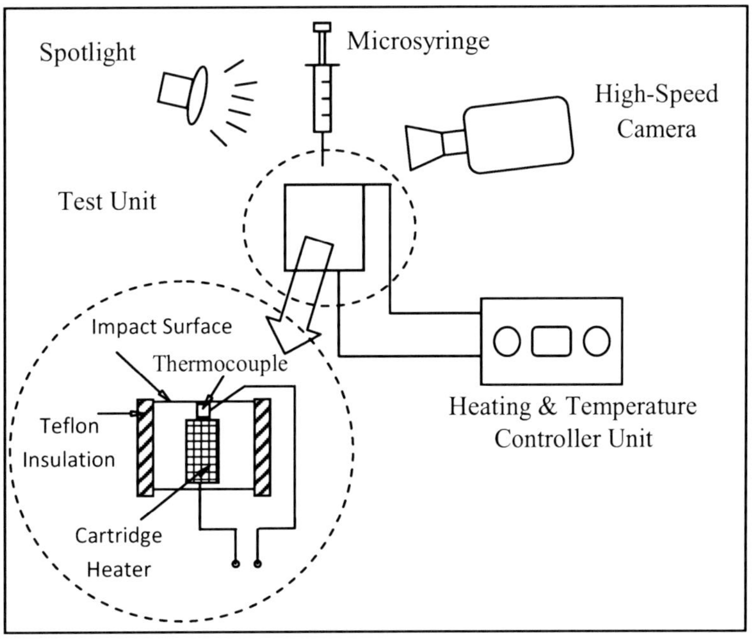

Figure 1: Schematic of Experimental Setup.

The microsyringe needle has a diameter of 0.79 mm and its tip was coated with hydrophobic wax, to allow for manual generation of uniform liquid droplets of 3 mm in diameter. The droplets fall under gravity to strike on the substrate surface. The test unit consists of a 50 mm-diameter aluminum cylinder with a height of 60 mm. The impact surface was heated by a 300 W cartridge heater (diameter = 6.35 mm and length = 50 mm) embedded at the bottom of the cylinder, 10 mm below the impacting surface. The desired surface temperature was maintained using a digital temperature controller to manipulate the power input to the cartridge heater in response to the thermocouple inserted just beneath the impact surface. The wall of the aluminum cylinder was insulated radially by a Teflon tube to reduce heat loss from the side to the surrounding. The heated aluminum cylinder was clamped onto a retort stand to allow for variation in inclination angle of the substrate surface. A high-speed camera (HSV-100, NAC), equipped with a 52-mm close-up macro lens was used to record the time history of the droplet impact and spread process. The lens was focused and zoomed to maximum magnification of the water droplet images. The film speed was set at 500 frames per second, and the droplet images were recorded on Super VHS videotapes. The images were exported frame by frame to a PC through a frame grabber, and then stored in the computer for analysis. Since the normal duration of carrying out droplet impact experiments was less than one second, the effect of heating due to thermal radiation from the spotlights is negligible.

2.2. Materials and Characterization

Test liquids are deionized water, glycerol, a mixture of these two liquids having a volumetric proportion of 25 % water and 75% glycerol, and Sodium Dodecyl Sulfate (SDS) surfactant with concentrations of 0.01%, 0.1% and 1% in deionized water. These fluids were used to examine the effects of surface tension and viscosity varying in the ranges of $21.8 < \gamma < 72.8 \times 10^{-3}$ N/m and $0.89 < \mu < 916.2$ mPa·s, respectively. The viscosity was measured using a "RV12 HAAKE Rotovico" viscometer. A "FTA 200" system (First Ten Angstroms), based on best curve-fitting to the Young-Laplace equation to a symmetric pendant drop, was used to determine the surface tension (against air) of these liquids. Such an instrument can give accuracy for surface tension of about 10^{-3} N/m. The measured values of surface tension, viscosity and density for each liquid sample are summarized in Table **1**.

Table 1: Measured values of surface tension (against air), viscosity, and density of fluids used in experiments.

	De-ionized water	DI water- glycerol mixture (1:3)	Glycerol	Surfactant concentration		
				0.01%	0.1%	1%
γ (N/m)	0.0725	0.0645	0.0635	0.071	0.046	0.022
μ (mPa·s)	0.89	39.5	816	0.89	0.89	0.89
ρ (kg/m^3)	999	1156	1242	999	999	999

The three kinds of substrates used in the experiments are silicon wafer, Teflon, and aluminum block. It is widely accepted that the wettability of surfaces can be evaluated by using contact angle measurements. The contact angles of these three substrates were also measured using the "FTA 200" system. Only advancing contact angles, corresponding to the situations when droplets are produced by injecting liquid, were recorded. In addition, a surface profilometer (Surftest 301, Mitutoyo) was used to examine the surface morphology of the substrates. Three different cross-sectional areas on the material surface were picked, and the average value of the three readings would be an approximate gauge for the substrate's surface roughness. Prior to the start of the experiment, all substrates were cleaned with acetone and de-ionized water. Contact angles obtained for de-ionized water together with the surface roughness of the substrates stated in terms of the maximum peak to valley value, R_t are listed in Table **2**. Contact angles of the other fluids on aluminum substrates are shown in Table **3**.

Table 2: Contact angle and surface roughness of various substrates.

Substrate	Contact angle of de-ionized water (°)	Surface roughness R_t (μm)
Silicon wafer	41.0	0.30
Teflon	81.0	3.83
Aluminum	63.4	11.77

Table 3: Contact angles of various fluids on aluminum substrates.

Name of fluids	Contact angle (°)
DI Water- Glycerol Mixture (1:3)	65.9
Glycerol	68.4
0.01% surfactant concentration	41.8
0.1% surfactant concentration	24.7

With careful experimentation, systematic uncertainty can be minimized. The accuracies of HAAKE viscometer and FTA 200 system are ±2% and ±3°, respectively. The accuracy of captured frame is determined by the resolution of images. The calibration analysis shows that the resolution of images is around 100 microns per pixel, giving rise to an accuracy of ±0.25 mm in droplet diameter measurement. Taking a typical spreading diameter of a droplet of about 5 mm, it can be estimated that the measurement error is within ±5%. The uncertainties of the calculated parameters are estimated based on the measurement accuracy. Under the assumption that the uncertainties of the measured data are independent and random with normal distributions, the uncertainties of Re, We and Oh numbers are estimated to be 5.5%, 7% and 8.5%, respectively.

3. RESULTS AND DISCUSSION

The diameters and heights of the deforming droplets were measured from the recorded images using digitizing software. The side view of the expanding and slipping droplet on the flat and inclined surfaces is schematically shown in Fig. **2** with the definitions of the spreading diameter, *d* and height, *h*. Graphical curves of d/D_o against time/ms were plotted to analyze the effect of different test conditions on the spread of the droplets. The diameter of a droplet was measured from the droplet images just before impact. It should be pointed out that the time at which the liquid droplet just touched the impact surface was set at $t = 0$. However, during the experiments, it was quite difficult to capture the image of a droplet just exactly coming into contact with the impact surface due to the limitation of the camera which was set at 2 ms interval for each frame. Therefore, a simple correction, based on kinetics of a free falling object, was made to 'adjust' the droplet so that the droplet would just strike on the impact surface at time zero.

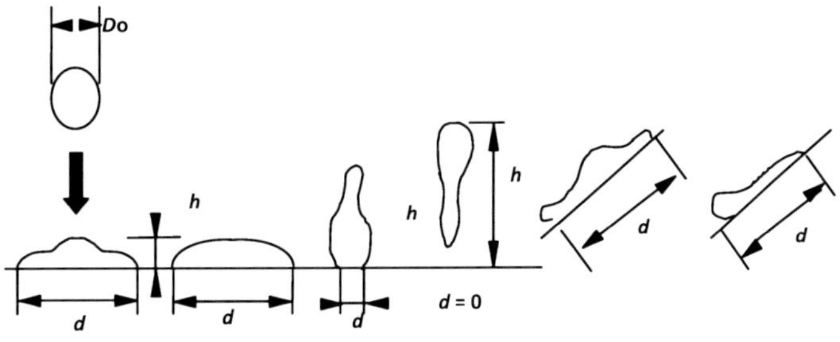

Figure 2: Schematic diagram for the definition of droplet diameter, *d* and height, *h*.

3.1. Effect of the Substrate Temperature

Fig. **3** shows sequential images of water droplets spreading on an aluminum substrate of various initial temperatures of 26°C, 150°C, and 240°C. In Fig. **4** the deformation, *viz.* the spreading and recoil of a droplet is presented quantitatively as the variation of dimensionless spreading diameter with time. When a droplet strikes on a solid surface under room temperature $T = 26$ °C, it spreads radially outward along the impact surface (cf. $t = 4$ ms), and reaches its maximum spread and minimum thickness (cf. $t = 8$ ms). During this period, the initial kinetic energy of a pre-impact droplet is in part irreversibly dissipated by viscous shear stress and in part converted into surface energy. Once the motion of the liquid-solid contact line becomes stationary (*i.e.*, the droplet possesses maximum surface energy and zero kinetic energy corresponding to its maximum spread), the droplet starts to recoil because of surface tension and the reflected waves from the solid surface, as evidenced by a decrease in the spread. Subsequently, the sessile droplet experiences damped oscillations on the substrate; visually it is observed that the oscillation in droplet height is more significant than that in droplet spread due to the restrained motion of the contact line on the substrate surface. At last, the droplet reaches its final shape, retaining thermodynamic equilibrium and hence static surface energy. The height to release the water droplets was kept at 2 cm above the substrates in these experiments, giving a droplet Weber number (defined as $We = \rho D_0 V_0^2 / \gamma$ where ρ is the liquid density, V_0 is the impact velocity, and γ is the liquid surface tension against air) of approximately 16.7. This value is far lower than the critical Weber number of 300 for occurrence of splashing obtained by Cossali *et al.* [21]. For the same reason, no rebound was observed in the experiments. Relevant studies were reported in the published literatures. However, it should be noted that the published study with the same surface characteristics, droplet initial velocity and time frame setting as the present work is not available. As a qualitative comparison, Table **4** shows the water droplets impacting on an aluminum surface (current research) and on a stainless-steel surface (Mao *et al.* [7]) under room temperature. It is found that the droplets possess similar morphological characteristics. At the selected time of t = 8 ms, both droplets reach the maximum spread and start spread-recoil oscillation till t = 30 ms, when the droplets approach to their final equilibrium state.

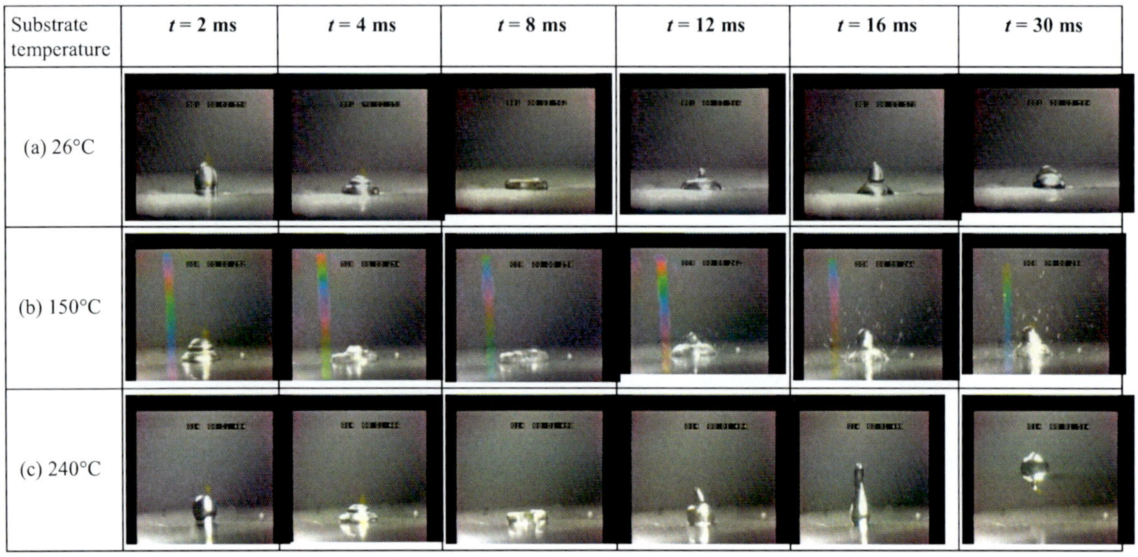

Figure 3: Water Droplets on a 0° Angle Aluminum Substrate of Different Temperatures (a) 26°C, (b) 150°C, and (c) 240°C.

Table 4: A qualitative comparison of water droplets impact on aluminum and stainless-steel surfaces under room temperature.

Time	DI water on an aluminum surface (Current study)	DI water on a stainless-steel surface (Mao *et al.* [7])
t = 8 ms		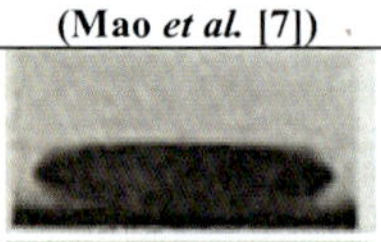
t = 12 ms		
t = 30 ms		

When the substrate temperature is at T = 150°C which is above the boiling point of water (T_s = 100°C), heat transfer has contributions from two mechanisms, *viz.* forced convection due to droplet spreading and nucleate boiling. Bubbles are formed from nucleation sites on the substrate and/or probably due to impurities in the liquid, immediately after the droplet impacts on the substrate (cf. $t = 4$ ms). Bubbles grow very fast and exhibit vibrancy. The miniaturization of droplet diameter occurs after time elapses about $t = 12$ ms. Furthermore, as can be seen from Fig. **4**, droplets at 150°C have a slightly higher fluctuating spread compared to droplets at lower temperature. The peak spread is also higher than at 26°C. The boiling and evaporation of the droplet at the edge may account for the higher peak spread upon impact and the subsequent fluctuations [4].

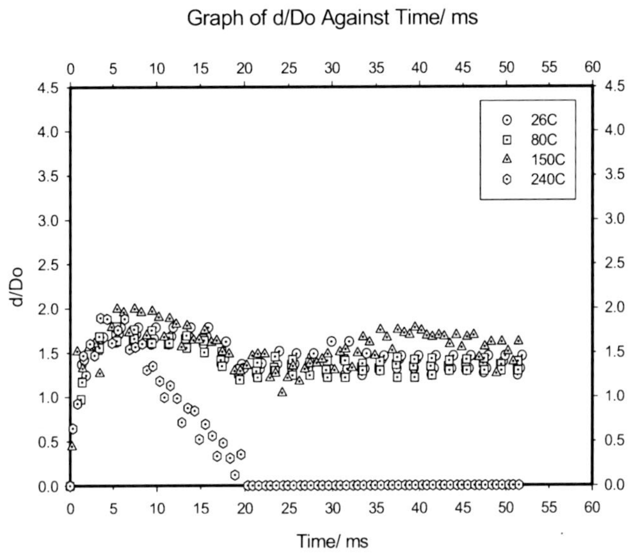

Figure 4: Spreading diameter of water droplets on a 0° angle aluminum substrate of different temperatures.

However, miniaturization disappears and no bubbles are present if the substrate temperature is $T = 240$ °C which is above the water's Leidenfrost temperature. Because of the occurrence of film boiling, the pressure of the vapor escaping from below the droplet is increasing and the droplet is not in real contact with the substrate surface. Compared to the other two cases, the kinetic energy of the droplet is not dissipated so much by the substrate due to friction resulted from the viscous and slippery interactions between the fluid and the substrate. Therefore, it can be seen that after rising to a peak spread upon impact, the droplet shrinks back and continuously elongates upwards. Eventually, the droplet rebounds and lifts itself away from the surface. Our findings of water droplets behavior in different heat transfer regimes are qualitatively in agreement with results reported by Bernardin *et al.* [15].

It is also interesting to note that before reaching their maximum, the droplet spread on the substrate exhibits almost the same shape and trend regardless of the substrate temperature. This suggests that the effect of the substrate temperature on the droplet spread at its initial stage is not important.

3.2. Effect of the Type of Liquid

As a comparison to water droplets, Fig. **5** displays a sequence of images showing the spreading of glycerol droplets on an aluminum substrate of various initial temperatures, 26°C, 150°C, and 240°C.

Compared to Fig. **3**, three different phenomena can be observed in Fig. **5**. Firstly, Glycerol droplets have a tendency to stick on the substrate surface compared to water droplets. The glycerol droplet undergoes lighter damped oscillation after impacting on the surface and it stabilizes quickly. This can be mainly attributed to viscous dissipation effects originated from high viscosity of the glycerol. Secondly, although bubble nucleation in glycerol droplets occurs at 150°C on the substrate, no miniaturization is present. Finally, for the case of high temperature $T = 240$°C, there is no rebounding of glycerol droplets. However, a cloud of wet vapor, which is formed as result of evaporation of the liquid from the puddle, is observed. It turns out that the heat transfer behavior of glycerol droplet impingement is quite different from that of water droplets. Hence detailed mapping of glycerol droplet impact and heat transfer regimes as a function of the substrate temperature is needed in our future studies.

To quantify the effect of liquid properties on the droplet dynamic behavior, the calculated Reynolds, Weber and Ohnesorge numbers for the droplets of DI water, DI Water-glycerol mixture and glycerol at room temperature are listed in Table **5**. Weber number represents the ratio of inertial to surface tension forces, while Ohnesorge number relates the ratio of viscous force to inertial and surface tension forces. It is obvious that the Ohnesorge number of glycerol is much higher than that of DI water, which indicates a greater influence of the viscosity of glycerol on the impact process. The relatively low Ohnesorge number and high Reynolds number of DI water imply that the inertial force is more dominant throughout the water droplet impacting process rather than the effects of surface tension and viscous forces. It is also seen that the surface tension plays the similar role in the impact process for three tested droplets, due to the relatively close Weber numbers.

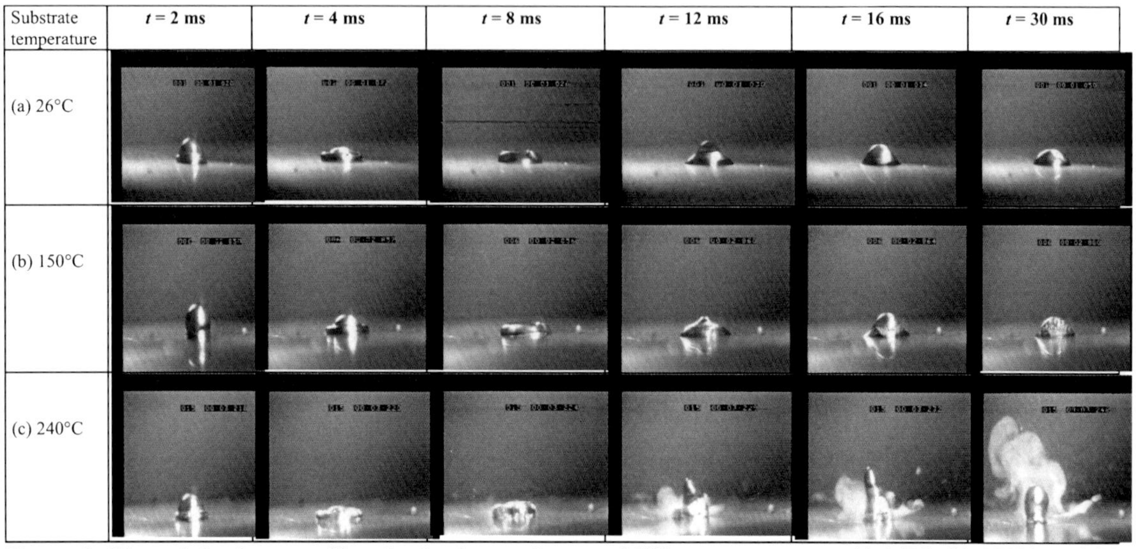

Figure 5: Glycerol droplets on a 0° angle aluminum substrate of different temperatures (a) 26°C, (b) 150°C, and (c) 240°C.

Table 5: Calculations of dimensionless parameters for different droplets.

Parameters	**DI water**	**DI water-glycerol mixture (1:3)**	**Glycerol**
Reynolds number, $\text{Re} = \frac{V_0 D_0}{\nu}$	2157	54.8	2.9
Weber number, $\text{We} = \frac{\rho D_0 V_0^2}{\gamma}$	16.7	22.0	23.5
Ohnesorge number, $\text{Oh} = \frac{\mu}{\sqrt{\rho \gamma D_0}}$	0.002	0.21	1.7

3.3. Effect of the Angle of Inclination

In many of the applications, the droplets usually impact on oblique surfaces. Fig. **6** illustrates a sequence of images showing the spreading of water droplets on an aluminum substrate with three different angles of inclination at room temperature of $T = 26^{\circ}$C. In all three cases, the water droplets were released from the same height above the impact point and hence have the same Weber number, $We = 16.7$. When the substrate surface is inclined, the contribution of the tangential gravitational component to the spreading of impacting droplets is discernible. After landing for a short while (*e.g.* $t = 4$ ms), the deforming droplet starts to deviate from its symmetric shape. The droplet is squeezed by gravity towards its leading edge, resulting in a rise in the front part. This is related to the feature of substrate advancing contact angle. The trailing edge is pulled by the droplet and hence becomes thinner, corresponding to the substrate receding contact angle. This asymmetric deforming structure together with momentum makes the entire droplet slide down the incline.

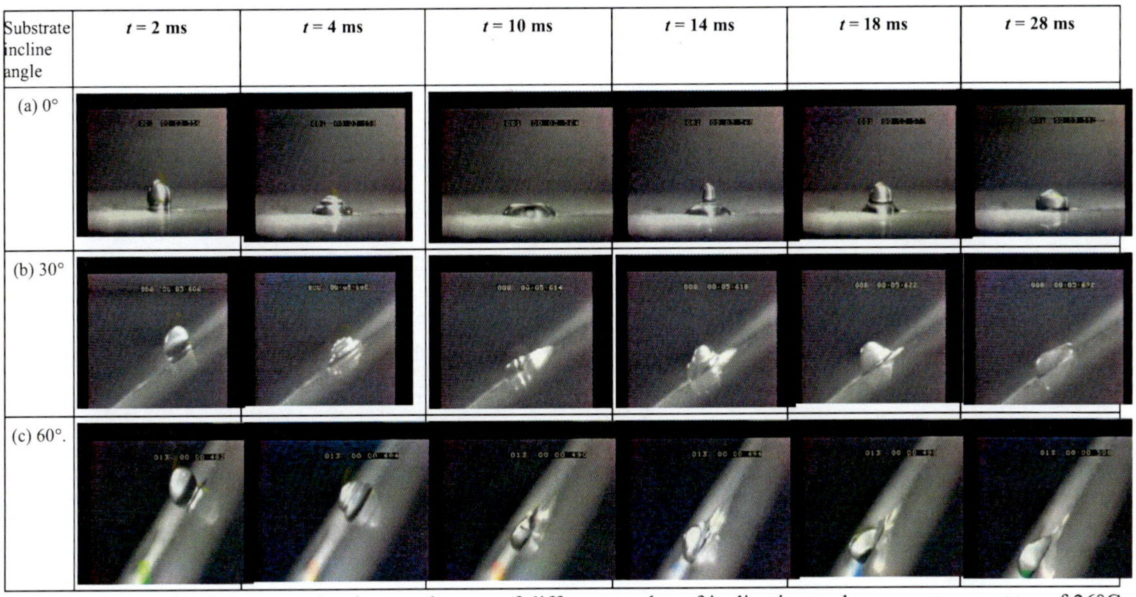

Figure 6: Water droplets on an aluminum substrate of different angles of inclination under room temperature of 26°C (a) 0°, (b) 30°, and (c) 60°.

Fig. **7** presents the quantitative droplet spread (normalized by the droplet initial diameter) versus time. It is observed that the spreading diameter for the inclined surfaces is lower than that for the horizontal surface during initial period, say $t < 8$ ms because of a decrease in the droplet impact momentum along the normal direction to the surface. However, afterwards, the spreading diameter of inclined surfaces quickly surpasses that of 0° angle of inclination due to the effect of gravity force along the tangential direction, *i.e.*, the droplet spreading direction. Such an effect not only causes the liquid film to be slippery, leading to larger wetting areas but also induces stronger oscillations. For instance, the spread of the droplet on the 60° angle of inclination is much larger and takes longer to stabilize. Other than gravitational contributions, the droplet spread on an inclined substrate exhibits a 3D effect which would cause stronger oscillations.

In addition, it should be pointed out that the drop size was kept unchanged throughout all the experiments reported here. Since the gravitational force is proportional to the cubic power of drop diameter, it is expected that the spreading of droplets of larger sizes will be affected more significantly by the angle of inclination.

3.4. Effect of the Substrate Wettability

Fig. **8** displays a sequence of images showing the effect of substrate surface wettability on water droplet impact process. The droplet Weber number is still kept as 16.7. At the initial stage (*i.e.*, $0 < t < 8$ ms) the droplet spreads on these three substrates exhibit almost the same shape regardless of the difference in surface wettability (*i.e.*, contact angle). This indicates that during this period, the deformation process is completely dominated by the inertial and viscous forces. However, from Fig. **9** that plots the normalized spread against time, it shows that the effect of surface wettability on the deformation process, though not significant, still exists. The maximum spread on the more wetting (*i.e.*, lower contact angle) silicon wafer is much larger than that on the less wetting surfaces of Teflon and aluminum. Moreover, it can also be seen that at the final equilibrium stage, the spread meniscus of water droplets corresponds to the static contact angles of these three substrates, showing that the spread increases with decreasing contact angle. The observation is supported by our previous report [22].

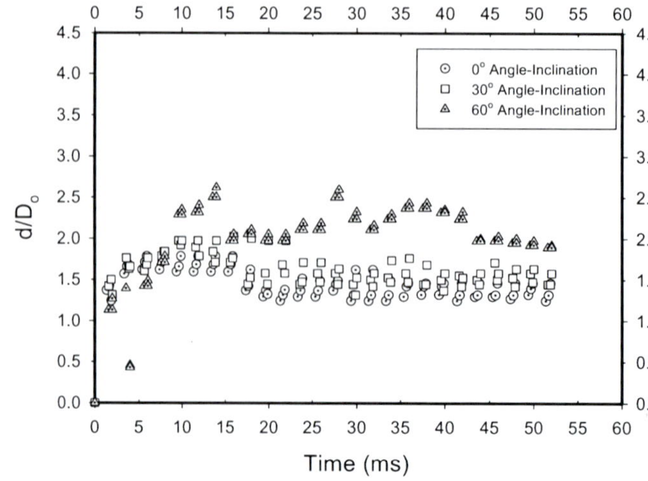

Figure 7: Spreading diameter of water droplets on an aluminum substrate at different angles of inclination.

It should also be noted although the contact angle for Teflon (θ = 81°) is very different from that for aluminum surface (θ = 63°), the final spreads of water droplets on these two substrates are about the same magnitude. This could be partially attributed to the effect of surface roughness. Such an effect causes contact angle hysteresis [16-17], and may even result in local change of equilibrium contact angle. It will also be interesting to examine the effect of surface roughness on droplet heat transfer, specifically boiling processes (*e.g.* bubble nucleation), which will be our next task. Nevertheless, the results presented amply demonstrate that the spread of droplets, or more precisely, the motion of three-phase contact line is dependent on the surface wettability, which in turn is related to the molecular interactions (*e.g.*, the van der Waals interactions) between the fluid and the solid.

Substrate type	t = 2 ms	t = 4 ms	t = 6 ms	t = 8 ms	t = 16 ms	t = 30 ms
(a) Aluminum						
(b) Silicon Wafer						
(c) Teflon						

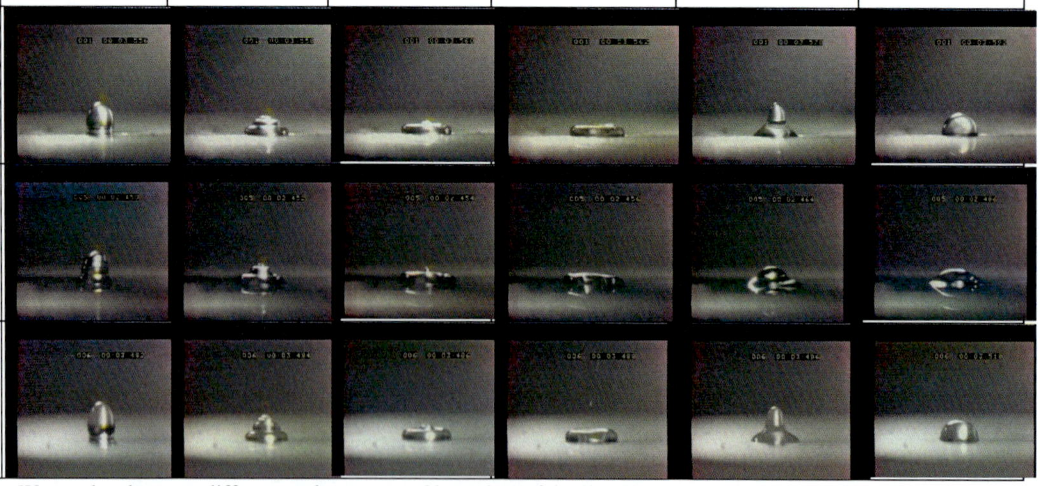

Figure 8: Water droplets on different substrates at 0° angles of inclination under room temperature of 26°C (a) aluminum, (b) silicon wafer, and (c) teflon.

3.5. Effect of the Surfactant

A surfactant (surface-active agent) is a soluble compound that is made up of two different components on a molecular level, one being hydrophilic and the other being hydrophobic. This dual nature greatly increases the surfactant's tendency to adsorb at interfaces. One of the effects of adding surfactants into water is to reduce the liquid surface tension. Since surface-active impurities are usually present even in a very clean system, it inevitable, to some extent, to have surfactants.

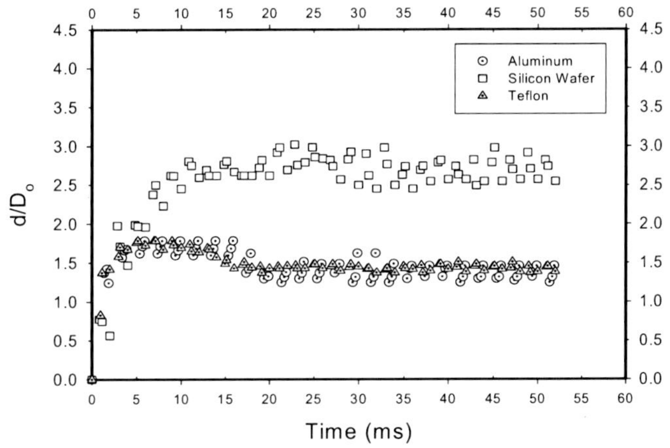

Figure 9: Spreading diameter of water droplets on 0° angle of inclination of different substrates.

Figs. **10** and **11** respectively show the normalized spread and height of impacting liquid droplets on an aluminum surface versus time for three different SDS surfactant concentrations: 0.01%, 0.1%, and 1% (w/w). According to the literature [23], the Critical Micellar Concentration (CMC) for SDS is 2.38 g/liter, corresponding to 0.24% (w/w). The equilibrium surface tension at CMC is 36.5×10^{-3} N/m. In principle, the surfactant effect belongs to surface thermodynamic category. As expected, such an effect is insignificant at initial spreading stage of impacting droplets. However, as seen from Fig. **10**, the concentration of SDS surfactant plays an important role in the droplet dynamic process, especially at the later stage. As shown in Table **3**, an increase in the SDS concentration causes a decrease in surface tension, and simultaneously leads to a reduction in the contact angle. Therefore, as the SDS concentration increases, the droplet wets the surface more readily, resulting in enhanced spreading.

Further, the results in Fig. **11** show that adding SDS surfactant causes a decrease in the intensity of oscillation and decay rates of oscillation increase and their frequencies decreases as surfactant concentration increases. It should also be noted that the characteristic time of the process of droplet impact and spread is within one second. This causes a strong deviation from thermodynamic equilibrium, and thereby the so-called dynamic surface tension may play some part. In addition, large deformation of droplet shape may also cause non-uniform distribution of surfactant molecules on the droplet surface, giving rise to the Marangoni effect. As a result, such Marangoni effect would exert tangential stresses on the droplet surface to stabilize interfacial waves, and thus damp the oscillation of droplets.

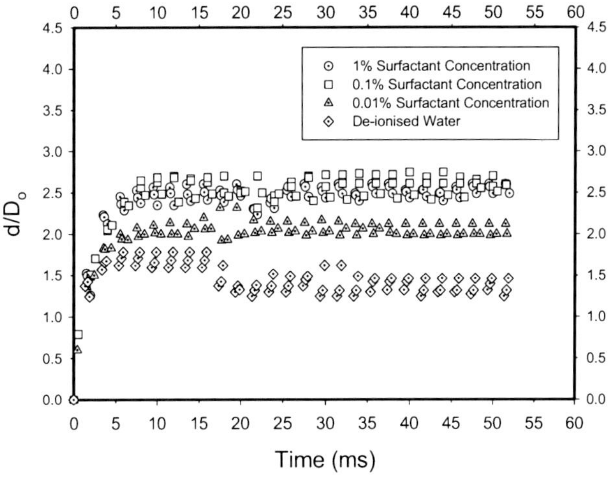

Figure 10: Spreading diameter of liquid droplets of different SDS concentrations on an aluminum substrate at 0° angle of inclination.

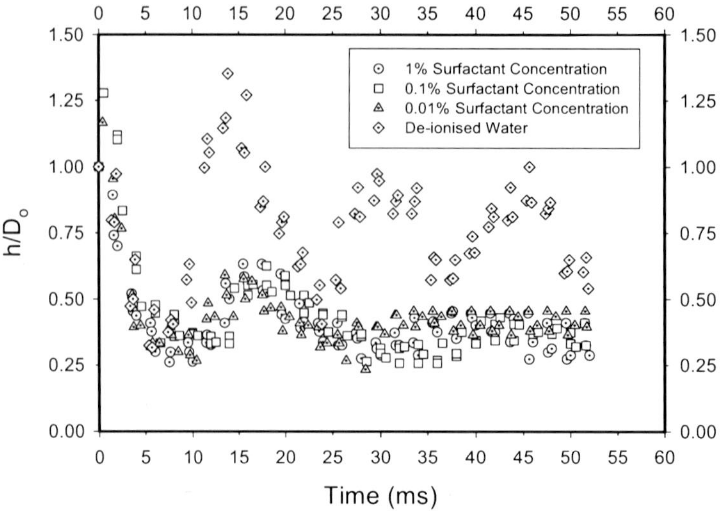

Figure 11: Height of liquid droplets of different SDS concentrations on a aluminum substrate at 0° angle of inclination.

4. SUMMARY

The effect of the various parameters on the behavior of a liquid droplet impacting and spreading on a heated solid surface can be summarized as follows:

1. At room temperature, the droplet, upon impacting on a substrate, exhibits spreading, attaining the maximum of deformation, recoiling, then experiences damped oscillations, and finally reaches equilibrium.

2. When the substrate temperature is above the boiling point of water but below its Leidenfrost temperature, bubbles grow very fast with miniaturization occurring at a later time. The water droplet has a slightly higher fluctuating spread and higher peak compared with that which impinges the substrate at room temperature.

3. Miniaturization disappears and no bubbles are present if the substrate temperature is above the water's Leidenfrost temperature due to the occurrence of film boiling. After rising to a peak spread upon impact, the droplet rebounds from the surface.

4. Prior to the droplet spread reaching its maximum size, the effect of the substrate temperature is insignificant.

5. Compared to a water droplet, the glycerol droplet undergoes lighter damped oscillation after impacting on the surface and it stabilizes quickly due to its high viscosity. No miniaturization is present at 150°C and there is no rebounding of the droplet even at 240 °C.

6. The spreading diameter for the inclined surfaces is lower than that for the horizontal surface during initial period but it quickly surpasses that of the horizontal surface and low angles of inclination due to the effect of gravity force along the tangential direction.

7. Droplet spread increases with higher wetting fluids with corresponding lower contact angles.

8. Droplet spread also increases with water of higher surfactant concentration due to the reduction in surface tension. Adding SDS surfactant causes a decrease in the intensity of oscillation. Decay rates of such oscillations increase while their frequencies decrease as surfactant concentration increases.

ACKNOWLEDGEMENT

The authors wish to express their appreciation to Mr. C.Y. Tan for his assistance in the experimental work.

REFERENCES

[1] L. C. Chow, M. S. Sehmbey, and M. R. Pais, "High heat flux spray cooling," *Ann. Rev. Heat Transfer*, vol. 8, pp. 291-317, 1997.

[2] T. L. Cox and S. C. Yao "Heat transfer of sprays of large water drops impacting on high temperature surfaces," *Trans. ASME J. Heat Transfer*, vol. 121, pp. 446-450, 1999.

[3] M. Ciofalo, I. D. Piazza, and V. Brucato, "Investigation of the cooling of hot walls by liquid water sprays," *Int. J. Heat Mass Transfer*, vol. 42, pp. 1157-1175, 1999.

[4] S. Chandra and C. T. Avedisian, "On the collision of a droplet with a solid surface," *Proc. Roy. Soc. Lond. A*, vol. 432, pp. 13-41, 1991.

[5] B. Scheler and D. W. Bousfield, "Newtonian drop impact with a solid surface," *AIChE J*, vol. 41, pp. 1357-1367, 1995.

[6] M. R. Pasandideh-Fard, Y. M. Qiao, S. Chandra, and J. Mostaghimi, "Capillary effects during droplet impact on a solid surface," *Phys Fluids*, Vol. 8, pp. 650-659, 1996.

[7] T. Mao, D. C. S. Kuhn, and H. Tran, "Spread and rebound of liquid droplets upon impact on flat surfaces," *AIChE J*, vol. 43, pp. 2169-2179, 1997.

[8] J. Lee, J. Kim, and K. T. Kiger, "Time- and space-resolution heat transfer characteristics of single droplet cooling using microscale heater arrays," *Int. J. Heat Fluid Flow*, vol. 22, pp. 188-200, 2001.

[9] G. Trapaga and J. Szekely, "Mathematical modeling of the isothermal impingement of liquid droplets in spraying processes," *Metall. Trans.*, vol. 22B, pp. 901-914, 1991.

[10] N. Hatta, H. Fujimoto, and H. Takuda, "Deformation process of a water droplet impinging on a solid surface," *Trans. ASME, J. Fluids Eng.*, vol. 117, pp. 394-401, 1995.

[11] Z. Zhao, D. Poulikakos, and J. Fukai, "Heat transfer and fluid dynamics during the collision of a liquid droplet on a substrate – I. Modeling," *Int. J. Heat Mass Transfer*, vol. 39, pp. 2771-2789, 1996.

[12] H. Fujimoto and N. Hatta, "Deformation and rebounding processes of a water droplet impinging on a flat surface above Leidenfrost temperature," *Trans. ASME, J. Fluids Eng.*, vol. 118, pp. 142-149, 1996.

[13] M. Reiber and A. Frohn, "A numerical study on the mechanism of splashing," *Int. J. Heat Fluid Flow*, vol. 20, pp. 455-461, 1999.

[14] M. Bussmann, S. D. Aziz, S. Chandra, and J. Mostaghimi, "Photographs and simulations of molten metal droplets landing on a solid surface," *Trans. ASME J. Heat Transfer*, vol. 122, pp. 422, 2000.

[15] J. Bernardin, C. J. Stebbins, and I. Mudawar, "Mapping of impact and heat transfer regimes of water drops impinging on a polished surface," *Int. J. Heat Mass Transfer*, Vol. 40, pp. 247-267, 1997.

[16] J. Bernardin, C. J. Stebbins, and I. Mudawar, "Effects of surface roughness on water drop impact history and heat transfer regimes," *Int. J. Heat Mass Transfer*, vol. 39, pp. 73-88, 1996.

[17] G. Palasantzas and J. Th. M. deHosson, "Wetting on rough surfaces," *Acta Mater.*, vol. 49, 3533-3538, 2001.

[18] Y. Hardalupas, A. M. K. P. Taylor, and J. H. Wilkins, "Experimental investigation of sub-millimeter droplet impingement onto spherical surfaces," *Int. J. Heat Fluid Flow*, vol. 20, pp. 477-485, 1999.

[19] D. Attinger, Z. Zhao, and D. Poulikakos, "An experimental study of molten microdroplet surface deposition and solidification: Transient behavior and wetting angle dynamics," *Trans. ASME J. Heat Transfer*, vol. 122, pp. 544-556, 2000.

[20] P. G. deGennes, "Wetting: statics and dynamics," *Rev. Modern Phys.*, vol. 57, pp. 827-863, 1985.

[21] G. E. Cossali, A. Coghe, and M. Marengo, "The impact of a single drop on a wetted solid surface," *Exp. Fluids*, vol. 22, pp. 463–472, 1997.

[22] C. Yang and K. C. Leong, "Influences of substrate wettability and liquid viscosity on isothermal spreading of liquid droplets on solid surfaces," *Exp. Fluids*, vol. 33, pp. 728-731, 2002.

[23] X. G. Zhang and O. R. Basaran, "Dynamic surface tension effects on impact of a drop with a solid surface," *J. Colloid Interface Sci.*, vol. 187, pp. 166-178, 1997.

CHAPTER 3

Numerical Simulation of Multiphase Flow and Interface Mass Transfer on Particle Scale

Chao Yang*, Zai-Sha Mao and Jie Chen

Institute of Process Engineering, Chinese Academy of Sciences, Beijing 100190, China

Abstract: More and more attentions have been paid to better understanding of the mechanisms of multiphase flow and interphase mass transfer on the mesoscale (particle scale), namely about solid particles, bubbles and drops. The behavior of particle swarms and effects of surfactant are subsequently studied to approximate real industrial processes. Besides, the sub-particle scale Marangoni effect, induced by interphase mass transfer, is also studied for seeking measures of mass transfer enhancement. The computational fluid dynamics and computational transport principles are being developed into reliable and efficient tools for improving the macroscopic performances of unit operations and process equipments.

Keywords: Particle (drop, bubble or solid particle), multiphase flow, mass transfer, numerical simulation, Marangoni effect, particle swarm.

1. INTRODUCTION

The flow of and mass transfer from/to drops, bubbles and solid particles are often observed in nature and various areas of engineering. Owing to the rapid progress of the computer techniques and numerical methods in fluid mechanics, the application of numerical simulation has recently become increasingly popular in understanding multiphase flow and transport on particle scale. In this chapter, this topic is discussed in detail in the following six sections. Firstly, theoretical basis and numerical methods frequently adopted are summarized. We have improved orthogonal boundary-fitted coordinate system based simulation and level set method to compute the motion and mass transfer of fluid particles (bubbles and drops), and also developed the mirror fluid method for motion of solid particles. Thereafter, the modified cell model is proposed to examining the flow and transport behavior of particle swarms. Then the validity of these methods is well manifested in contrast with the reported experimental data respectively in sections 4 and 5. In consideration of trace quantities of surfactants unavoidable in most industrial extraction equipments, the study on the motion and mass transfer of a solute to/from a single drop with an adsorbed surfactant on the interface is expounded to better understanding of liquid extraction processes and for scientific design of equipments. In the last section, the Marangoni effect is mathematically formulated and numerically simulated to shed light to the mechanisms based on which the interphase interaction may be either promoted or suppressed.

2. THEORETICAL BASIS

The mathematical formulation of two-phase particle flow may be exemplified using two fluid systems in which a liquid drop or a gas bubble moves in another continuous liquid as it follows in this section. So the flow field and solute transport in both fluid phases must be formulated with the first principles of fluid mechanics and transport phenomena. When a solid particle is involved, the flow in the solid domain is usually not necessary and the particle is tracked mechanically as a rigid body. The mass transfer inside a solid particle is sometimes necessary for example in a porous particle, but the macroscopic fluid flow is generally absent. In this context, the two-phase flow with a solid particle is a simplified case of general two-fluid two-phase systems.

2.1. FLUID MECHANICS

2.1.1. Governing Equations

The motion of a particle (drop, bubble or solid particle) under gravity through an immiscible continuous fluid

***Address correspondence to Chao Yang:** Institute of Process Engineering, Chinese Academy of Sciences, Beijing 100190, China; Tel.: +86-10-62554558. Fax: +86-10-62561822; E-mail: chaoyang@home.ipe.ac.cn

Lixin Cheng and Dieter Mewes (Eds)

phase is considered with the following assumptions: (1) the fluid is viscous and incompressible; (2) the physical properties of the fluid and particle are constant; (3) the two-phase flow is axisymmetric or two-dimensional; (4) the flow is laminar. The flow in each fluid phase is governed by the continuity and Navier-Stokes equations:

$$\nabla \cdot \mathbf{u} = 0 \tag{1}$$

$$\rho\left(\frac{\partial \mathbf{u}}{\partial t} + \mathbf{u} \cdot \nabla \mathbf{u}\right) = -\nabla p + \rho \mathbf{g} + \nabla \cdot \boldsymbol{\tau} + \mathbf{S} \tag{2}$$

where $\boldsymbol{\tau}$ is the stress tensor defined as:

$$\boldsymbol{\tau} = \mu(\nabla \mathbf{u} + (\nabla \mathbf{u})^{\mathrm{T}}) \tag{3}$$

and the source term **S** is differently formulated under different conditions.

Boundary conditions for the governing equations are considered with the existence of an interface. For a bubble or a drop, the normal velocity in each phase is equal at the interface. For a solid particle, the normal velocity from the continuous phase cannot pierce into the particle, and the tangential velocity at the solid surface should satisfy the "no-slip" condition.

2.1.2. Forces on Single Particles

Many terminologies are used in describing the interactions between the particles and the surrounding continuous fluid phase, such as drag, lift, virtual mass force, Basset force, turbulent dispersion force, lubrication force, *etc.* (Fig. **1**) [1]. Some of these forces have sound physical basis, but some does not; some forces have been correlated with simple formulae with sufficient accuracy, but some demands further attention. These interphase forces appear in the governing equations as body forces in the Eulerian-Lagrangian model.

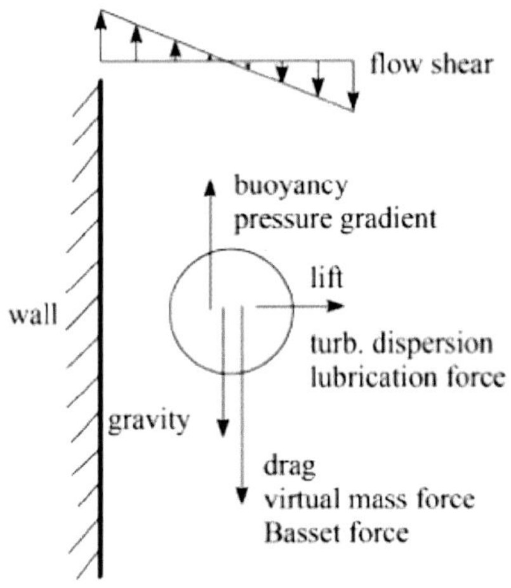

Figure 1: Forces arising from the interaction of a particle with a continuous fluid phase.

2.1.2.1. Drag Force

A moving particle is subjected to drag force from the surrounding fluid, which is usually calculated by:

$$F_{\mathrm{D}} = -\frac{1}{2} C_{\mathrm{D}} \rho_{\mathrm{c}} \left(\frac{\pi d_{\mathrm{p}}^2}{4}\right) |\mathbf{u}_{\mathrm{p}} - \mathbf{u}_{\mathrm{c}}| (\mathbf{u}_{\mathrm{p}} - \mathbf{u}_{\mathrm{c}}) \tag{4}$$

In fact, this equation is also the definition of the drag coefficient, C_{D}, though it applies both to a particle in steady and unsteady motion. However, the term of drag force is generally referred to as that in steady motion. A single solid particle in steady motion is the most thoroughly studied case.

For the case of creeping flow (particle Reynolds number $Re_\mathrm{p} \ll 1$) [2], a spherical particle (including bubble and drop) of axial symmetry, the drag coefficient is analytically derived from the so-called Hadamard-Rybczynski solution of the flow field:

$$C_\mathrm{D} = -\frac{8}{Re_\mathrm{p}}\frac{2+3\lambda}{1+\lambda}, \lambda = \frac{\mu_\mathrm{p}}{\mu_\mathrm{c}} \tag{5}$$

For a solid sphere, $\lambda \to \infty$, $C_\mathrm{D} = 24/Re_\mathrm{p}$; for a bubble, $\lambda \to 0$, $C_\mathrm{D} = 16/Re_\mathrm{p}$; the drag coefficient of a drop lies in between.

When Re_p is greater than 0.1, the assumption of creeping flow is no more valid, no analytical solution is available for the flow field around an axisymmetric particle. The only way is the experimental measurement and the subsequent empirical correlation. For a solid sphere, a simple and popularly used is the Schiller and Naumann correlation [3]:

$$C_\mathrm{D} = \begin{cases} \frac{24}{\mathrm{Re_p}}\left(1 + 0.15\mathrm{Re_p^{0.687}}\right) (\mathrm{Re_p} < 1000) \\ 0.44, (\mathrm{Re_p} \geq 1000) \end{cases} \tag{6}$$

Many works have been devoted to improve it by piece-wise correlation, but they make little difference when used for simulating the solid particles laden engineering multiphase flows, particularly as the time-averaged flow fields are the main concern.

For gas bubbles, it is recognized that the drag coefficient is a function of Reynolds number and bubble shape, and the latter is in turn dependent on Re and other parameters of gas-liquid systems. Thus, a correlation valid under a wide range of regimes has been proposed by Tomiyama [4]:

$$C_\mathrm{D} = \max\left\{\min\left[\frac{24}{Re_\mathrm{p}}\left(1 + 0.15Re_\mathrm{p}^{0.687}\right), \frac{72}{Re_\mathrm{p}}\right], \frac{8}{3}\frac{Eo}{Eo+4}\right\} \tag{7}$$

for slightly contaminated systems, and

$$C_\mathrm{D} = \max\left\{\frac{24}{Re_\mathrm{p}}\left(1 + 0.15Re_\mathrm{p}^{0.687}\right), \frac{8}{3}\left[\frac{Eo}{Eo+4}\right]\right\} \tag{8}$$

for contaminated systems ($10^{-2} < Eo < 10^3$, $10^{-14} < Mo < 10^7$). The situation is the same: the effects of many other parameters on C_D have not been examined to get generally accepted results.

Similarly, the drag exerted on single drops is also dependent on drop shape. When $Re_\mathrm{p} < 600$ and surface tension is relatively large, a drop keeps to be spherical or elliptic. When Re_p in the range from 600 to 900, drop velocity reaches the maximum and C_D goes to the minimum. Further increase of Re_p induces greater change in shape and the drop becomes pulsating, and C_D increases sharply along with Re_p. When $Re_\mathrm{p} > 1000 \sim 3000$, a drop would break up due to the hydrodynamic instability.

Hu and Kintner [5] proposed a correlation based on a large amount of experimental data:

$$C_\mathrm{D}WeE^{0.15} = \begin{cases} \frac{4}{3}\left(\frac{Re_\mathrm{p}}{E^{0.15}} + 0.75\right)^{1.275}, 2 < C_\mathrm{D}WeE^{0.15} < 70 \\ 0.045\left(\frac{Re_\mathrm{p}}{E^{0.15}} + 0.75\right)^{2.37}, C_\mathrm{D}WeE^{0.15} \geq 70 \end{cases} \tag{9}$$

with $E = \rho_\mathrm{c}^2\sigma^3/\mu_\mathrm{c}^4 g\Delta\rho$, and the Weber number $We = u^2 d_\mathrm{p}\rho_\mathrm{c}/\sigma$.

In summary, the knowledge on single spheres in a quiescent fluid is relatively satisfactory, but other situations (turbulence, shear, rotation, oscillation and unsteady motion), particularly for bubbles and drops, have not been understood sufficiently.

2.1.2.2. Unsteady Forces

Added mass force (or virtual mass force) and Basset force appear when a particle is in rectilinear unsteady motion. These two forces are clarified as ones with sound physical meanings. If a particle is accelerating in an ideal fluid, extra force is needed to move the fluid around the particle, and such a force may be calculated by:

$$\boldsymbol{F}_{\mathrm{A}} = -C_{\mathrm{A}} V_{\mathrm{p}} \rho_{\mathrm{c}} \frac{\mathrm{D}\mathbf{u}_{\mathrm{slip}}}{\mathrm{D}t} \quad (10)$$

where $\mathbf{u}_{\mathrm{slip}}$ is the slip velocity of the particle and V_{p} the particle volume. This equation is also the definition of the added mass force coefficient C_{A}. For bubbles with ρ_{b} far less than ρ_{c}, $\boldsymbol{F}_{\mathrm{A}}$ is much greater than the force to accelerate the bubble mass and becomes not negligible.

History force (also called the Basset force) is a force related with the history of acceleration of a particle. As compared to a steadily moving particle, an accelerating one at the same slip velocity has thinner momentum boundary layer, which is equivalent to larger friction at the particle surface. It got such a name for Basset in 1988 got a force expressed by the following integral exerted on a solid particle in sinusoidal oscillation in a creeping flow [6]:

$$\boldsymbol{F}_{\mathrm{H}} = -\frac{3d_{\mathrm{p}}^{2}\mu_{\mathrm{c}}}{2}\sqrt{\frac{\pi}{\nu}}\int_{0}^{t}\frac{\mathrm{d}\mathbf{u}_{\mathrm{p}}/\mathrm{d}\tau}{\sqrt{t-\tau}}\mathrm{d}\tau \quad (11)$$

This force is essentially part of the difference of that to drive an accelerating sphere minus the drag at steady state motion.

Although the extension [7],

$$\boldsymbol{F}_{\mathrm{H}} = -\frac{3d_{\mathrm{p}}^{2}\mu_{\mathrm{c}}}{2}\sqrt{\frac{\pi}{\nu_{\mathrm{c}}}}\left(\frac{\mathbf{u}_{\mathrm{slip,0}}}{\sqrt{t}} + \int_{0}^{t}\frac{\mathrm{d}\mathbf{u}_{\mathrm{p}}/\mathrm{d}\tau}{\sqrt{t-\tau}}\mathrm{d}\tau\right) \quad (12)$$

is often used in simulation of single particle motion, many efforts were devoted to developing the equations of $\boldsymbol{F}_{\mathrm{H}}$ valid for more general cases, such as non-oscillating motion and the particle with finite Re, *etc.*

2.1.2.3. Lift Force

Lift force is referred to that exerted on a particle in the direction perpendicular to the particle motion path. Two mechanisms would create the lateral force: the shear in the flow induces the Saffman force, and the rotation of a particle leads to the Magnus force. Such distinction seems to be of theoretical sense, because a particle will rotate in shear flow so that two forces appear together for freely moving particles. The lift force is crucial for obtaining correct spatial distribution of phase holdups, for example, the radial profile of gas holdup in a bubble column.

Drew and Lahey [8] gave a general expression for the Saffman force in an ideal fluid:

$$\boldsymbol{F}_{\mathrm{L}} = \mathrm{C}_{\mathrm{L}} V_{\mathrm{p}} \rho_{\mathrm{c}} (\mathbf{u}_{\mathrm{p}} - \mathbf{u}_{\mathrm{c}}) \times (\nabla \times \mathbf{u}_{\mathrm{c}}) \quad (13)$$

Many factors influence the lift force coefficient C_{L}, and no general and reliable correlation on C_{L} is available. Moreover, the correlations reported in the literature give wide scattered values of C_{L}. So some authors ignored the lift force as compared with the drag, and sometimes the lift force coefficient is even used as an adjustable parameter to match gas holdup prediction with experimental data.

Besides these forces as already mentioned, some other forces are used to describe the interaction between the dispersed and continuous phases, about which argument exists on their physical soundness. For example, the turbulent dispersion force is used to describe the spread of bubbles due to the effect of flow turbulence. This force is partly physical in the sense that turbulent vortices exert impact onto a particle and push it to move in the fluid. The wall (lubrication) force is conceived as a resistance to the motion of a particle when approaching a solid wall to a certain distance, the flow of a continuous phase fluid in the narrow gap between the particle and solid wall creates a net force away from the wall. Different expressions for lubrication force were developed and tested, but no generally accepted one is available. These two forces sometimes are used to compensate the lateral lift forces so that the gas holdup profiles with either wall peak or core peak can be resulted from numerical simulation to match experimental data. Obviously, the physical mechanisms underlying these forces need further thorough investigation. In general, it is better to model these mechanisms based on their physical essences instead of using analog and imitation, so that accurate constitutive equations can be resulted with less empirical constants and wider applicability.

2.1.3. Drag Coefficient

The rectilinear motion of a particle is described by Newton's second law:

$$m_{\mathrm{p}}\frac{\mathrm{d}u_{\mathrm{p}}}{\mathrm{d}t}=F_{\mathrm{B}}-(F_{\mathrm{D}}+F_{\mathrm{A}}+F_{\mathrm{H}}) \tag{14}$$

where $m_{\mathrm{p}}=\rho_{\mathrm{p}}V_{\mathrm{p}}$ represents the mass of a particle. The forces on the right are buoyancy F_{B}, drag force F_{D}, added mass force F_{A}, and history force F_{H}, respectively.

One method to numerically simulate the motion of a single particle is to correlate every single force with experimental data and compute particle trajectory. Nevertheless, it is difficult even impossible to separate the forces due to the nonlinearity of particle motions. Therefore, the total force usually is correlated by experimental measurements. In different states of particle motion, the mechanisms and magnitudes of forces would be multifarious. Although this method is also not very easy, it is possible to simplify the solution of single particle motion. In studying the motion of single bubbles in a high viscosity liquid, Zhang *et al.* [9] made a successful trial to correlate the total force instead of the individual forces based on the concept of linear splitting of the total force. In their work, the drag force, added mass force, and history force are considered as one resistance, namely the total drag force, and the total drag coefficient C_{DA} is defined as:

$$C_{\mathrm{DA}}=\frac{F_{\mathrm{D}}+F_{\mathrm{A}}+F_{\mathrm{H}}}{\frac{1}{2}\rho_{\mathrm{C}}u_{\mathrm{p}}^{2}(\frac{1}{4}\pi d_{\mathrm{e}}^{2})}=\frac{4}{3}d_{\mathrm{e}}\frac{(\rho_{\mathrm{C}}-\rho_{\mathrm{p}})g-\rho_{\mathrm{p}}\frac{\mathrm{d}u_{\mathrm{p}}}{\mathrm{d}t}}{\rho_{\mathrm{C}}u_{\mathrm{p}}^{2}} \tag{15}$$

when the bubble rises steadily ($\mathrm{d}u_{\mathrm{b}}/\mathrm{d}t=0$), the conventional drag coefficient C_{D} results:

$$C_{\mathrm{D}}=\frac{4}{3}d_{\mathrm{e}}\frac{(\rho_{\mathrm{C}}-\rho_{\mathrm{p}})g}{\rho_{\mathrm{C}}U_{\mathrm{T}}^{2}} \tag{16}$$

Via dimensional analysis, it can be shown that the drag coefficient is a function of Reynolds number and non-Newtonian parameters. The latter depends on the choice of a specific rheological model.

Using the modified Reynolds number Re_{MC}, the correlation also covers accelerating bubbles:

$$C_{\mathrm{DA}}=\frac{16}{Re_{\mathrm{MC}}}\left(1+0.12{Re_{\mathrm{MC}}}^{0.6}\right)\left(1+0.196Ac^{0.767}Ar^{0.381}\right) \tag{17}$$

where the dimensionless group are:

$$Ac=\frac{d_{\mathrm{p}}a}{u_{\mathrm{p}}^{2}},\ Ar=\frac{d_{\mathrm{p}}^{3}\rho_{\mathrm{C}}^{2}g}{\mu^{2}} \tag{18}$$

Instead of splitting the total force exerted on a single particle into drag, unsteady force and lateral lift, the above alternative to correlate the total force which drives a particle move transiently through the continuous phase seems to be successful. Studies on modeling total drag coefficient will be represented particularly in section 4.1.2.

2.2. Mass Transfer

2.2.1. Convective Diffusion Equation

For the case with constant physical properties of both liquid phases, including that on the interface, the solution for mass transfer would be decoupled from the problem of fluid flow. Thus the information of the flow field, required for solution of convective diffusion problem, can be provided directly from the numerical simulation of steady-state fluid flow.

In general, the transient mass transfer to/from a drop (or a bubble) is governed by the convective diffusion equation in the vector form:

$$\frac{\partial c}{\partial t} + \mathrm{u} \cdot \nabla c = D\nabla^2 c \tag{19}$$

subject to two interfacial conditions:

$$D_1 \frac{\partial c_1}{\partial \mathrm{n}_1} = D_2 \frac{\partial c_2}{\partial \mathrm{n}_2} \text{ (flux continuity at the interface)} \tag{20}$$

$$c_2 = mc_1 \text{ (interfacial dissolution equilibtium)} \tag{21}$$

In the above equations, subscript 1 indicates the continuous phase and 2 for the dispersed phase. The solution of Eq. (19) is relied on the resolved fluid flow both in the dispersed and the continuous phase as addressed by Li and Mao [10].

2.2.2. Mass Transfer Coefficient

In accordance with Fick's first law, for steady mass transfer local diffusive flux across the interface may be defined and expressed with the local mass transfer coefficient, k_{loc}, as:

$$N_{\mathrm{loc}} = -D_1 \frac{\partial c_1}{\partial \mathrm{n}_1} = k_{\mathrm{loc}} \left(c_1^{\infty} - \frac{\bar{c}_2}{m} \right) \tag{22}$$

where the remote boundary concentration c_1^{∞}, and the only available measurement of the bubble/drop concentration $\bar{c}_2$ (the average over the whole drop, take drop as a example) is used to define the driving force and mass transfer coefficient. The latter may be expressed in terms of dimensionless concentration gradient as:

$$k_{\mathrm{loc}} = -\frac{D_1}{\left(c_1^{\infty} - \frac{\bar{c}_2}{m} \right)} \frac{\partial c_1}{\partial \mathrm{n}_1} \tag{23}$$

Then, the local Sherwood number is:

$$Sh_{\mathrm{loc}} = \frac{2Rk_{\mathrm{loc}}}{D_1} = -\frac{2R}{\left(c_1^{\infty} - \frac{\bar{c}_2}{m} \right)} \frac{\partial c_1}{\partial \mathrm{n}_1} \tag{24}$$

and the drop area averaged Sh_{od} reads:

$$Sh_{\mathrm{od}} = \frac{\oint Sh_{\mathrm{loc}} \mathrm{d}s}{\oint \mathrm{d}s} \tag{25}$$

On the other hand, the overall mass transfer coefficient k_{od} may be evaluated from the overall solute conservation based on the drop as follows:

$$k_{\mathrm{od}}(c_2^* - \bar{c}_2)S = V_{\mathrm{d}} \frac{\mathrm{d}\bar{c}_2}{\mathrm{d}t} \tag{26}$$

where $\bar{c}_2$ is the average concentration of the drop at any time instant, which is almost the only available measures of solute concentration of drops in conventional experiments. If the time interval $t_{out} - t_{in}$ is chosen small enough, k_{od}, S and V_d may be taken approximately as constants. Then integrate the above equation gives.

$$k_{od} = -\frac{V_d}{S}\frac{1}{t_{out}-t_{in}}\ln\left(\frac{c_2^*-\bar{c}_{2,out}}{c_2^*-\bar{c}_{2,in}}\right) \tag{27}$$

where V_d and S are the volume and surface area of the drop, and for a spherical drop $V_d/S = d/6$. The corresponding Sherwood number is:

$$Sh_{od} = \frac{d}{D}k_{od} \tag{28}$$

Mathematical functions for physical models are the foundation of numerical simulation. Based on those principal theories, two approaches may be adopted for resolving the fluid flow and mass transfer in two fluid phases coupled by the boundary conditions. They are the boundary-fitted orthogonal coordinate frame method and the level set method. In addition to this, mirror fluid method and cell model are developed for solid particle and particle swarms respectively. Specific methods will be expounded in next section. However, no matter whichever numerical approach is used for simulating dispersed two-phase flows, the need for accurate constitutive equations of interphase forces is inevitable, which are applicable to a wide range covering turbulent, accelerating motion subject to breakage and coalescence of fluid particles. Therefore, more systematic efforts in experimental, theoretical and simulation studies are necessary.

3. OVERVIEW OF NUMERICAL METHODS

3.1. Orthogonal Boundary-Fitted Coordinate System (OBFCS) and Coordinate Transformation

For the free-boundary problem of a buoyancy-driven particle, it is beneficial to use an orthogonal Boundary-Fitted Coordinate System (OBFCS) so as to enforce the boundary conditions more accurately at the surface of the deformed particle. In order to ensure that the two coordinate grids, inside and outside the particle, match up exactly at the free surface, the strong constraint method is used for the outer domain and the weak constraint method for the inner domain. For an axisymmetrical flow, it is convenient to use the common cylindrical frame. When the coordinates are nondimensionalized with the volume-equivalent radius, R, of a particle, the two-dimensional sectional plane (x, y) through the symmetry axis can be mapped orthogonally to the computational plain (ξ, η) by the covariant Laplace equations:

$$\begin{cases} \frac{\partial}{\partial\xi}\left(f(\xi,\eta)\frac{\partial x}{\partial\xi}\right) + \frac{\partial}{\partial\eta}\left(\frac{1}{f(\xi,\eta)}\frac{\partial x}{\partial\eta}\right) = 0 \\ \frac{\partial}{\partial\xi}\left(f(\xi,\eta)\frac{\partial y}{\partial\xi}\right) + \frac{\partial}{\partial\eta}\left(\frac{1}{f(\xi,\eta)}\frac{\partial y}{\partial\eta}\right) = 0 \end{cases} \tag{29}$$

where $f(\xi,\eta)$ is called the distortion function defined as $f(\xi,\eta) = h_\eta/h_\xi$, representing the ratio of scale factors of h_η to h_ξ, which are defined by

$$h_\xi = \sqrt{\left(\frac{\partial x}{\partial\xi}\right)^2 + \left(\frac{\partial y}{\partial\xi}\right)^2},\ h_\eta = \sqrt{\left(\frac{\partial x}{\partial\eta}\right)^2 + \left(\frac{\partial y}{\partial\eta}\right)^2} \tag{30}$$

A unit square in the computational plane corresponds to the physical external and internal domains as indicated in Fig. **2** [10]:

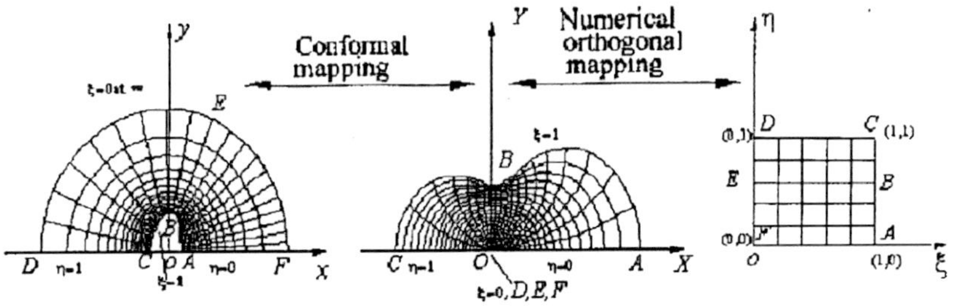

Figure 2(a): The correspondence of external physical domain (x, y) of a particle to the auxiliary domain (X, Y) and the computational domain (ξ, η).

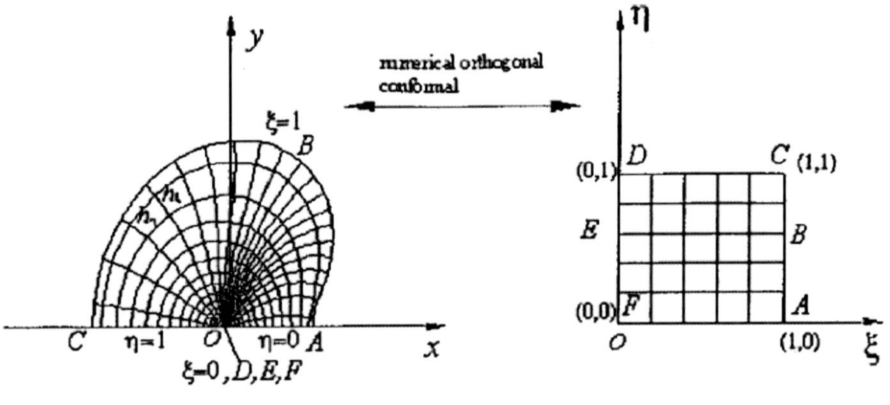

Figure 2(b): The correspondence of internal physical domain (x, y) of a particle to the computational domain (ξ, η).

In choosing the distortion function for implementing orthogonal mapping, Ryskin & Leal [11, 12] and Dandy & Leal [13] chose an infinite region for the exterior phase and specified the distortion function:

$$f_1(\xi_1, \eta_1) = \pi\, \xi_1(1 - 0.5\cos\pi\eta_1) \tag{31}$$

where subscript 1 denotes the outer continuous phase.

Li *et al.* [14] noted that the distortion function becomes vanishingly small at the far infinity and makes it difficult to enforce the remote boundary conditions accurately at $\xi_1 = 0$ (Fig. **2a**). This shortcoming can be overcome by choosing a large enough exterior region around a solid sphere and using a new form of distortion function:

$$f_1(\xi_1, \eta_1) = \frac{\pi}{\epsilon}(1 - 0.5\cos\pi\eta_1) \tag{32}$$

called the finite domain distortion function. Here $\epsilon = 2.5$ is a parameter for specifying the position of the remote boundary, and f is the same order of unity everywhere in the exterior domain. This choice of the distortion function has successfully been applied in a series of our numerical work [14-17].

For the interior phase, the distortion function provided by Ryskin and Leal [12] is used:

$$f_2(\xi_2, \eta_2) = \pi\, \xi_2(1 - 0.5\cos\pi\eta_2) \tag{33}$$

3.1.1. Stream Function-Vorticity Formulation

The motion of a single particle in an infinite fluid medium in a two-dimensional coordinate system may be described by a set of partial differential equations of stream function ψ and vorticity ω in a sectional plane (x, y) passing through the axis of symmetry. The governing equations are:

$$v_1 L_1^2(y_1\omega_1) - \frac{1}{h_{\xi_1}h_{\eta_1}}\left[\frac{\partial\psi_1}{\partial\xi_1}\frac{\partial}{\partial\eta_1}\left(\frac{\omega_1}{y_1}\right) - \frac{\partial\psi_1}{\partial\eta_1}\frac{\partial}{\partial\xi_1}\left(\frac{\omega_1}{y_1}\right)\right] = \frac{\partial\omega_1}{\partial t} \quad (34)$$

$$L_1^2\psi_1 + \omega_1 = 0 \quad (35)$$

$$v_2 L_2^2(y_2\omega_2) + \frac{1}{h_{\xi_2}h_{\eta_2}}\left[\frac{\partial\psi_2}{\partial\xi_2}\frac{\partial}{\partial\eta_2}\left(\frac{\omega_2}{y_2}\right) - \frac{\partial\psi_2}{\partial\eta_2}\frac{\partial}{\partial\xi_1}\left(\frac{\omega_2}{y_2}\right)\right] = \frac{\partial\omega_2}{\partial t} \quad (36)$$

$$L_2^2\psi_2 + \omega_2 = 0 \quad (37)$$

where subscript 1 denotes the outer continuous phase, 2 denotes the particle, and the differential operator is:

$$L^2 = \frac{1}{h_\xi h_\eta}\left[\frac{\partial}{\partial\xi}\left(\frac{f}{y}\frac{\partial}{\partial\xi}\right) + \frac{\partial}{\partial\eta}\left(\frac{1}{fy}\frac{\partial}{\partial\eta}\right)\right]. \quad (38)$$

When being transformed in terms of the following non-dimensional physical variables (Ω and Ψ) defined by:

$$\omega_1 = \frac{U_T}{R}\Omega_1, \omega_2 = \frac{U_T}{R}\Omega_2, \quad (39)$$

$$\psi_1 = R^2U_T\Psi_1, \psi_2 = R^2U_T\Psi_2, \quad (40)$$

the governing equations become:

$$L_1^2(Y_1\Omega_1) - \frac{Re_1}{2}\frac{1}{H_{\xi_1}H_{\eta_1}}\left[\frac{\partial\Psi_1}{\partial\xi_1}\frac{\partial}{\partial\eta_1}\left(\frac{\Omega_1}{Y_1}\right) - \frac{\partial\Psi_1}{\partial\eta_1}\frac{\partial}{\partial\xi_1}\left(\frac{\Omega_1}{Y_1}\right)\right] = \frac{Re_1}{2}\frac{\partial\Omega_1}{\partial\theta} \quad (41)$$

$$L_1^2\Psi_1 + \Omega_1 = 0 \quad (42)$$

$$L_2^2(Y_2\Omega_2) + \frac{Re_2}{2}\frac{1}{H_{\xi_2}H_{\eta_2}}\left[\frac{\partial\Psi_2}{\partial\xi_2}\frac{\partial}{\partial\eta_2}\left(\frac{\Omega_2}{Y_2}\right) - \frac{\partial\Psi_2}{\partial\eta_2}\frac{\partial}{\partial\xi_1}\left(\frac{\Omega_2}{Y_2}\right)\right] = \frac{\partial\Omega_2}{\partial\theta} \quad (43)$$

$$L_2^2\Psi_2 + \Omega_2 = 0 \quad (44)$$

$$L^2 = \frac{1}{H_\xi H_\eta}\left[\frac{\partial}{\partial\xi}\left(\frac{f}{Y}\frac{\partial}{\partial\xi}\right) + \frac{\partial}{\partial\eta}\left(\frac{1}{fY}\frac{\partial}{\partial\eta}\right)\right] \quad (45)$$

where $\theta = tU_T/R$ and the relevant dimensionless parameters are:

$$Re_1 = \frac{2RU_T\rho_1}{\mu_1}, Re_2 = \frac{\zeta}{\lambda}Re_1, \lambda = \frac{\mu_2}{\mu_1}, \zeta = \frac{\rho_2}{\rho_1} \quad (46)$$

The physical velocity components are related to the stream function by

External domain (left-handed frame) $U_{\xi_1} = -\frac{1}{Y_1H_{\eta_1}}\frac{\partial\Psi_1}{\partial\eta_1}, U_{\eta_1} = \frac{1}{Y_1H_{\xi_1}}\frac{\partial\Psi_1}{\partial\xi_1}$ (47)

Internal domain (right-handed frame) $U_{\xi_2} = \frac{1}{Y_2H_{\eta_2}}\frac{\partial\Psi_2}{\partial\eta_2}, U_{\eta_2} = -\frac{1}{Y_2H_{\xi_2}}\frac{\partial\Psi_2}{\partial\xi_2}$ (48)

3.1.2. Convective Transport Equation

For an axisymmetric drop moving in an infinite immiscible stagnant liquid, the motion and mass transfer may be formulated in a two-dimensional orthogonal curvilinear coordinate system. The external and internal domains of a drop in the physical plane (x, y) may be transformed respectively into a unit square in the computational plane (ξ, η), as shown in Figs. **1** and **2**. This task of grid generation has been resolved by Ryskin and Leal [11]. In our case, the expanded form of Eq. (19) in such an orthogonal grid reads

$$\frac{\partial c}{\partial t}+\frac{u_\xi}{h_\xi}\frac{\partial c}{\partial \xi}+\frac{u_\eta}{h_\eta}\frac{\partial c}{\partial \eta}=\frac{D}{h_\xi h_\eta y}\left[\frac{\partial}{\partial \xi}\left(\frac{h_\eta y}{h_\xi}\frac{\partial c}{\partial \xi}\right)+\frac{\partial}{\partial \eta}\left(\frac{h_\xi y}{h_\eta}\frac{\partial c}{\partial \eta}\right)\right] \quad (49)$$

In the present numerical procedure, the governing equations are non-dimensionalized and the following non-dimensional variables and groups are defined for this purpose:

$$\boldsymbol{C_1}=\frac{c_1}{c_1^{\infty}}, \boldsymbol{\Psi_1}=\frac{\psi_1}{R^2U}, \boldsymbol{H_{\xi_1}}=\frac{h_{\xi_1}}{R}, \boldsymbol{H_{\eta_1}}=\frac{h_{\eta_1}}{R}, \boldsymbol{Y_1}=\frac{y_1}{R}, \boldsymbol{Pe_1}=\frac{2RU}{D_1} \quad (50)$$

Among them Pe is the Peclet number, symbolizing the relative strength of convection to molecular diffusion. With subscripts 1 inserted into Eq. (49) to index the external domain and Eq. (47) incorporated, the non-dimensional convective diffusion equation for the external region is:

$$\left(\frac{Pe_1}{2}H_{\xi_1}H_{\eta_1}Y_1\right)\frac{\partial C_1}{\partial \theta}+\frac{Pe_1}{2}\left[\frac{\partial}{\partial \xi_1}\left(-\frac{\partial \Psi_1}{\partial \eta_1}C_1\right)+\frac{\partial}{\partial \eta_1}\left(\frac{\partial \Psi_1}{\partial \xi_1}C_1\right)\right]=\left[\frac{\partial}{\partial \xi_1}\left(f_1 Y_1\frac{\partial C_1}{\partial \xi_1}\right)+\frac{\partial}{\partial \eta_1}\left(\frac{Y_1}{f_1}\frac{\partial C_1}{\partial \eta_1}\right)\right] \quad (51)$$

The governing equation for the interior of the drop is denoted with subscript 1 replaced by 2 in the above formulation.

It is difficult to construct the orthogonal curvilinear coordinates for complicated and seriously deformed interfaces and the orthogonal boundary-fitted coordinate system is usually used to calculate a buoyancy-driven drop with slightly deformed interface at low Reynolds numbers. In order to calculate a seriously deformed interface at high Reynolds numbers, an improved level set method is proposed by our series works [18-20].

3.1.3. Numerical Solution Procedure

The numerical simulation proceeds in the following steps: (1) Numerical simulation of steady fluid flow inside and outside the particle is carried out by solving Eqs. (41)-(44) with pertinent boundary conditions. This can be either done by numerical solution of steady Navier-Stokes equations with necessary under-relaxation, or by solution of time-dependent Navier-Stokes equations for a long enough time so that the flow approaches the steady state. (2) Utilizing the solved velocity field (or the stream function) as the starting initial flow conditions, simultaneous solution of transient mass transfer of a solute into the particle and the fluid flow under influence of the changing surface tension due to non-uniform concentration profile on the particle surface. This stage of simulation is carried out in the real-time domain. Exact solution of the flow and concentration fields must be achieved for each time step, so that the correct mass transfer Sh_{oc} can be evaluated by Eq. (25). In each time step, it is necessary to solve the stream function Ψ, the vorticity Ω and the concentration of solvent C and to refresh the surface parameters involved in boundary conditions in a few iterations so that these fields become compatible with one another at the end of a time step.

3.2. Level Set Method

Since Osher and Sethian [21] introduced the level set method for modeling fronts, this method has been applied in many fields with the particular advantages in easy and accurate capture of interface in a fixed Euler grid. By this method, the topology of the interface and its change are fully described by the zero set of the level set function, which is initially defined in the whole flow field and then advanced along with the fluid flow. The method is stable and surface geometric parameters such as curvature become easy to be calculated, and three-dimensional calculation problems also become easy to be implemented. The major shortcoming of the level set approach is less accurate treatment of interface compared with the boundary-fitted coordinate method.

In our study, numerical simulation of a single drop in a liquid-liquid extraction system is performed in an axisymmetric cylindrical reference frame based on the drop. The motion of a drop with a finite degree of deformation is coupled with simultaneous mass transfer for considering the effect of unsteady-state effects.

3.2.1. Level Set Method for Fluid Flow

In a two-dimensional coordinate system, mass and momentum conservation with the level set approach incorporated [18] are written in terms of dimensionless variables as:

$$\frac{\partial u'}{\partial X} + \frac{1}{r}\frac{\partial}{\partial Y}(rv') = 0 \tag{52}$$

$$\frac{\partial}{\partial \theta}(\rho u') + \frac{\partial}{\partial X}\left(\rho u'u' - \frac{\mu}{Re}\frac{\partial u'}{\partial X}\right) + \frac{1}{r}\frac{\partial}{\partial Y}\left(r\rho v'u' - r\frac{\mu}{Re}\frac{\partial u'}{\partial Y}\right) = -\frac{\partial p'}{\partial X} + \frac{1}{Fr}\rho g_X - \frac{1}{We}\kappa(\phi)\delta_\varepsilon(\phi)\frac{\partial \phi}{\partial X} + \frac{1}{Re}\frac{\partial}{\partial X}\left(\mu\frac{\partial u'}{\partial X}\right) + \frac{1}{Re}\frac{1}{r}\frac{\partial}{\partial Y}\left(r\mu\frac{\partial v'}{\partial X}\right) \tag{53}$$

$$\frac{\partial}{\partial \theta}(\rho v') + \frac{\partial}{\partial X}\left(\rho u'v' - \frac{\mu}{Re}\frac{\partial v'}{\partial X}\right) + \frac{1}{r}\frac{\partial}{\partial Y}\left(r\rho v'v' - r\frac{\mu}{Re}\frac{\partial v'}{\partial Y}\right) = -\frac{\partial p'}{\partial Y} + \frac{1}{Fr}\rho g_Y - \frac{1}{We}\kappa(\phi)\delta_\varepsilon(\phi)\frac{\partial \phi}{\partial Y} + \frac{1}{Re}\frac{\partial}{\partial X}\left(\mu\frac{\partial u'}{\partial Y}\right) + \frac{1}{Re}\frac{1}{r}\frac{\partial}{\partial X}\left(r\mu\frac{\partial v'}{\partial Y}\right) - \left\{\frac{2}{Re}\frac{\mu v'}{r^2}\right\} \tag{54}$$

where $r \equiv 1$ for Cartesian coordinates, $r \equiv Y$ for cylindrical coordinates and curly brackets indicate the term presents only in cylindrical coordinates. The dimensionless groups Re, Fr and We are Reynolds, Froude and Weber numbers, respectively,

$$Re \equiv \frac{\rho LV}{\mu}, Fr \equiv \frac{V^2}{Lg}, We = \frac{L\rho_1 V^2}{\sigma} \tag{55}$$

where the characteristic length L is chosen as the initial diameter of a spherical particle (d) and the corresponding reference velocity as $V = \sqrt{dg}$.

The level set function ϕ is introduced into the formulation of multiphase flow and mass transfer systems to define and capture the interface between two fluids. The interface is defined as the zero set of the level set function ϕ, which is defined as the signed algebraic distance of a node to the interface, being positive in the continuous fluid phase and negative in the drop.

The following equation is used to advance the level set function exactly as the drop moves:

$$\frac{\partial \phi}{\partial \theta} + \frac{\partial}{\partial X}(u'\phi) + \frac{1}{r}\frac{\partial}{\partial Y}(rv'\phi) = 0 \tag{56}$$

$\kappa(\phi)$ is the curvature of drop surface defined as:

$$\kappa(\phi) = -\nabla \cdot \mathbf{n} = -\nabla \cdot \left(\frac{\nabla \phi}{|\nabla \phi|}\right) \tag{57}$$

where $\mathbf{n}$ is the unit vector normal to the interface pointing towards the continuous phase and $\delta_\varepsilon(\phi)$ is the regularized delta function defined as:

$$\delta_\varepsilon(\phi) = \begin{cases} \frac{1}{2\varepsilon}\left(1 + \cos\left(\frac{\pi\phi}{\varepsilon}\right)\right) \text{ if } |\phi| < \varepsilon \\ 0 \text{ otherwise} \end{cases} \tag{58}$$

where ε prescribes the finite "half thickness" of the interface. Usually we take $\varepsilon = 1.5\Delta x$, where Δx is the dimensionless uniform mesh size near the interface. $H_\varepsilon(\phi)$ is the regularized Heaviside function expressed as:

$$H_\varepsilon(\phi) = \begin{cases} 0 \text{ if } \phi < -\varepsilon \\ \frac{1}{2}\left(1 + \frac{\phi}{\varepsilon} + \frac{\sin\left(\frac{\pi\phi}{\varepsilon}\right)}{\pi}\right) \text{ if } |\phi| \leq \varepsilon \\ 1 \text{ if } \phi > \varepsilon \end{cases} \tag{59}$$

for defining the corresponding regularized (smoothed) density function ρ and the regularized viscosity μ as:

$$\rho_\varepsilon(\phi) = {}^{\rho_2}/_{\rho_1} + (1 - {}^{\rho_2}/_{\rho_1})H_\varepsilon(\phi) \tag{60}$$

$$\mu_\varepsilon(\phi) = {}^{\mu_2}/_{\mu_1} + (1 - {}^{\mu_2}/_{\mu_1})H_\varepsilon(\phi) \tag{61}$$

The subscripts 1 and 2 denote the continuous phase and drop respectively.

Generally, ϕ will no longer be a distance function (*i.e.*,$|\nabla\phi| \neq 1$) after some iterations, even if Eq. (56) moves the interface ($\phi = 0$) at the correct velocity. Maintaining ϕ as a distance function is essential for providing the interface with an invariant width and a sound basis for estimating the surface curvature. This problem can be solved by adopting a reinitialization method proposed by Sussman *et al.* [22] to solve the following initial problem to steady state in a virtual time domain:

$$\frac{\partial\phi}{\partial\tau} = \mathrm{sgn}(\phi_0)(1 - |\nabla\phi|) \tag{62}$$

$$\phi(x,0) = \phi_0(x) \tag{63}$$

where τ is the virtual time for reinitialization, $\phi_0(x)$ the level set function at any computational instant, and $\mathrm{sgn}(\phi_0)$ the sign function needed for enforcing $\nabla\phi = 1$. Eq. (62) has the property that ϕ remains unchanged at the interface, therefore, the zero level set of ϕ_0 and ϕ is the same. Away from the interface ϕ will converge to $\nabla\phi = 1$, *i.e.*, the actual distance function.

The following perturbed Hamilton-Jacobi equation was introduced by Chang *et al.*[23, 24] to guarantee mass conservation:

$$\frac{\partial\phi}{\partial\tau} + \big(A_0 - A(\tau)\big)(-Q + \kappa(\phi))|\nabla\phi| = 0 \tag{64}$$

where A_0 denotes the initial total mass of both fluids at $\tau = 0$ and $A(\tau)$ the total mass corresponding to the level set function $\phi(\tau)$. The parameter Q is a positive constant and is usually set to be 1.

Yang and Mao [18] found that a considerable amount of mass in the fluid particle vanished gradually although an excellent conservation of total mass with time was observed with the reinitialization by Eq. (64). Therefore, the definition of $A(\tau)$ was modified to

$$A(\tau) = \sum_{\phi_{ij}\le\varepsilon} \rho_\varepsilon(\phi_{ij})R_j\Delta x\Delta y \tag{65}$$

where $\phi_{ij} \le \varepsilon$ denotes the nodes in the fluid particle and the interface has a virtual thickness of 2ε, *i.e.*, $A(\tau)$ in Eq. (65) is taken as the mass of drop/bubble instead of the total mass as in Eq. (64). The improved reinitialization procedure can maintain the level set function as a distance function and guarantee the drop mass conservation.

Since the fluids are incompressible, mass conservation is equivalent to volume conservation. Furthermore, a volume-amending method [20] was proposed, in which we reinitialize the level set function ϕ to an exact signed distance function by solving Eq. (62) firstly and then the position of the interface is corrected according to the loss or gain of the particle (bubble/drop) volume ($V = \sum_{\phi\le 0} H(\phi)R\Delta x\Delta y$):

$$\Delta V = \frac{V(t) - V_0}{V_0} \tag{66}$$

Assuming the bubble/drop is spherical, the increment of the particle radius may be expressed as:

$$\Delta R = R - R_0 = ((1 + \Delta V)^{1/\alpha} - 1)R_0 \tag{67}$$

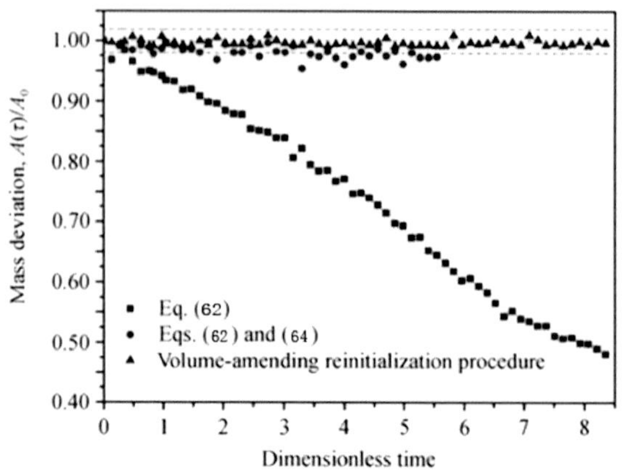

Figure 3: Volume deviation with time during the numerical simulation for the interactive behavior of two bubbles.

where α is the dimension of the system simulated: $\alpha = 2$ for an axisymmetric fluid particle and $\alpha = 3$ for a three-dimensional case. Since the volume increase of the particle corresponds to that of the distance from the interface to the particle center, the correction to the level set function ϕ may be taken in proportion to ΔR. Thus, a relationship is established between the increment of the level set function ϕ and the loss or gain of the particle (bubble/drop) volume, so that the correction to the level set function ϕ may be expressed as:

$$\boldsymbol{\delta\phi = \beta((1+\Delta V)^{1/\alpha} - 1)} \tag{68}$$

where β is an empirical coefficient, large β may cause the reinitialization divergent, and a small one leads to low computational efficiency. β is usually chosen between 0.01-0.1 by experience. Thus, the volume-amending equation can be written as:

$$\boldsymbol{\phi = \phi_0 + \delta\phi = \phi_0 + \beta((1+\Delta V)^{1/\alpha} - 1)} \tag{69}$$

After volume-amending reinitialization procedure, mass conservation has been well satisfied, as shown in Fig. **3** [20].

3.2.2. Level Set Method for Mass Transfer

In the following, the "one-fluid" formulation coupled with the level set function for interphase mass transfer is derived. When the distribution coefficient of a solute, m, is unity, the solute concentration across the interface is continuous. The only difficulty is the discontinuity of molecular diffusion coefficients, which can easily be handled by the level set function with the same smoothed Heaviside function as used in Eqs. (60) and (61). Thus, Eq. (19) for two phases may be expediently solved in a single domain by the level set method similar to the solution of multiphase flow.

However, more general case is that m is not equal to unity. In this case some measures must be taken to make the concentration field become continuous across the interface, just as the continuity of fluid velocity at the interface. For this purpose, the concentration transformation method, such as $\hat{c}_1 = c_1\sqrt{m}$ and $\hat{c}_2 = c_2/\sqrt{m}$, is proposed [19]. Condition (21) is turned into $\hat{c}_2 = \hat{c}_1$ at the interface, and $\hat{c}_1$ and $\hat{c}_2$ become continuous in the whole domain. Using these definitions, the transformed Eq. (19) remains in the following form:

$$\boldsymbol{\frac{\partial \hat{c}_i}{\partial t} + \mathbf{u}\cdot\nabla\hat{c}_i = D_i\nabla^2\hat{c}_i, i = 1, 2} \tag{70}$$

where $\hat{c}_i$ denotes $\hat{c}_1$ or $\hat{c}_2$, and D_i represents the corresponding D_1 or D_2.

The difficulty that remains is only condition (20). When the transformations are utilized, Eq. (20) becomes:

$$\frac{D_1}{\sqrt{m}}\frac{\partial \hat{c}_1}{\partial n_1} = \sqrt{m}D_2\frac{\partial \hat{c}_2}{\partial n_2} \tag{71}$$

Therefore, at the interface, which is of finite thickness, the diffusion coefficients in both phases should be locally replaced by $D_1/\sqrt{m}$ and $\sqrt{m}D_2$ to satisfy the original condition of mass flux continuity. This would make the diffusivity in the interface region different from that in the bulk domain and may result in unacceptable errors. In our studies, a simple transformation is applied to make the diffusivity being equal throughout a fluid phase. For the continuous and dispersed phases, Eq. (70) can be rewritten separately as:

$$\frac{\partial \hat{c}_1}{\partial(\sqrt{m}t)} + \frac{1}{\sqrt{m}}\mathbf{u}\cdot\nabla\hat{c}_1 = \frac{1}{\sqrt{m}}D_1\nabla^2\hat{c}_1 \tag{72}$$

$$\frac{\partial \hat{c}_2}{\partial(\frac{1}{\sqrt{m}}t)} + \sqrt{m}\mathbf{u}\cdot\nabla\hat{c}_2 = \sqrt{m}D_2\nabla^2\hat{c}_2 \tag{73}$$

and then can be put into a common equation over the whole domain in the form analogous to the momentum equation:

$$\frac{\partial \hat{c}}{\partial \hat{t}} + \hat{\mathbf{u}}\cdot\nabla\hat{c} = \nabla\cdot(\hat{D}\nabla\hat{c}) \tag{74}$$

through different definitions of $\hat{t}$, $\hat{D}$ and $\hat{\mathbf{u}}$ by the regularized Heaviside function $H_\varepsilon(\phi)$:

$$\hat{t}(\phi) = \begin{cases}\sqrt{m}t, \text{if } \phi \geq 0\\ \frac{1}{\sqrt{m}}t, \text{if } \phi < 0\end{cases} \tag{75}$$

$$\hat{D}(\phi) = \sqrt{m}D_2 + (\frac{1}{\sqrt{m}}D_1 - \sqrt{m}D_2)H_\varepsilon(\phi) \tag{76}$$

$$\hat{\mathbf{u}}_\varepsilon(\phi) = \sqrt{m}\mathbf{u} + (\frac{1}{\sqrt{m}}\mathbf{u} - \sqrt{m}\mathbf{u})H_\varepsilon(\phi) \tag{77}$$

where $\hat{\mathbf{u}}$ should be the velocity field in a frame of reference moving with the drop.

The governing equations for interphase mass transfer are nondimensionalized and the expanded expressions in a two- dimensional axisymmetric coordinate system are:

$$\frac{\partial C}{\partial \theta} + u'\frac{\partial C}{\partial X} + v'\frac{\partial C}{\partial Y} = \frac{1}{Pe_1}\left[\frac{\partial}{\partial X}\left(D\frac{\partial C}{\partial X}\right) + \frac{1}{Y}\frac{\partial}{\partial Y}\left(rD\frac{\partial C}{\partial Y}\right)\right] \tag{78}$$

where C is the dimensionless concentration based on the reference concentration, using c_1^∞ for solute transportation from the continuous phase to the drop and the dimensionless group Pe_1 is the Peclet number denoted as:

$$Pe_1 \equiv \frac{LV}{D_1} \tag{79}$$

where D_1 is chosen as the characteristic molecular diffusivity.

Besides, the other transformation forms of concentration, such as $\hat{c}_1 = mc_1$ and $\hat{c}_2 = c_2$, or $\hat{c}_1 = c_1$ and $\hat{c}_2 = c_2/m$ have also been tested. The overall interphase mass transfer can be solved in the similar way, and completely identical results were obtained. Therefore, the transformation form has no effect on the numerical solutions of interphase mass transfer.

3.2.3. Numerical Solution Procedure

The main calculation steps of level set algorithm are outlined as follows: (1) initialize the flow field (**u** and p) with the quiescent drop/bubble, physical parameters (density, viscosity, surface tension and molecular diffusivity of solute in each phase), concentration, and ϕ as the signed normal distance to the interface, (2) compute the velocity and pressure to steady state and update the level set function by the SIMPLEC algorithm, (3) compute the unsteady concentration field over the whole domain with the steady state flow field, and calculate the corresponding overall mass transfer coefficients and Sherwood numbers.

3.3. Mirror Fluid Method

The mirror fluid method [25] was proposed for simulating solid-fluid two-phase flow. The mirror fluid method has been implemented to compute the motion of a rigid spherical or elliptic particle in a Newtonian fluid for the purpose of method validation.

The basic idea of the mirror fluid method is to take the whole domain as a Eulerian one for the fluid with a Lagrangian subdomain embedded; *i.e.*, a solid-fluid flow problem with geometric complexity is resolved on a simple domain, including a mirror fluid domain which is originally occupied by the solid. In this aspect, the mirror fluid method and the fictitious-domain method have the same advantages that a fixed and regular mesh can be applied for the entire computation without repeated remeshing or using boundary-fitted coordinates, and a simple finite-difference method can be used in the mirror fluid method instead of the finite-element method and a complicated weak formulation [26, 27]. The no-slip boundary condition is enforced implicitly on solid-fluid surface segments by mirror relations. Therefore, to make a surface segment subjected to a zero net force, the key is to guarantee the shear rates across the surface having the same magnitude but with opposite direction. This may be implemented by taking the inside flow in the subdomain occupied by the solid particle (*i.e.*, the mirror fluid domain) as the flipped mirror image of the outside flow in the real fluid phase at the same surface segment or, in other words, by rotating the outside flow field and pressure field by 180° around the surface segment (as shown in Fig. **4** [25]).

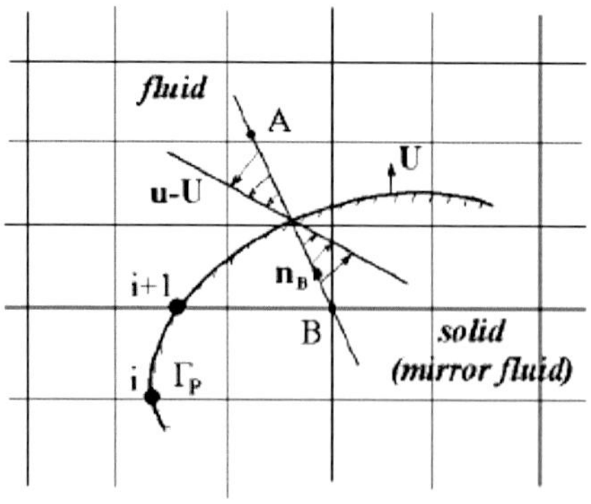

Figure 4. Schematic diagram of the solid-fluid interface for mirror fluid method.

In the mirror fluid method, the flow parameters (**u** and p) are defined by the mirror relations. Thus, the key step of the mirror fluid method is to accurately specify the flow parameters inside the mirror fluid domain, in which the variable value at each node corresponds to that of a point in real fluid by the mirror relations. The following formulations are derived only in two-dimensional Cartesian or cylindrical coordinates for a single solid particle moving in a fluid as a typical example, while this is easy to be extended to a three-dimensional space and straightforward to multiple-particle-fluid systems.

As depicted in Fig. **4**, we can find the mirror location of $\mathrm{A}(x_\mathrm{A}, y_\mathrm{A})$ in the real fluid region, corresponding to node $\mathrm{B}(x_\mathrm{B}, y_\mathrm{B})$ in the mirror fluid domain in terms of a distance function defined as in the level set approach [18, 22]. The signed algebraic distance function denoted as ϕ, being positive in the continuous fluid phase, negative in the solid phase, and zero at the solid-fluid interface, to facilitate the mirror calculations. After the particle is advanced by a time step, ϕ can be redesignated with the known center position of the particle. The unit normal vector to the interface is denoted as:

$$\mathbf{n} = \frac{\nabla\phi}{|\nabla\phi|} \tag{80}$$

In two-dimensional Cartesian or cylindrical coordinates, the above equation is expanded as:

$$\mathbf{n} = \begin{pmatrix} n_x \\ n_y \end{pmatrix} = \begin{pmatrix} \dfrac{\partial\phi/\partial x}{\sqrt{(\partial\phi/\partial x)^2 + (\partial\phi/\partial y)^2}} \\ \dfrac{\partial\phi/\partial y}{\sqrt{(\partial\phi/\partial x)^2 + (\partial\phi/\partial y)^2}} \end{pmatrix} \tag{81}$$

and can be calculated at the nodes immediately close to the solid-fluid interface. So the unit normal vector passing through B to the interface is thus calculated:

$$\mathbf{n}_B = \begin{pmatrix} n_{xB} \\ n_{yB} \end{pmatrix} = \left(\frac{\nabla\phi}{|\nabla\phi|}\right)_B \tag{82}$$

In Fig. **4**, the straight line along $\mathbf{n}_B$ passing through B is

$$\frac{x - x_B}{n_{xB}} = \frac{y - y_B}{n_{yB}} \tag{83}$$

So the coordinates of A mirrored with B can be decided by solving the following equation set of Eqs. (84) and (85) coupled with the constraint in Eq. (86):

$$\frac{x_A - x_B}{n_{xB}} = \frac{y_A - y_B}{n_{yB}} \tag{84}$$

$$(x_A - x_B)^2 + (y_A - y_B)^2 = (2\phi_B)^2 \tag{85}$$

$$\phi_A\phi_B \leq 0 \text{ (if } \phi_B \leq 0, \text{then } \phi_B \geq 0) \tag{86}$$

As shown in Fig. **5** [25], the points denoted by solid circle symbols (A) are one-to-one mirror images of the nodes in a solid particle denoted by open circle symbols (B). Then the fictitious velocity and pressure of node B in the mirror fluid are obtained easily:

$$\mathbf{u}_B = -(\mathbf{u}_A - \mathbf{U}) + \mathbf{U} = 2\mathbf{U} - \mathbf{u}_A \tag{87}$$

$$p_B = p_A \tag{88}$$

Be aware that for a rotating solid particle, the velocity at the particle-fluid interface (*i.e.*, $\mathbf{U}$) should be replaced by $\mathbf{U} + \omega \times (\mathbf{x} - \mathbf{x}_p)$. Such specification ensures that the shear and normal stresses on the two sides of the solid-fluid interface are with the same magnitude but opposite direction. The density and viscosity of the mirror fluid are designated as simply equal to those of the real fluid.

By the mirror fluid method, the main calculation steps of solid-fluid flows are summarized as follows: (1) Initialize the flow field ($\mathbf{u}$, $\mathbf{U}$, p) , $\mathbf{a}$, ω, physical parameters (density and viscosity), and ϕ as the signed normal distance to the interface. (2) Assume the velocity of the solid particle be $\mathbf{U}$ at the end of a time increment Δt: $\mathbf{U} = \mathbf{U}_0 + \mathbf{a}_0\Delta t$, where $\mathbf{U}_0$ and $\mathbf{a}_0$ are the velocity and acceleration of the solid particle at the end of the former time step. (3) Calculate the location (A) in the real fluid mirrored with any node (B) in the subdomain occupied by the solid particle through solving the equation set of Eqs. (84) and (85). (4) Solve the mass and momentum conservation equations of fluid in the whole domain by the SIMPLE algorithm with the interface boundary condition implicitly enforced by the mirror fluid method. (5) Calculate the total drag force exerted by the liquid on the particle and the body force by the external field. Then calculate the acceleration **a** by solving Newton's second law. (6) Estimate the particle velocity $\tilde{\mathbf{U}}$ at the end of current time step according to the new acceleration value: $\tilde{\mathbf{U}} = \mathbf{U}_0 + \frac{1}{2}(\mathbf{a}_0 + \mathbf{a})\Delta t$. (7) Compare $\tilde{\mathbf{U}}$ and **U**. If $\tilde{\mathbf{U}} = \mathbf{U}$ (meaning $\mathbf{a}_0 = \mathbf{a}$), and go on to the next time step of the simulation. If not equal, go back to step (2) with suitably adjusted $\mathbf{a}_0$. (8) Move the particle center to a new position with the converged surface velocities and respecify ϕ of fluid and solid particle. Keep the **u** and p values for nodes in the fluid and discard these in the particle according to the renewed signed distance function ϕ. (9) Let $\mathbf{U}_0 = \mathbf{U}$, $\mathbf{u}_0 = \mathbf{u}$, $\mathbf{a}_0 = \mathbf{a}$, and $\omega_0 = \omega$, and repeat steps 2–8 for the next time step.

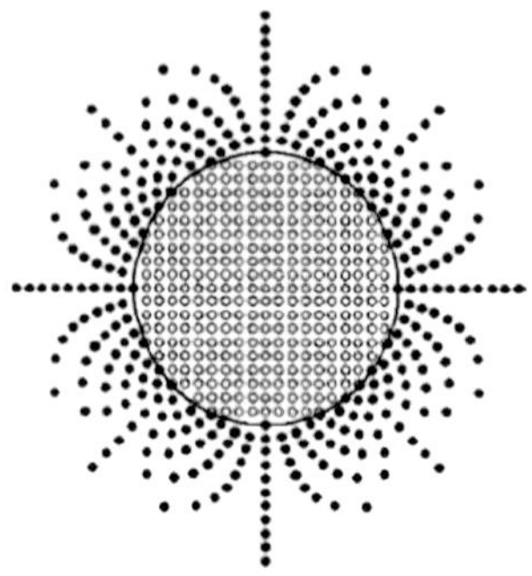

Figure 5. Sketch map of a typical mirror relation between the nodes in a solid particle and the corresponding mirrored nodes in a real fluid.

3.4. Cell Model

The cell model developed since 1950s is a useful tool for exploring the behavior of particle assemblages. The multiphase dispersion is visualized as an assemblage of identical unit cells, in which a typical particle is located at the cell center and surrounded by its share of the continuous phase (Fig. **6**). Thus, only one cell is necessarily picked up and studied as the representative of particles in an assemblage. The cell model was first proposed for creeping flow through porous medium as encountered in studies of particle sedimentation, permeability of porous medium and suspension viscosity [28]. The Darcy law of porous medium may be interpreted by this unit cell technique. Happel [29] first studied the fluid flow through multi-particle systems by the cell model. It is reasonably successful in predicting the pressure drop at low particle Reynolds numbers in packed and fluidized beds. Since the importance of permeating flow for petroleum production, the cell model has been extended to examine the flow behavior of the non-Newtonian fluid in porous medium.

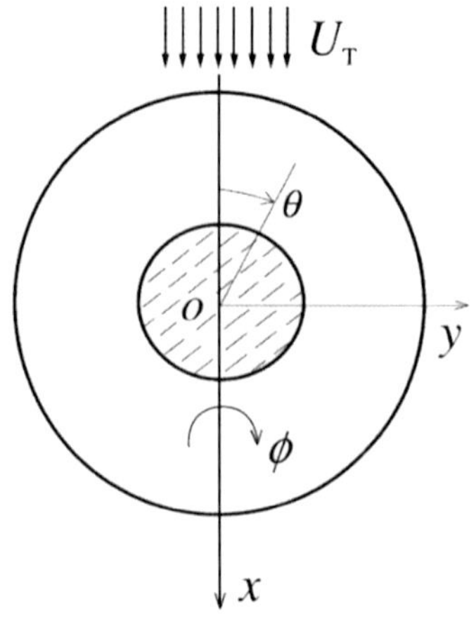

Figure 6: Sketch of a spherical unit cell with the reference coordinate system.

3.4.1. Formulation

In a typical cell, a spherical particle is fixed at the center of the cell as depicted in Fig. 7. To resolve the fluid flow relative to the center particle, the Navier–Stokes and continuity equations are solved in an axisymmetric coordinate system. The axisymmetric assumption facilitates recasting the Navier–Stokes equations in the stream function-vorticity form. Therefore, the flow in the cell may be described by a set of partial differential equations in terms of stream function ψ and vorticity ω in a sectional plane (x, y) passing through the axis of symmetry as indicated in Fig. **6** for a solid particle cell model. The nondimensional governing equations are listed as Eqs. (41) and (42) [12,15]. If bubble or drop swarms are dealt with, the fluid flow inside the center fluid sphere must also be considered by solving Eqs. (43) and (44) in a couple way.

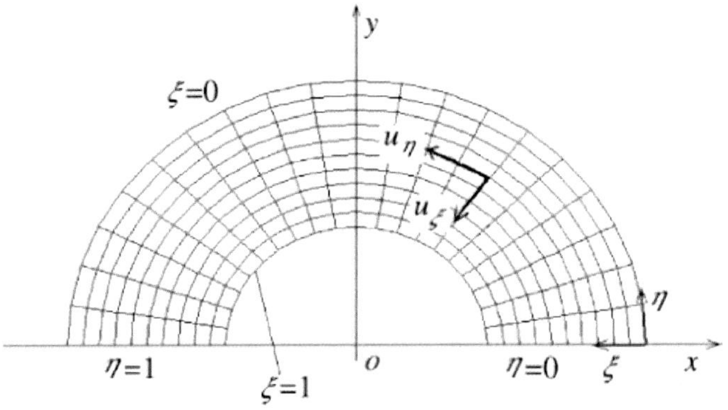

Figure 7. Grid for numerical solution, physical reference frame and computational coordinate system for an assemblage of solid spheres.

To resolve the external mass transfer to the particle, the convective diffusion equation is solved in an axisymmetric orthogonal coordinate system based on the information of fluid flow from the numerical solution already obtained. For the mathematical formulation of viscous flow in such an assemblage, it is customary to adopt the following assumptions: (1) the mass transfer is at steady-state and (2) the physical properties are not altered by mass transfer. These assumptions decouple the mass transfer and fluid flow problems. The non-dimensional governing equation of mass transfer is expanded in the computational plane (ξ, η) as Eq. (51). Meanwhile, the mass transfer coefficient can be expressed with Eqs. (27) and (28).

3.4.2. Outer Boundary Conditions

Specification of stream function and vorticity at the outer cell boundary is the key to successful cell model simulation. For the boundary distribution of stream function, the uniform flow is usually assumed *a priori* when a particle is fixed at the cell center. For the case of particle settling, no fluid exchange between cells is enforced to set the boundary values of the stream function. For the second boundary condition, Happel [29] used the shear-free boundary condition when the cell model was first proposed and Kuwabara [30] later adopted the zero vorticity boundary condition instead.

a. Model H

In Happel's original free surface cell model [29], zero shear stress (no momentum exchange) and zero radial velocity (no fluid exchange) are designated, eliminating all interactions between cells. This is termed as the case of "sedimentation". Here the solid sphere at the center of the cell moves downward in a stationary spherical envelope [28]. In the present notation, the employed boundary conditions are:

$$u_\xi = 0 \text{ (zero normal velocity)} \tag{89}$$

$$\tau_{\xi\eta} = 0 \text{ (no shear boundary)} \tag{90}$$

These conditions were also used by Mohan and Raguraman [32].

In the moving reference frame with a cell-centered sphere, the sphere is stationary and the fluid flows in the x-direction at the terminal velocity. The dimensionless value is unity at the outer cell boundary. For this case, the boundary condition (89) is replaced by specifying the stream function on the outer cell boundary as

$$\psi = \frac{y^2}{2} \text{ (uniform external flow)} \tag{91}$$

In this case, the physical picture viewed from the particle's reference frame is the same as the case of "permeation". Here, fluid flows through the interstitial space between particles [28]. El-Kaissy and Homsy [33] also used the outer cell boundary condition set given as Eqs. (89) and (91). Gu *et al.* [34] employed this set of boundary conditions in their study on the flow of a non-Newtonian fluid through a particle assemblage.

b. Model K

As to the outer cell boundary conditions, Kuwabara [30] proposed zero velocity boundary combined with the uniform external flow condition of Eq.(91):

$$\omega = 0 \text{ (zero velocity boundary)} \tag{92}$$

which was accepted by LeClair and Hamielec [35, 36], Tal(Thau) and Sirignano [37] and Tal(Thau) *et al.*[38], but El-Kaissy and Homsy [33] along with Mohan and Raguraman [32] argued that the foregoing set was conceptually erroneous. However, the comparison between experimental data and predictions obtained from the zero vorticity model is inconclusive regarding which theoretical cell model agrees better with the physical situation in particle assemblages. Model K boundary conditions are the most easily enforced for the system in the stream function-vorticity formulation since boundary values are set explicitly.

3.4.3. Improved Cell Models

Either of these above two conditions seeks to isolate a cell from its neighbors, and the particle interactions in suspension are reflected by the constraints due to the domain geometry on the fluid flow in a cell. It seems that the zero vorticity boundary condition forces a more stringent constraint on the inter-cell interaction [31]. The numerical implementation [31] of outer cell boundary conditions used in the present study covers shape of the cell, non-uniform flow at outer boundary, fore-aft symmetry of flow at outer boundary and maximum value of stream function.

A cylindrical cell model was validated by Tal(Thau) *et al.* [37, 38], but the universality of the cylindrical cell is not convincingly justified. This is far from the reality in a more randomly and densely packed bed or in a distended particle-liquid system, where a particle feels the presence of neighboring ones with equal probability in all spatial directions in a statistical sense. In view of this requirement, it is believed that the hexagonal lattice represents the physical situation more realistically. However, mathematical solution of fluid flow of a hexagonal cell is quite difficult with a spherical particle at the center. Therefore, a spherical cell is adopted as the suitable choice in our work [31].

In a symbolic triangular array of spherical cells (Fig. **8**), liquid may flow through between the cells along certain by-passing channels to experience the least resistance. Hence, the uniformity at the outer boundary should be relaxed to allow the liquid flow along a preferential route as exemplified by line *a-b-c*. It is expected that allowing non-uniform liquid flow would result in a lower drag coefficient than that when uniform liquid flow is enforced.

We conceive that all the cells are identical, but not isolated. Since the cells are presumed identical. It is inferred that the liquid flowing out of a cell at a downstream outer boundary point A in Fig. **8(a)** enters another cell at a point corresponding to point B on the upstream half surface with the same velocity vector.

This requires that point B be in symmetry to A with respect to the horizontal plane $x = 0$. The fore-aft symmetry of flow at the outer boundary is adopted in our work [31].

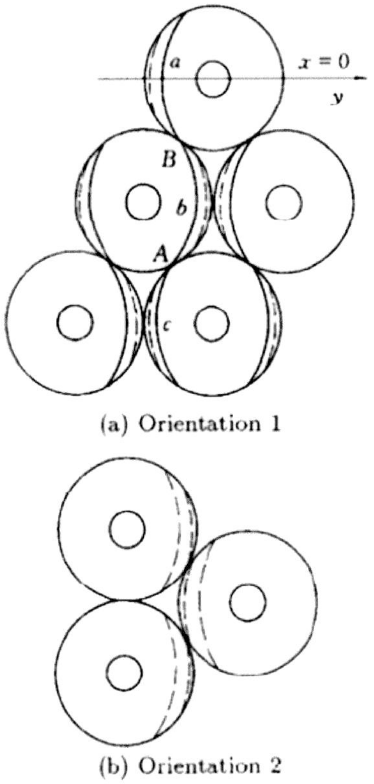

(a) Orientation 1

(b) Orientation 2

Figure 8: Formation of non-uniform liquid flow close to the outer cell boundary.

The conventional way to specify the uniform flow at the outer boundary implies that the maximum value of stream function at the equator of the outer cell surface is

$$\psi_{max} = \frac{1}{2}R_0^2 \text{ (maximum } \psi \text{ at the cell outer equator)} \quad (93)$$

The conception implied in Eq. (93) is questionable. A reasonable sectional area of flow must be figured out and assigned to each cell. It is believed that regular tetragonal packing of spheres (or its equivalent hexagonal lattice) is the closest approximation to the actual random arrangement in a particulate suspension, because the tetragonal packing has the highest degree of spatial symmetry. We figured out the parameter should be 0.4512 and proposed two modified cell models as follows [31].

a. Model M1

The suitability of the above outer cell boundary conditions was addressed and several modifications were proposed [31]. Notably, non-uniform external flow is allowed and only the maximum value of the stream function (the total fluid flow shared by a cell), ψ_{max}, is specified as in Eq. (95). The boundary values of ψ are evaluated from the other boundary condition (free surface) in conjunction with the internal fluid flow. Thus, the flow channeling occurs more readily in this model as compared with Model H [31]. The boundary conditions are:

$$\tau_{\xi\eta} = 0 \text{ (free shear boundary, for designating } \psi \text{ and } \omega \text{ at the outer boundary)} \quad (94)$$

$$\psi_{max} = 0.4512R_0^2 \text{ (maximum } \psi \text{ at the cell outer equator)} \quad (95)$$

b. Model M2

The proposed second version of the modified model added an additional constraint on the relation between entering and exiting flows [31]:

$$\omega(x) = \omega(-x) \text{ (fore − aft symmetry of boundary vorticity)} \quad (96)$$

Eq. (96), used in conjunction with Eqs. (94) and (95), suggests that the exiting flow is exactly the entering flow to a downstream neighbor cell. It should be noted that these three conditions do not overspecify the mathematical model since Eq. (95) is the condition for a boundary point only.

Nowadays, the dynamics of individual particle and particle swarms have been studied extensively using various Computational Fluid Dynamics (CFD) approaches. In a certain sense, the developments of computational approaches facilitate fundamentally analog computation. Consulted our work, four methods have been introduced with their relative merits. On the basis of the previous works, we have proposed some modified measurements to perfect the methods for more complicated conditions. These modified methods have been applied to compute the behavior of a single particle and particle swarms and the related resulted will be reported in following sections.

4. BUOYANCY-DRIVEN MOTION AND MASS TRANSFER OF SINGLE PARTICLES

4.1. Drop and Bubble Motion

4.1.1. Simulation by Level Set Method

As we have described in Section 3.2, the level set method has many advantages. This approach allows for large surface tension and jump in density and viscosity across the interface without reconstructing the numerical grid. Unfortunately, the algorithm for the level set function is in a non-conservative form. The volume loss (or gain) of the bubble/drop would occur as the computational time elapses to result in the deviation from the correct interface position and the distortion of the flow field. Up to now, three kinds of correcting methods have been reported for improving the mass conservation [39]. First of all, a variety of modified reinitialization procedures [23, 40-43] were adopted extensively. Secondly, several ''hybrid'' methods have been developed, by combining the level set approach with the volume-tracking [44-46] or front-tracking [47, 48] methods. Lastly, the mass conservation in the level set method can be significantly improved by reducing spatial discretization errors. The use of AMR (adaptive mesh refinement) for better resolution of interface structures has been attempted to demonstrate improvements in mass conservation [41, 44, 48-50]. A volume-amending was proposed by Li *et al.* [20] to preserve mass (volume) conservation in the level set method, which can be accomplished without significant complication in coding or increase of computational burden.

4.1.1.1. Motion of a Single Bubble in Liquid

In the past decades, many experimental studies have been focused on the motion of bubbles and drops [18, 51-57]. This section is focused on the determination of the fluid particle shape and terminal velocity at steady state motion. We simulate the single axisymmetric 2D air bubbles rising respectively in an aqueous solution of sugar with high or low viscosity (Figs. **9** and **10** [20]). The predicted bubble terminal velocities are 0.181 and 0.317 m/s, which agree with the experimental data of 0.190 and 0.306 m/s, respectively [58]. The steady shapes of spherical cap shown in Figs. **9** and **10** also agree well with the experimental visualization [58].

4.1.1.2. Interactional Motion of Two Bubbles

It has been shown that the volume conservation can be satisfied in a two-dimensional computational system by the level set approach with the volume-amending method incorporated as it has been done by the level set approach with other reinitialization procedures incorporated [20]. However, it is a greater challenge to be satisfied in the fully three-dimensional problems [51]. Wang *et al.* [58] have successfully simulated the interactional motion of two three-dimensional buoyancy-driven deformable bubbles. In order to verify the present method to be effective in fully three-dimensional complex free-surface problems, the interactive motion of a pair of bubbles that are not vertically in line is tested. It involves complex topological deformation of the interfaces as shown by the experimental results [59]. In the simulation by Oka and Ishii [51], mass conservation was well kept but it was very luxurious to use three reinitialization procedures simultaneously.

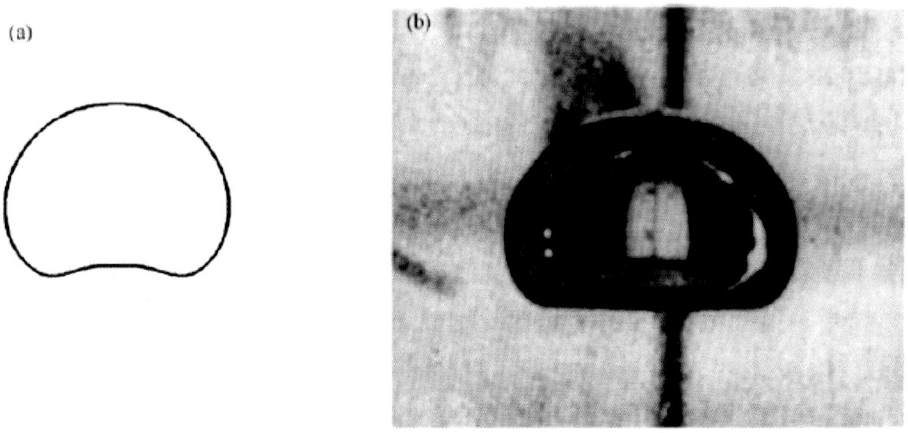

(a) Calculated shape (b) Measured shape (Bhaga and Weber, Figure 3(b))[24]

Figure 9: Comparison of the shape of a 9.3 cm^3 bubble in a sugar solution with high viscosity at steady state. $R = 0.013$ m, $\rho_l = 1390\ \mathrm{kg \cdot m^{-3}}$, $\rho_g = 1.226\ \mathrm{kg \cdot m^{-3}}$, $\mu_l = 2.786\ \mathrm{Pa \cdot s}$, $\mu_g = 1.78 \times 10^{-5}\ \mathrm{Pa \cdot s}$, $\sigma = 0.0794\ \mathrm{N \cdot m^{-1}}$.

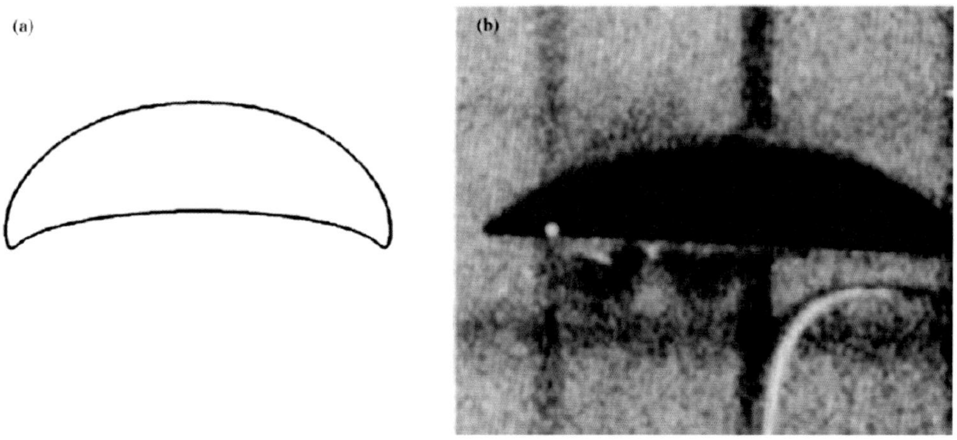

(a) Calculated shape (b) Measured shape (Bhaga and Weber, Figure 3(g))[24]

Figure 10: Comparison of the shape of a 9.3 cm^3 bubble in a sugar solution with low viscosity at steady state. $R = 0.013$ m, $\rho_l = 1320\ \mathrm{kg \cdot m^{-3}}$, $\rho_g = 1.226\ \mathrm{kg \cdot m^{-3}}$, $\mu_l = 0.525\ \mathrm{Pa \cdot s}$, $\mu_g = 1.78 \times 10^{-5}\ \mathrm{Pa \cdot s}$, $\sigma = 0.0754\ \mathrm{N \cdot m^{-1}}$.

Two spherical bubbles with the radius ratio $R_1/R_2 = 1.5$ are initially at rest in the vertical center plane of the computational box. Most of the conditions are set just as those in Manga and Stone's experiments [59] other than the distance between two bubbles, such as $We \approx 50$ and $Re \approx 5 \times 10^{-3}$. Some of the important stages of the predicted interactive behavior are demonstrated in Fig. **11** [20]. The two bubbles deform considerably while rising, but the deformations of two bubbles are quite different, which resemble the experimental result. The upper bubble deforms towards the steady-state shape of a single bubble in free rise. The trailing bubble deforms a little. The lower part of upper bubble becomes concave while that of the trailing one becomes protuberant. In the following time, the distance between two bubbles turns shorter and shorter. The trailing bubble is entrained into the wake of the preceding one and then they integrate together. The numerical result is in qualitatively agreement with the experimental data [59].

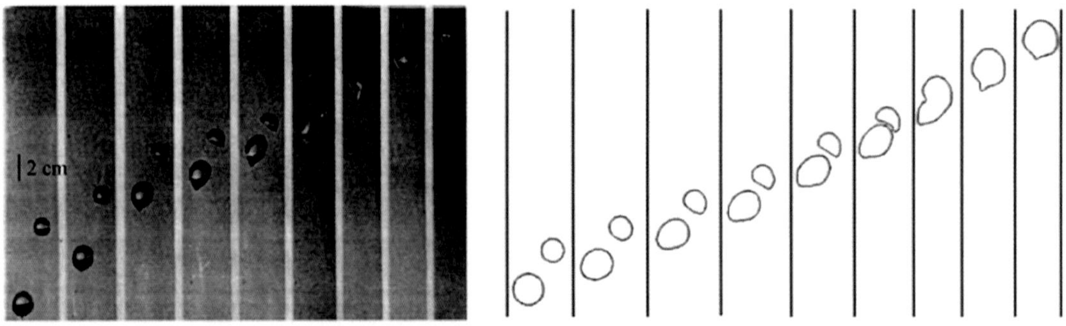

(a) Experimental result of Manga and Stone (b) Numerical result

Figure 11: Experimental and numerical results for interactive behavior of two bubbles rising in corn syrup (Photographs and simulated results are shown at 10 s intervals).

4.1.1.3. Parasitic Flow of a Neutrally Buoyant Drop

Parasitic flows may occur in the numerical simulation of incompressible multiphase flow with level set method due to errors in the calculation of surface tension terms, specifically for the curvature and unit normal vector. A simple weighted integration method based on the level set method is proposed by Wang *et al.* [60], in which the contribution of not only the center node but also the rest area of a control volume to the calculation of surface tension is considered in a balanced manner. The weighted integration method (WIM) is more consistent with the concept of a banded interface in the level set method. It is applied to the temporal evolution of a two-dimensional neutrally buoyant liquid drop and a buoyancy driven deformable bubble in an immiscible fluid for the validation of WIM. Fig. **12** [60] shows that the parasitic flows are evidently suppressed by the weighted integration method.

4.1.2. Unsteady Motion of a Bubble

Steady motion and mass transfer of fluid particles are numerically simulated in an orthogonal boundary-fitted coordinate system. In fact, our studies on mass transfer computation are usually based on steady motion, which will be expounded at the latter of this section. Therefore, this section will focus on the unsteady motion of a fluid particle.

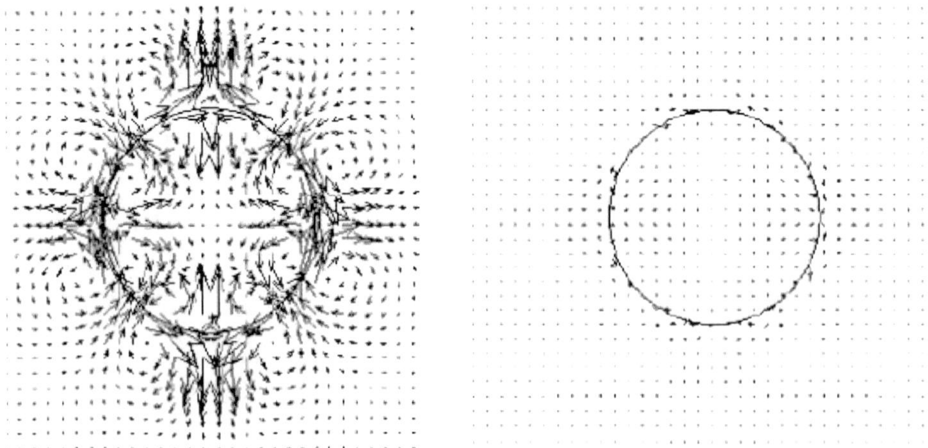

Figure 12: Spurious currents (with the same velocity scales) at the 1000th time steps for a stationary, neutrally buoyant drop with the conventional continuum surface force method (left hand side) and WIM (right hand side) respectively ($\varepsilon = \Delta x$, $Oh = 2.88 \times 10^{-2}$, $\rho_b/\rho_c = 1$, $\mu_b/\mu_c = 1$, $\sigma = 0.02\ \mathrm{N \cdot m^{-1}}$, $R = 0.01$ mm).

In the past, fewer attempts were made on the unsteady forces on bubbles or particles in Newtonian fluids [61-67] and particles in viscoelastic fluids [68-73] and shear thinning fluids [74, 75]. The added mass force is important for a bubble in unsteady motion when it accelerates or decelerates (*e.g.*, near the nozzle or in a turbulent flow with swirl). Without the added mass force included in mathematical formulations, the bubble acceleration reaches an unrealistic value because of the relatively strong buoyant forces and the small mass of the bubbles [76, 77]. The history force (also called the Basset force) represents the fading memory effect of the relative acceleration between the bubble and the surrounding liquid. The history force formulation involves a history integral which makes its estimation computationally intensive, and it is usually neglected for the liquid of low viscosity [78] because it is difficult to measure directly the added mass force and the history force exerted on a bubble.

The motion of a single bubble rising unsteadily and steadily in a quiescent non-Newtonian liquid (Table **1**) has been investigated experimentally. By using a Charged Coupled Device (CCD) camera to follow the rising bubble, the sequences of the recorded frames are digitized and analyzed using image analysis software, and the measurements of the acceleration and steady motion of bubbles are obtained. Using the experimental data, we propose an empirical correlation to predict the total drag coefficient calculated from the accelerating motion to the steady motion with the added mass force and history force included.

The rectilinear motion of a bubble is described by Newton's second law as Eq.(14). The forces include buoyancy F_{B}, drag force F_{D}, added mass force F_{A}, and history force F_{H}. The drag force, added mass force, and history force are summed as a total drag force, and the total drag coefficient C_{DA} is defined as Eq.(15). When the particle rises steadily ($du_{\mathrm{p}}/dt = 0$), the conventional drag coefficient C_{D} is defined as Eq.(16). The drag coefficient is showed as a function of the Reynolds number and the non-Newtonian parameters, which depends on the choice of a specific rheological model.

Table 1: Rheological and physical properties of different solutions.

Solution	C_w(wt %)	μ_0 (mPa · s)	λ(s)	$n(-)$	$\varepsilon(\%)^a$	K (mPa · s^n)	$b(-)$	$\varepsilon(\%)^a$
CMC	0.25	13.17	93.8	0.98	0.5	11.8	0.99	1.06
	0.5	34.76	32.08	0.96	0.51	29.5	0.95	0.9
	0.75	57.93	0.376	0.95	1.18	61.9	0.93	4.2
	1.0	122.02	0.207	0.90	0.94	138.3	0.90	3.12
xanthan gum	0.036	25.65	0.505	0.63	1.13	30.1	0.65	5.01
	0.06	115.49	0.984	0.49	1.21	112.6	0.50	5.56
HEC	0.3	47.0	0.062	0.79	1.83	63.5	0.86	3.47
	0.5	315.9	0.261	0.68	0.84	376.0	0.77	2.89
PAAS	0.01	17.94	1.36	0.53	2.73	51.8	0.52	10.9
	0.03	162.73	0.522	0.37	2.53	218.2	0.41	12.48

[a]ε: Relative fitting errors.

CMC-aqueous carboxymethyl cellulose sodium salt; HEC-hydroxylethyl cellulose; PAAS-sodium polyacrylate.

It has been found that the three-parameter Carreau viscosity equation:

$$\mu_a(\dot{\gamma}) = \mu_0[1 + (\varsigma\dot{\gamma})^2]^{(n-1)/2} \quad \textbf{(97)}$$

provided an adequate representation of the shear viscosity of the fluids of interest. Here μ_0 is the zero-shear viscosity, ς is a characteristic time of the fluid, and n is the slope of the shear stress-shear rate curve in the shear-thinning region. This model reduces to the Newtonian relation when $\varsigma = 0$ or $n = 1$.

Since the fluid viscosity varies as a function of the shear rate, the first problem in developing the drag curve is the definition of Reynolds number since it contains both fluid viscosity and bubble velocity. Re_{M} is defined by introducing $2U_{\mathrm{T}}/d_{\mathrm{e}}$ as a representative shear rate over the system:

$$Re_{\mathrm{M}} = \frac{\rho_{\mathrm{l}} d_{\mathrm{e}} U_{\mathrm{T}}}{\mu_0/\{1+[\varsigma(2U_{\mathrm{T}}/d_{\mathrm{e}})]^2\}^{(1-n)/2}} = \frac{\rho_{\mathrm{l}} d_{\mathrm{e}} U_{\mathrm{T}}}{Q\mu_0} \text{ (for shear thinning liquids)} \quad (98)$$

with

$$Q = \frac{1}{\{1+[\varsigma(2U_{\mathrm{T}}/d_{\mathrm{e}})]^2\}^{(1-n)/2}} \quad (99)$$

where μ_0 means the effective viscosity over the system. If $Q = 1$, the bubble motion can be regarded as a Newtonian fluid case.

For the PAAS solutions showing shear-thinning and viscoelastic behavior, a modified Reynolds number based on the zero shear viscosity (Re_0), De, and n is introduced:

$$Re_{\mathrm{C}} = Re_0(1 + De^{n-1}) \text{ (for shear thinning viscoelastic liquids)} \quad (100)$$

The Deborah number, De, and the Reynolds number, Re_0 , based on the zero shear viscosity, are given by $De = \varsigma U_T/d_{\mathrm{e}}$ and $Re_0 = \rho_{\mathrm{l}} d_{\mathrm{e}} U_T/\mu_0$.

Fig. **13** [79] shows the results calculated using the combination of modified Reynolds number Re_{MC} for the non-Newtonian liquids. On the basis of the 28 experimental data for bubble steady motion, a simple correlation is proposed in the form of:

$$C_{\mathrm{D}} = \frac{16}{Re_{\mathrm{MC}}}(1 + 0.12 Re_{\mathrm{MC}}{}^{0.6}) \quad (101)$$

The value of the drag coefficient in Eq. (101) converges to $C_{\mathrm{D}} = 16/Re$ at very low values of Reynolds number (*i.e.*, reaches the values predicted by the Hadamard-Ribczynky model [80] of a bubble with a shear-free interface). It can be seen that Eq. (101) shows good agreement with the experiments with the average deviation of 15.9%.

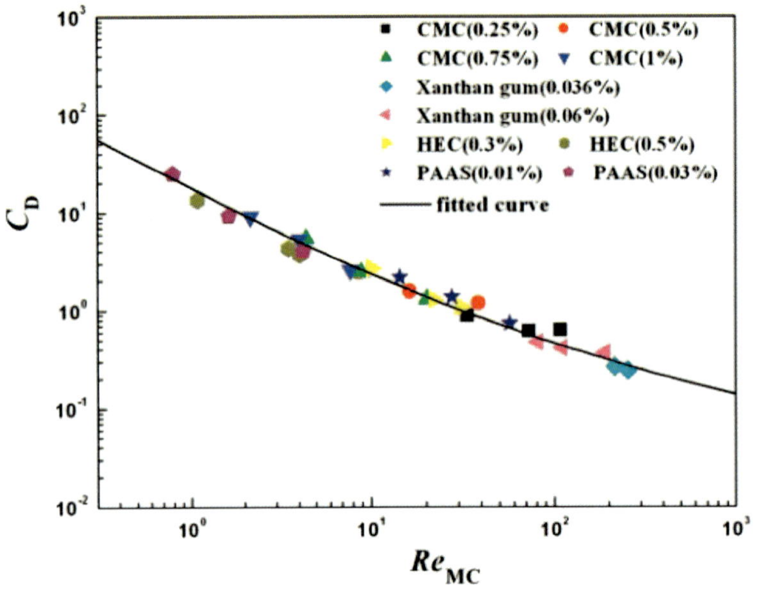

Figure 13: Relationship between C_{D} and Re_{MC} in steady rising.

In the presence of acceleration, the total drag coefficient depends additionally on acceleration as quantified by the acceleration number Ac proposed by our previous work [9]. In this section, only shear thinning non-Newtonian liquids are treated and the PAAS solution is excluded for its viscoelastic behavior. By fitting the measured values of C_{DA}, $Ac(= d_{\mathrm{e}}a/u_{\mathrm{b}}^2)$, and $Ar(= d_{\mathrm{e}}^3\rho_{\mathrm{l}}^2 g/[\mu_a(\dot{\gamma})]^2)$ with the least-squares method, the following correlation for the total drag coefficient results on the basis of 246 data (146 of steady motion and 100 in transient motion), as indicated in section 2.1.3:

$$C_{\mathrm{DA}} = \frac{16}{Re_{\mathrm{MC}}}\left(1 + 0.12{Re_{\mathrm{MC}}}^{0.6}\right)\left(1 + 0.196Ac^{0.767}Ar^{0.381}\right) \qquad (102)$$

The average relative deviation between the predicted and measured total drag coefficient is about 25.4%, which is acceptable for the large range of rheological parameters involved in the experiments. The validity of Eq. (102) to single bubbles in shear thinning non-Newtonian liquids is also shown by plotting $C_{\mathrm{DA}}/(1 + 0.196Ac^{0.767}Ar^{0.381})$ of various systems against Re_{M} in Fig. **14** [79]. These data for non-Newtonian liquids cover a range of flow index s from 0.37 to 0.97 corresponding to the Reynolds numbers from 0.15 to 261. In these wide ranges, the present correlation approximates and applies to both accelerating and steady motion of rising bubbles.

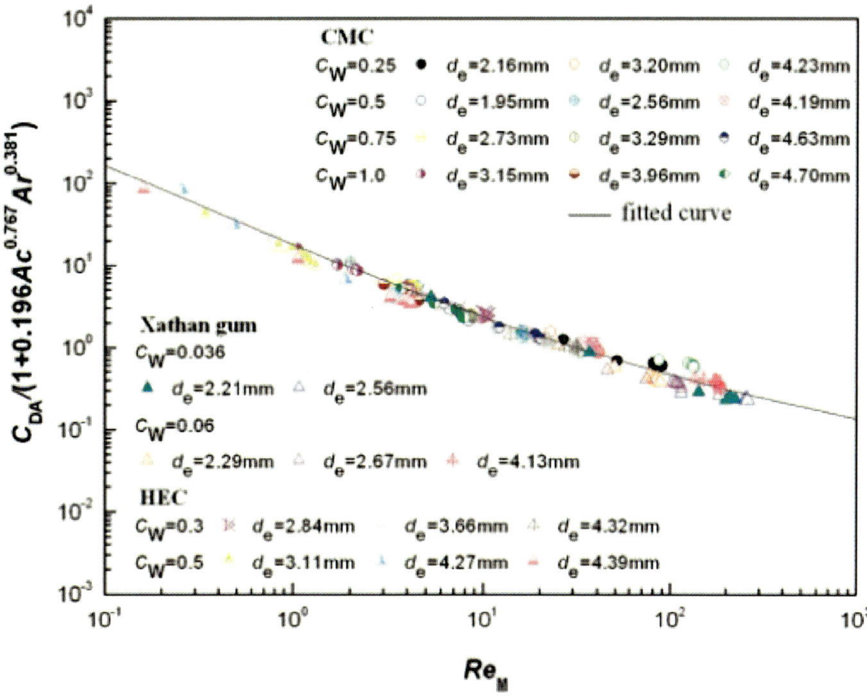

Figure 14: Correlation of transient C_{DA} against Re_{M}.

4.1.3. Simulation of Bubble Motion in a Non-Newtonian Fluid

It is highly important to obtain the knowledge of bubble behavior not only in a Newtonian continuous phase but also in a non-Newtonian one, because the investigation on bubble motion provides useful and essential information for realizing optimal process design and operation. It is well known that a large number of investigations concerning various aspects of the bubble or drop motion in non-Newtonian fluids have been reported theoretically and experimentally in the past. Many of them have been well summarized and reviewed by Chhabra [81, 82] and Kulkarni and Joshi [83].

The motion of a single bubble rising freely through shear thinning fluids represented by the Carreau model of Eq. (97) is computed numerically using a level set method for tracking the bubble interface. The changes in the local viscosity field around a bubble rising in shear thinning non-Newtonian fluids are examined.

Real density and viscosity ratios of gas and liquid are adopted when simulating the shear thinning non-Newtonian two-phase flow systems.

The terminal rising velocity of air bubbles in five non-Newtonian liquids (Tables **2** and **3**) with shear thinning property are determined and these bubbles are simulated. Table **3** shows the numerical and experimental results for a single bubble rising freely in pure shear thinning non-Newtonian solutions. The predicted terminal velocities are in close agreement with the present experimental values with the relative deviation below 10%. The satisfactory agreement indicates the validity of numerical procedure for the numerical simulation of a single bubble rising in a shear thinning non-Newtonian continuous phase. In these solutions the apparent viscosity around the bubble decreases due to the shear, so the bubble can rise more and more easily. It is interesting to note that the zero-shear viscosity of CMC2 and XG is roughly equal (Table **2**), but U_T of the bubble in XG is much larger than that in CMC2 since XG has stronger shear thinning property and the liquid around the bubble thus becomes less viscous. Hence, understanding the viscosity distribution over the whole domain is important to elucidate the bubble behaviors in shear thinning non-Newtonian liquids.

Table 2: Parameters for Carreau model and physical properties of liquids.

Liquid	μ_0 (Pa s)	λ (s)	n	ρ_l (kg m^{-3})	σ (mN m^{-1})
CMC1	0.0377	0.512	0.89	1000.0	72
CMC2	0.122	0.207	0.90	1000.0	72
HEC1	0.047	0.0624	0.79	1000.0	72
HEC2	0.315	0.260	0.68	1000.0	72
XG	0.115	0.984	0.49	1000.0	72

Table 3: Comparisons between experimental and numerical results for a single bubble rising in Carreau shear thinning liquids.

Liquid	d_e (mm)	$U_{T,exp}$ (cm s^{-1})	$U_{T,cal}$ (cm s^{-1})	$Re_{M,exp}$	$Re_{M,cal}$	D_{ev}^* %
CMC1	4.19	21.5	20.87	121.27	116.29	2.9
CMC2	3.96	9.88	9.39	4.05	3.832	4.4
HEC1	2.84	11.88	11.6	10.19	9.91	2.4
HEC2	3.11	5.25	4.74	1.041	0.91	9.7
XG	2.29	26.5	24.6	84.07	75.15	7.16

$^*D_{ev} = |U_{cal} - U_{exp}|/U_{exp} \times 100.$

The shear thinning liquids whose properties are shown in Table **4** are tested in order to elucidate the shear thinning effect on bubble rising. The numerical results of Re_M are also listed in Table **4**. The gas-related physical properties of the systems are $\rho_g = 1.2$ kg m^{-3}, $\eta_g = 1.8 \times 10^{-5}$Pa s, and $\sigma = 7.2 \times 10^{-2}$ N m^{-1}.

Table 4: Carreau model parameters and numerical results of bubbles with $d_e = 1$ cm.

System	μ_0 (Pa s)	λ (s)	n	U_T (cm s^{-1})	$2\lambda U_T/d_e$	Re_M
1	0.5	-	1.0	11.2	-	2.24
2	0.5	1.00	0.8	16.5	33.0	6.65
3	0.5	1.00	0.5	22.7	45.4	30.69
4	0.5	1.00	0.4	23.9	47.8	48.8
5	0.1	1.00	0.8	25.3	50.6	55.5
6	0.5	1.00	0.3	25.6	51.2	80.85

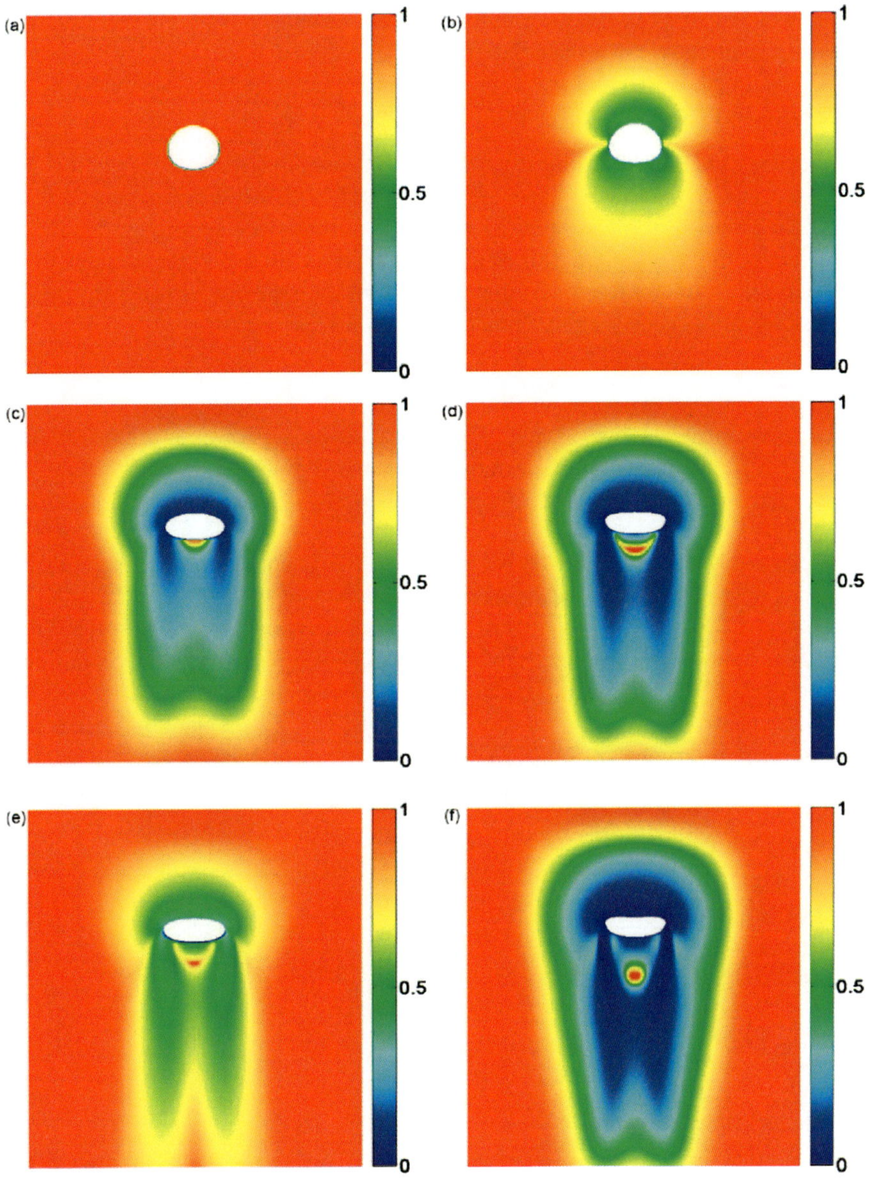

Figure 15: Viscosity distribution and shape of a bubble rising in a solution. (a) System 1: Newtonian fluid, $Re_M = 2.24$, $\mu_0 = 0.5$ Pa s. (b) System 2: Carreau shear thinning fluid, $Re_M = 6.65$ $\mu_0 = 0.5$ Pa s. (c) System 3: Carreau shear thinning fluid, $Re_M = 30.69$, $\mu_0 = 0.5$ Pa s. (d) System 4: Carreau shear thinning fluid, $Re_M = 48.8$, $\mu_0 = 0.5$ Pa s. (e) System 5: Carreau shear thinning fluid, $Re_M = 55.5$, $\mu_0 = 0.5$ Pa s. (f) System 6: Carreau shear thinning fluid, $Re_M = 80.85$, $\mu_0 = 0.5$ Pa s.

Fig. **15** [84] shows the viscosity distribution (non-dimensionalized by μ_0) around a bubble and the real bubble shapes for systems 1 through 6. The color bar shown at the right of the figures depicts the range of reduced apparent viscosity around the bubble. System 1 corresponds to the bubble rising in a Newtonian fluid with the same zero shear rate viscosity and the viscosity around the bubble is constant. As for systems 2 through 6, the viscosity around the bubble varies in correspondence to the bubble shape and the shear thinning property of the liquid, and the degree of the decrease in viscosity is the largest in close vicinity to the bubble. The change in viscosity becomes gradually drastic from system 3 to system 6, showing that a confined region with high viscosity exists in the wake of these bubbles and it becomes finally detached from the bubble rear surface. As seen in Table **4**, it is obvious that U_T (or Re_M) increases gradually due to the stronger shear thinning effect. The bubble can rise faster due to the large extent of decrease in viscosity around the bubble. Obviously, the bubble shape in the shear thinning non-Newtonian fluid differs much from that in the Newtonian case. As the shear thinning effect becomes intensive gradually, the bubble takes the more oblate shape, with the front surface flatter than the rear part.

4.2. Mass Transfer to/From a Drop

The convective diffusion equation for a single particle can be numerically solved when the fluid dynamic is resolved. Numerical simulation of interphase mass transfer in a boundary-fitted orthogonal reference frame has been successfully adopted for mass transfer to/from a steady or accelerating drop in solvent extraction systems. The boundary-fitted orthogonal coordinate method is used for the numerical simulation of interphase mass transfer at low Reynolds numbers. The level set approach is extended to interphase mass transfer, but a variable transform and higher order numerical scheme are required.

4.2.1. Mass Transfer Dominated by External Resistance

In a boundary-fitted orthogonal coordinate system, the general governing Eqs. (41)-(44) for steady and transient axisymmetrical cases of mass transfer in the external region around single buoyancy-driven drops in steady motion under low or intermediate Reynolds numbers are numerically solved by the control volume formulation.

For the situation of mass transfer dominated by the external resistance, the partition coefficient of solute between the drop and the continuous phase is rather large, so that the solute concentration in the drop and at the surface can be reasonably assumed to be zero during the extraction. Considering that the concentration of solute at the remote boundary far away from the drop may be maintained at constant c^{∞}, the mass transfer to/from the drop will eventually approach a steady state as the mass transfer proceeds. If only the steady solution is of interest, the partial derivative with respect to time could be dropped from Eq. (51) and the initial condition becomes irrelevant. Thus, the following initial and boundary condition may be assumed for mass transfer from the continuous phase to drops (mass transfer direction c → d):

$$C(\xi,\eta,0) = 0 \text{ at } \theta = 0 \tag{103}$$

$$C(1,\eta,\theta) = 0 \text{ at the drop surface, } \xi = 1 \tag{104}$$

$$C(0,\eta,\theta) = 1 \text{ at the remote boundary, } \xi = 0 \tag{105}$$

$$\frac{\partial C}{\partial \eta} = 0 \text{ on the axis of symmetry, } \eta = 0, \eta = 1 \tag{106}$$

As compared with the empirical correlation of external mass transfer for liquid drops with intermediate Reynolds numbers developed for the case of $Re > 70$ and relative viscosity $\lambda < 2$ by Harper and Moore[85],

$$Sh = \frac{2}{\sqrt{\pi}}\left[1 - \frac{1}{Re^{0.5}}(2.89 + 2.15\lambda^{0.64})\right]^{0.5} Pe^{0.5} \tag{107}$$

good agreement in the range of $0.1 < Pe < 2000$ is evidenced in Fig. **16** [16]. When Pe is very large, the transport is dominated by convection, and this may lead to very large concentration gradient at the surface. Since the mesh density is limited at the interface, the large gradient of concentration as $D \to 0$ is not accounted for with sufficient accuracy, and hence, the discrepancy becomes obvious at high Pe region.

It is easy to investigate the effect of relevant physical parameters on the average Sh by means of numerical experimentation. Fig. **17** shows that the average Sh increases along with Pe at different Reynolds numbers [16]. When Pe is small, the mass transfer is dominated by molecular diffusion, and Sh is hence very close to the lower limit of $Sh = 2$. However, Re plays a significant role in promoting the rate of mass transfer when Pe is large, since the convective transport controls the rate process in this case. It should be kept in mind that when interpreting mass transfer for a given extraction system, Sc is predescribed as a constant, where Re and Pe $(= ReSc)$ work in a synchronous way, and higher Re indicates stronger convection and higher mass transfer rate.

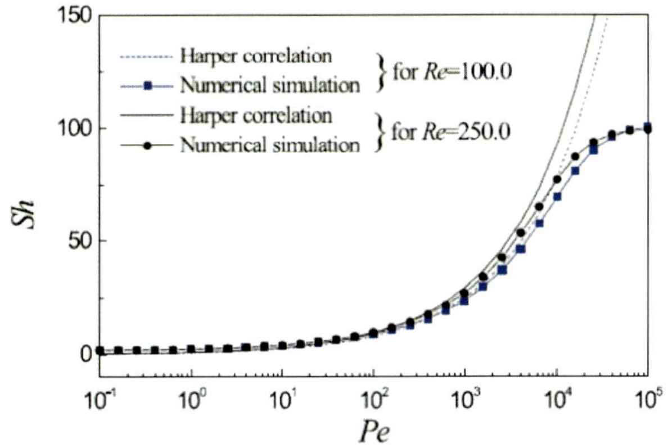

Figure 16: Comparison of numerical prediction and empirical correlation by Harper and Moore[85].

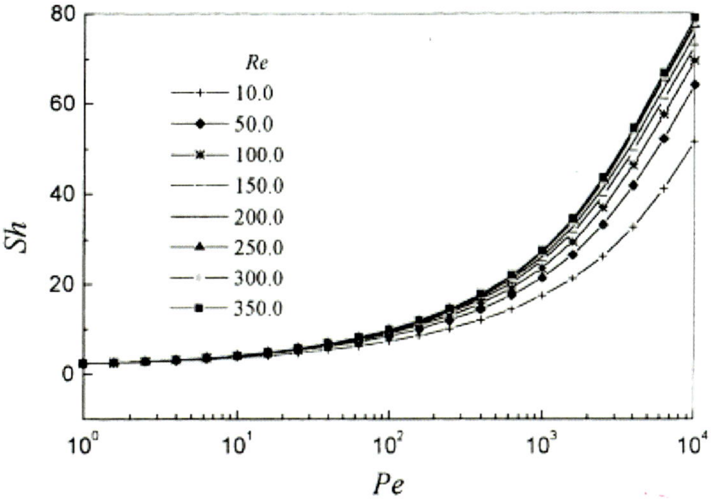

Figure 17: Sh for single drops in laminar flow regime as a function of Pe with different Reynolds numbers (We = 4.0, = 5.0, ζ = 2.0).

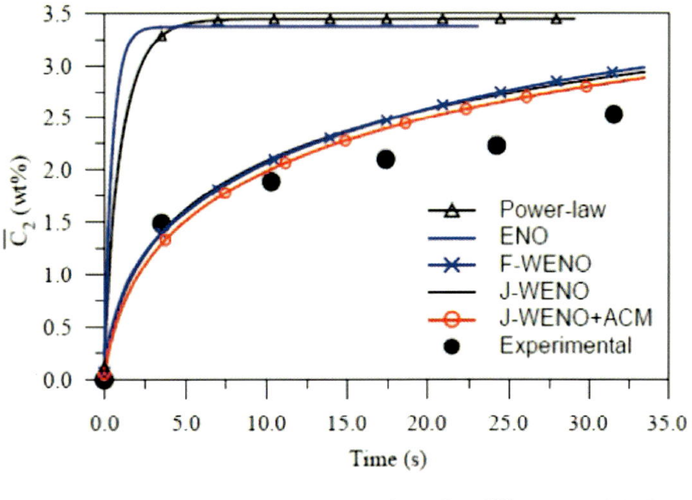

Figure 18: Comparison of predicted average drop concentrations for different spatial discretization schemes with experimental data.

4.2.2. Mass Transfer for a Steady Drop with Resistance in Both Phases

When the level set approach is applied for simulating the interphase mass transfer of single drops in an immiscible liquid with resistance in both phases, the concentration transformations in Section 3 must be adopted to impose the flux continuity and interfacial dissolution equilibrium conditions on the interface accurately if the distribution coefficient (m) is not equal to unity. The solution of convective diffusion equation (78) for interphase mass transfer is decoupled with the momentum equations.

Based on the comparison of different spatial discretization schemes such as the power-law, second-order ENO (essentially nonoscillatory), fifth-order WENO (weighted essentially nonoscillatory) and WENO coupled with ACM (artificial compression method) technique for computing the governing equations of mass transfer, the fifth-order WENO scheme is found to perform better and is adopted for the present simulation including the solution of the evolution and reinitialization equations of level set function, as shown in Fig. **18** [19].

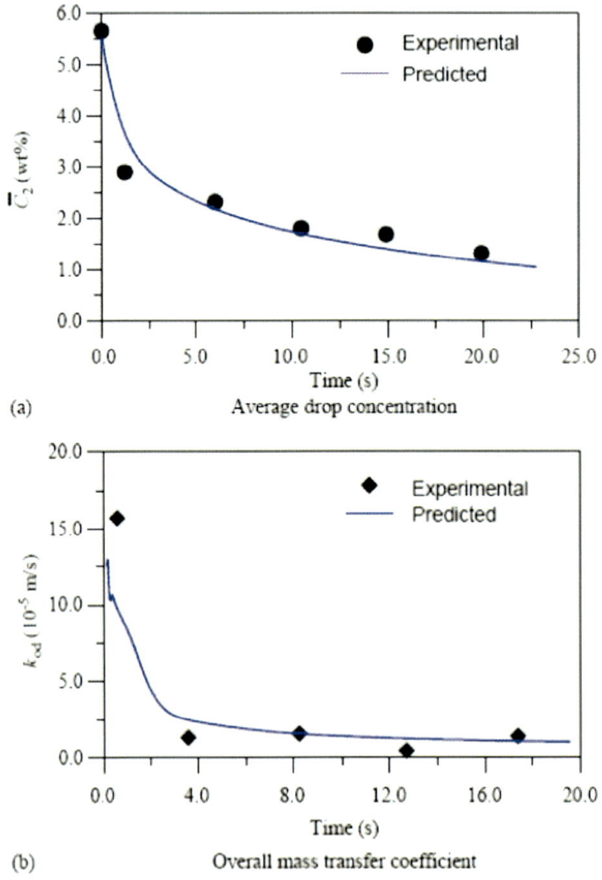

Figure 19: Predicted and experimental concentrations and overall mass transfer coefficients as a function of mass transfer time for serial No.BD16-1: (a) average drop concentration, (b) overall mass transfer coefficient (d = 1.69 mm, $\rho_1 = 988.2\ \mathrm{kg \cdot m^{-3}}$, $\rho_2 = 841.1\ \mathrm{kg \cdot m^{-3}}$, $\mu_1 = 1.44 \times 10^{-3}\ \mathrm{Pa \cdot s}$, $\mu_2 = 3.34 \times 10^{-3}\ \mathrm{Pa \cdot s}$, $\sigma = 1.00 \times 10^{-3}\ \mathrm{N \cdot m^{-1}}$, $D_1 = 5.5 \times 10^{-10}\ \mathrm{m^2 \cdot s^{-1}}$, $D_2 = 2.1 \times 10^{-10}\ \mathrm{m^2 \cdot s^{-1}}$, $m = 1.16$, $C_1^0 = 0$, $C_2^0 = 5.66\%$; solute transfer direction: d → c).

The Reynolds number and time history of concentration and mass transfer coefficient are well predicted by this algorithm as verified against the reported experimental data of the n-butanol–succinic acid–water system recommended by EFCE (the European Confederation of Chemical Engineering). As shown in Fig. **19** [19], for example, the predicted drop average concentrations and overall mass transfer coefficients for mass transfer direction from the drop to the continuous phase is in reasonably good agreement with the corresponding experimental data. In Table **5**, although better agreement between computational and experimental Reynolds numbers is observed, most predicted overall mass transfer coefficients, including the results in a boundary-fitted orthogonal coordinate system, are slightly higher than the experiments. Li *et al.* [86] conjectured the experimental system might not be very pure, though serious care was taken in conducting experiments. Despite the fact that the boundary-fitted orthogonal coordinate method offers more accurate numerical results, it is very difficult to construct orthogonal curvilinear coordinates for highly deformed drops and be extended to the simulation of three-dimensional problems.

Table 5: Comparison of predict and experimental Reynolds numbers and average overall mass transfer coefficients for *n*-butanol-succinic acid-water system (the same conditions as in Fig. **19**).

No.	Re_{exp}	Re_{pre}	$\bar{k}_{od,exp}$ $\times 10^5$ (m/s)	$\bar{k}_{od,pre}$ $\times 10^5$ (m/s) [19]	$\bar{k}_{od,pre}$ $\times 10^5$ (m/s) [86]
BC4-1	18.9	19.0	0.449	0.709	0.370
BC9-1	37.7	38.4	0.589	1.126	0.807
BC12-1	49.5	48.5	0.897	1.400	1.278
BC16-1	61.9	59.6	2.566	1.900	2.130
BD4-1	21.2	18.8	0.374	0.728	0.408
BD9-1	39.0	37.3	0.752	1.275	0.907
BD12-1	48.1	50.6	0.661	1.543	1.137
BD16-1	55.2	56.9	1.108	1.739	2.340

For two limiting cases of mass transfer into or out of single drops with resistance dominated only in the stagnant continuous phase or in the stagnant spherical drop, the numerical results are almost identical to the corresponding analytical solutions [19]. The good agreements between the results of numerical simulation and those from experiments and theoretical analysis for the transient interphase mass transfer of a drop also confirm the correct jump conditions at the interface are satisfied by the necessary transformation making the concentration continuous. These numerical tests indicate the present level set approach is simple, robust and effective in simulating the interphase mass transfer of single drops.

4.2.3. Mass Transfer to/From An Unsteady Drop

The main difficulty in numerical simulation is that the motion of a deformed drop with simultaneous mass transfer must be solved with the unknown shape of free surface [87]. Numerical method must ensure accurate estimation of the interfacial position and the concentration so that the mass transfer calculation remains accurate. We use the level set method to simulate the interphase mass transfer to/from a drop in unsteady motion.

Fig. **20** [88] shows the predicted solute distribution in both phases and the velocity vector field relative to the drop at selected dimensionless time for Case A in Table **6**. The results show the drop is gradually deformed into a spheroid from its initial spherical shape, and a toroidal recirculation is formed within the drop. In the center part of the drop the concentration of the solute progressively increases as it is swept to the downstream direction, suggesting that convective mass transfer does play an important role. The convection at the drop nose promotes the transport of solute to the drop.

Fig. **21** [88] shows the predicted solute distribution in both phases and the velocity vector field relative to the drop motion at different dimensionless time for Case B (Table **6**, liquid-liquid system, the direction of mass transfer: d → c). The results show that in the center part of the drop the solute concentration gradually decreases as it is swept by the recirculation towards the drop surface and then towards the rear of the drop where it enters the continuous phase as a plume. The plume is more distinct at early time when the concentration of the drop is the highest.

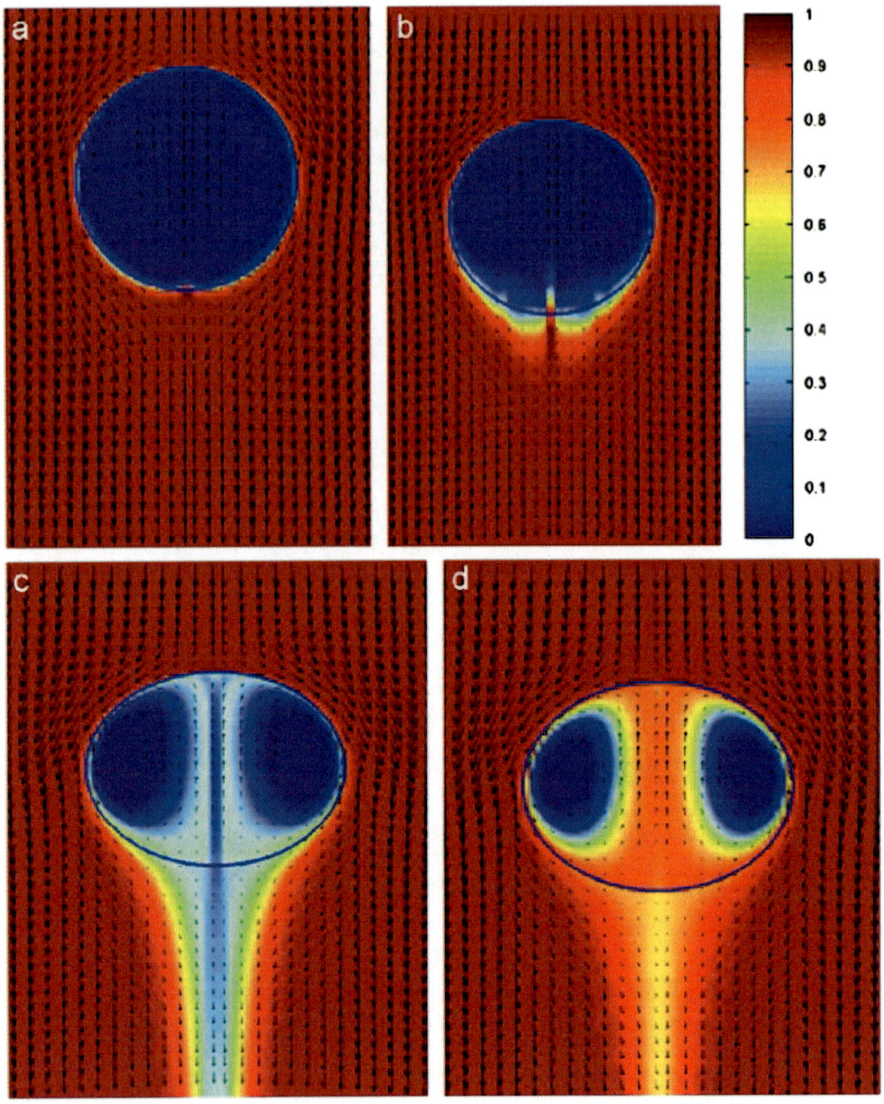

Figure 20: Predicted fractional solute concentration distribution and velocity vector field relative to the motion of the drop in Case A (Table **6**, $c_{c,0} = 2.92\text{wt}\%, c_{d,0} = 0, R = 0.52\text{mm}$) at various dimensionless time: (a) $t = 0.02$ s, (b) $t = 0.04$ s, (c) $t = 0.13$ s, and (d) $t = 0.49$ s.

4.3. Simulation of Solid Particle Motion by Mirror Fluid Method

The mirror fluid method is proposed for simulating solid-fluid two-phase flow [25]. In this method the entire computational domain is taken as a fixed Eulerian one, suitable flow parameters are assigned to the solid particle domain by mirror relations to ensure the surface segment subjected to zero net shear and normal forces so that the fluid-solid boundary conditions enforced implicitly. Meanwhile, the motion of solid particles is calculated using a Lagrangian numerical scheme. This algorithm has the advantages of solving multiphase flow by a set of coupled equations, eliminating smear of density and other variables profiles across the interfaces, using a simple finite-difference method for discretization and adopting a fixed Cartesian grid without need of remeshing.

The SIMPLE scheme and the control volume formulation are used for solving the governing equations for the motion of a rigid spherical or elliptic particle in a fluid. The Reynolds number, drag coefficient, and wake length of real spherical particles are well predicted as verified against the acknowledged experimental data (as shown in Figs. **22-23** [25]). The orientation angle and trajectory of an ellipse settling in a fluid are also well predicted as compared with the reported simulation results (as shown in Fig. **24** [25]). These numerical tests indicate that the mirror fluid method is simple, robust, and effective in simulating real solid-fluid flows.

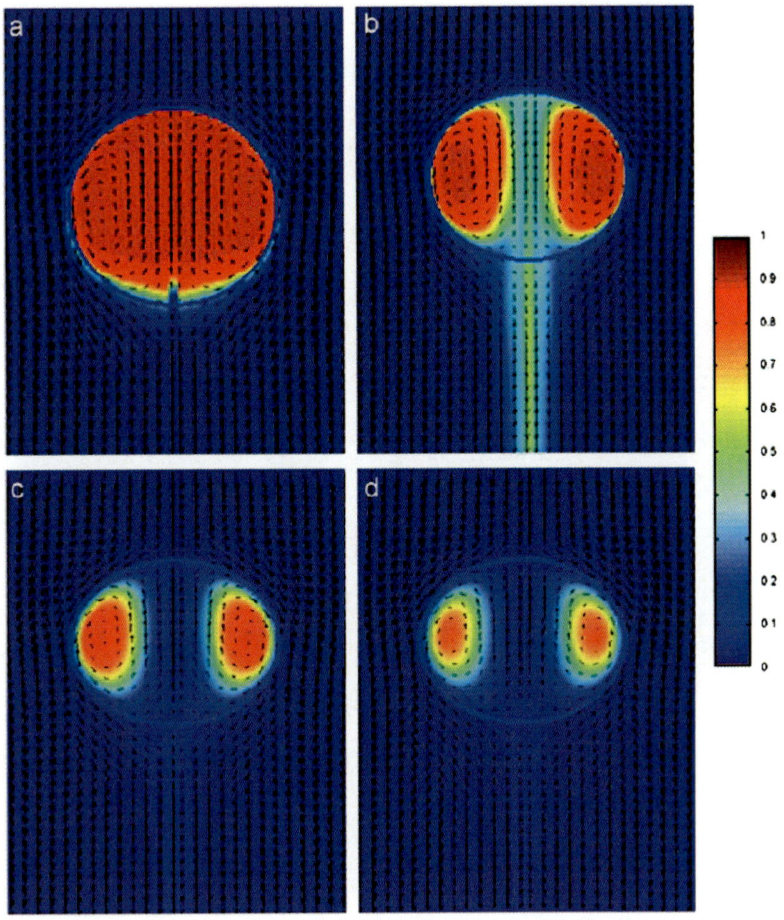

Figure 21: Predicted fractional solute concentration distribution and velocity vector field relative to the motion of the drop in Case B ($c_{c,0} = 0, c_{d,0} = 4.79\text{wt}\%, R = 0.59\text{mm}$) at various dimensionless time: (a) $t = 0.04$ s, (b) $t = 0.14$ s, (c) $t = 1.39$ s, and (d) $t = 2.10$ s.

Table 6: Relative deviations of predicted and experimental average overall mass transfer coefficients.

Case	System continuous phase /solute/disperse phase	Experimental average $k_{od} \times 10^5 (\text{m/s})$	Predicted average $k_{od} \times 10^5 (\text{m/s})$	Deviation (%)
A	water/succinic acid/ butanol	4.57	2.16	16.41
		3.23	2.75	11.99
		2.90	3.67	5.17
		1.99	3.82	7.87
B	butanol/ succinic acid/water	3.23	4.31	25.53
		2.81	3.51	19.94
		2.59	2.80	7.5
		2.54	2.53	0.39
		2.36	2.31	2.12

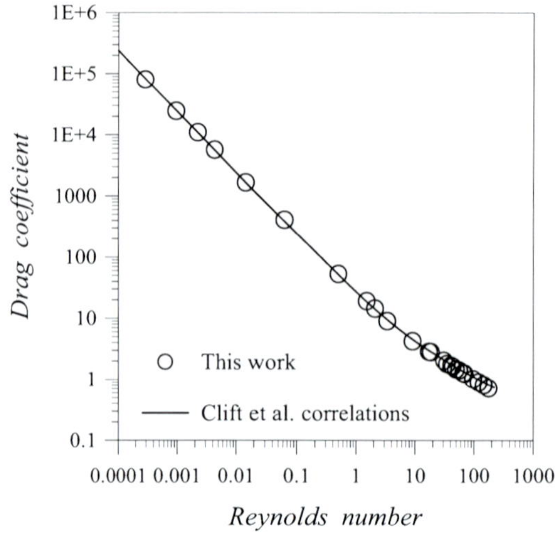

Figure 22: Plot of the drag coefficient against Reynolds number of a falling spherical particle ($R = 0.21$ mm, $\rho_l = 1000$ kg m^{-3}, $\rho_s = 2000$ kg m^{-3}, $\mu_l = 1.0 \times 10^{-3}$ Pa s).

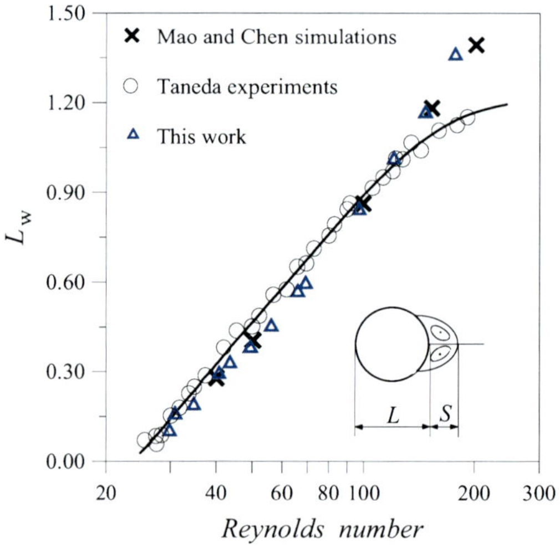

Figure 23: Comparison of predicted wake length with experimental data ($L_w = S/L$).

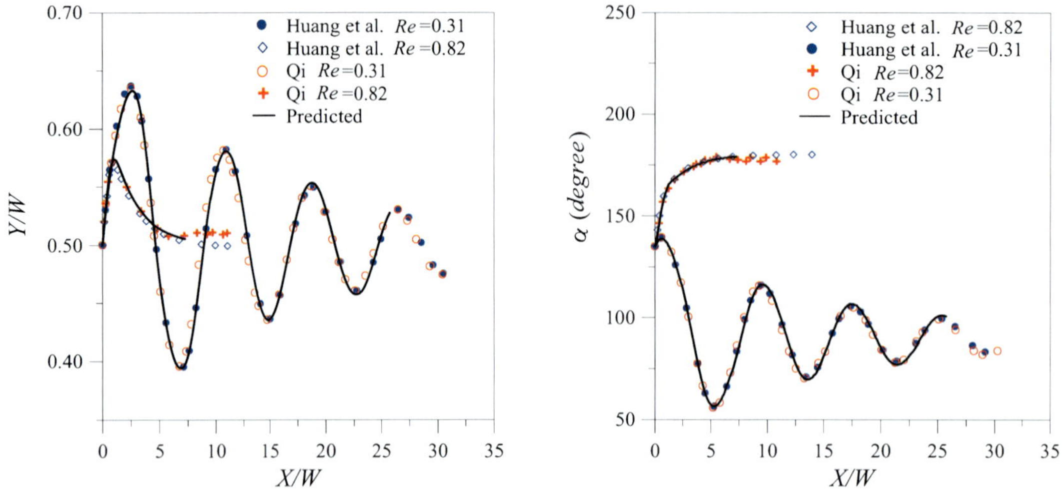

Figure 24: Comparison of the predicted trajectory Y/W and orientation angle α of an ellipse with the results of Huang *et al.* [89] and Qi [90] (X is the displacement at the gravity or vertical direction and Y at the horizontal direction).

Numerical simulation for a simple single particle is a foundation research, but has much significance for a real process. From these results, OBFCS and level set method are validated well with the experimental data, but OBFCS can satisfy mass conservation commendably while level set method is superior on dealing with deformable interfaces. Also, the mirror fluid method is effective in simulating real solid-fluid flows and can be extended to other more challenging particle flow problems with less restriction, especially when complicated and irregular solid-fluid interfaces are involved.

5. EFFECT OF SURFACTANT ON DROP MOTION BEHAVIOR

In most industrial extraction equipments, unavoidable trace quantities of surface-active contaminants may have profound effect on the behavior of drops and bubbles. Contaminant can eliminate internal circulation, significantly increase the drag and reduce mass transfer rate. More accurate account of the effect of surfactant can be resolved by the principles of computational fluid dynamics and transport. Such numerical approach is applied to the motion and mass transfer of a single liquid drop and the drop behavior is interpreted with reasonable precision. A numerical investigation of the flow field inside and around a deformed drop translating in another quiescent liquid contaminated by surfactant soluble in the continuous phase but not soluble in the dispersed phase is presented in this section.

5.1. Effect of Surfactant on Drop Motion

5.1.1. Governing Equations for Fluid Flow

The motion of a single drop in an infinite fluid medium may be described by a set of partial differential equations of stream function ψ and vorticity ω in a sectional plane (x, y) passing through the axis of symmetry. The nondimensional governing equations are Eqs. (41)-(44) [13].The physical velocity components are related to the stream function by Eqs. (47) and (48). The relevant nondimensional physical variables and parameters are defined by Eqs. (39), (40) and (46).

When a surfactant is present, the boundary conditions at the contaminated interface become different from those at a clean interface depicted by Dandy and Leal [13]. The shear and normal stress balance at the contaminated interface should be modified to take into account the effect of variable interfacial tension. The boundary conditions for Eqs. (41)-(44) become the following:

On the axis of symmetry ($\eta_1 = \eta_2 = 0, \eta_1 = \eta_2 = 1$):

$$\boldsymbol{\Psi}_1'' = \boldsymbol{\Psi}_2 = \mathbf{0}, \boldsymbol{\Omega}_1 = \boldsymbol{\Omega}_2 = \mathbf{0} \quad (108)$$

At the far domain boundary ($\xi_1 = 0$):

$$\boldsymbol{\Psi}_1'' = \boldsymbol{\Psi}_2 = \mathbf{0}, \boldsymbol{\Omega}_1 = \boldsymbol{\Omega}_2 = \mathbf{0} \quad (109)$$

At the drop surface ($\xi_1 = \xi_2 = 1$):

$$\boldsymbol{\Psi}_1'' = \boldsymbol{\Psi}_2 = \mathbf{0} \text{ (impermeable interface)} \quad (110)$$

$$\boldsymbol{U}_{\eta_1} = \boldsymbol{U}_{\eta_2} \text{ (kinematic continuity)} \quad (111)$$

$$\Delta\boldsymbol{\tau}_{\xi\eta} = \boldsymbol{\Omega}_1 - 2\boldsymbol{\kappa}'_{(\eta)}\boldsymbol{U}_{\eta_1} - \boldsymbol{\lambda}\left(\boldsymbol{\Omega}_2 - 2\boldsymbol{\kappa}'_{(\eta)}\boldsymbol{U}_{\eta_2}\right) = \frac{1}{CaH_{\eta_1}}\frac{\partial\sigma'}{\partial\eta_1} \text{ (shear stress balance)} \quad (112)$$

$$\Delta\boldsymbol{\tau}_{\xi\xi} = \frac{4\sigma'}{We}\left(\boldsymbol{\kappa}'_{(\xi)} + \boldsymbol{\kappa}'_{(\eta)}\right) \text{ (nomal stress balance)} \quad (113)$$

where Δ denotes the difference between tensor components in two bulk phase. It is more convenient to use the modified external stream function Ψ_1'':

$$\Psi_1'' = \Psi_1 - \frac{1}{2}Y_1^2(1 - \xi_1^3) \tag{114}$$

for this makes Ψ_1'' subject to a homogeneous boundary condition, Eq. (109).

In boundary conditions (112) and (113), three dimensionless parameters are introduced:

$$Ca = \frac{\mu_1 U_T}{\sigma_0}, We = \frac{2R\rho_1 U_T^2}{\sigma_0}, \sigma' = \frac{\sigma}{\sigma_0} \tag{115}$$

and it is evident that Eqs. (112) and (113) contain the changing surface tension σ' along the interface; thus the mass transfer of surfactant is involved with the solution of fluid flow, resulting in the coupling between them. In the above, the nondimensional surface tension, σ', appears as a variable and can be related to the local concentration of an absorbed surfactant by a suitable state equation. The simplest "ideal gas" equation of state for the adsorbed surfactant:

$$\sigma = \sigma_0 - RT\Gamma \tag{116}$$

can be used in cases with low surfactant concentration, where σ_0 is the surface tension of the clean interface, R is the universal gas constant, T is the absolute temperature, and Γ is the surfactant adsorption on the interface. This model is valid as long as the average distance between surfactant molecules at the interface remains large compared with the molecular size. Using viscous stress to scale the surface tension, the dimensionless form of the ideal gas law is:

$$\sigma' Ca^{-1} = Ca^{-1} - Ma\Gamma' \tag{117}$$

The relevant nondimensional parameter is defined by:

$$Ma = \frac{RT\Gamma_\infty}{\mu_1 U_T}, \Gamma' = \frac{\Gamma}{\Gamma_\infty} \tag{118}$$

When Γ is high, the Frumkin equation

$$\sigma = \sigma_0 + RT\Gamma_\infty \ln\left(1 - \frac{\Gamma}{\Gamma_\infty}\right) \tag{119}$$

should be used, where $\Gamma_\infty$is the maximum concentration of the absorbed surfactant. In the limit of $\Gamma \ll \Gamma_\infty$, the Frumkin equation reduces to the ideal gas equation of state. Similar to Eq. (116), Eq. (119) is nondimensionalized to:

$$\sigma' Ca^{-1} = Ca^{-1} + Ma\ln(1 - \Gamma') \tag{120}$$

Differentiating Eq. (120) and substituting into Eq. (112), the shear stress balance at the interface can be related to the surface concentration, which presents the first factor of coupling between the momentum balance and mass transport of surfactant:

$$\Delta\tau_{\xi\eta} = \Omega_1 - 2\kappa'_{(\eta)}U_{\eta_1} - \lambda\left(\Omega_2 - 2\kappa'_{(\eta)}U_{\eta_2}\right) = \frac{Ma}{(1-\Gamma')H_{\eta_1}}\frac{\partial\Gamma'}{\partial\eta_1} \tag{121}$$

This equation is the expanded form of interfacial shear stress balance derived for a general curvilinear orthogonal system. For a spherical drop in the spherical coordinate system, it reduces to:

$$\Delta\tau_{\xi\eta} = \frac{Ma}{(1-\Gamma')}\frac{\partial\Gamma'}{\partial\theta} \tag{122}$$

as used in Bel Fdhila and Duineveld [91]. Eq. (121) means that the variable surface tension at the interface, which creates additional tangential force to retard or accelerate the interfacial flow, must be compensated by the unequal shear stress at two sides of the interface, which are usually equal when the surface tension is a constant in pure systems.

5.1.2. Transient Convective Diffusion

It is assumed that the surfactant is soluble only in the continuous phase, and therefore, the mass transfer process occurs only in the continuous phase and on the interface. The mass transport outside the drop is depicted by a transient convective–diffusion equation in the vector form, like Eq. (19), which can be expanded into the form Eq. (49) when the orthogonal curvilinear coordinate system is applied.

For the mass transfer process in the continuous phase, the concentration of surfactant at the remote infinity c^{∞} is assumed to be constant, and used as the scaling factor for the concentration field. The boundary conditions at the interface need some interpretation. Since diffusion within the droplet is absent, the gradient of convective–diffusion flux of surfactant along the interface, as well as the flux of net sorption onto the interface, is equal to the diffusion flux from the continuous phase. Based on the physical notion, the boundary conditions of the transport equation can be written as:

$$\nabla_s \cdot (u_s \Gamma) - D_s \nabla_s^2 \Gamma = -D[\mathbf{n} \cdot \nabla c]_s \tag{123}$$

In this equation, D_s is the surface diffusion coefficient, Γ is the surface concentration of surfactant, ∇_s is the surface counterpart of operator ∇, $\mathbf{n}$ is the normal unit vector pointing into the droplet, u_s is the convective velocity at the interface, and ∇c is the concentration gradient of surfactant in the bulk. In the present numerical procedure, the boundary condition is also nondimensionalized to:

$$\nabla_s' \cdot (U_s \Gamma') - \frac{1}{Pe_s} \nabla_s'^2 \Gamma' = -\left[\mathbf{n} \cdot \nabla' \left(\frac{C}{Pe}\right)\right]_s \tag{124}$$

and the following nondimensional variables and groups are defined for this purpose:

$$U_s = \frac{u_s}{U_T}, \Gamma' = \frac{\Gamma}{\Gamma_\infty}, Pe_s = \frac{dU_T}{D_s}, Pe = \frac{U_T \Gamma_\infty}{Dc^\infty} \tag{125}$$

In our numerical work, the expanded form of Eq. (125) in the orthogonal curvilinear coordinate system reads:

$$-\frac{Pe_s}{Pe} f_1 Y_1 \frac{\partial C}{\partial \xi_1}\bigg|_{\xi_1=1} + \frac{Pe_s}{2}\frac{\partial (U_s Y_1 \Gamma')}{\partial \eta_1} = \frac{\partial}{\partial \eta_1}\left(\frac{Y_1}{H_{\eta_1}}\frac{\partial \Gamma'}{\partial \eta_1}\right) \tag{126}$$

Considering that the sorption of the surfactant at the interface reaches steady state and obeys Langmuir's kinetic law, the boundary condition at interface can also be formulated as:

$$-D[\mathbf{n} \cdot \nabla c]_s = Bc_s(\Gamma_\infty - \Gamma) - A\Gamma \tag{127}$$

In this equation, B and A are the adsorption and desorption rate constants respectively, and c_s is the sublayer concentration of surfactant immediately at the interface in the continuous phase. Using the nondimensional variables and groups:

$$Bi = \frac{d\alpha}{U_T}, M = \frac{\beta c_s}{\alpha}, C_s = \frac{c_s}{c_\infty} \tag{128}$$

where M is the dimensionless bulk concentration, Eq. (127) can be nondimensionized to:

$$-\left[\mathbf{n} \cdot \nabla \left(\frac{C}{Pe}\right)\right]_s = Bi(MC_s(1-\Gamma') - \Gamma') \tag{129}$$

Similarly in the external orthogonal grid, the expanded form of Eq. (129) reads:

$$-\frac{1}{Peh_{\xi_1}}\frac{\partial C}{\partial \xi_1}\bigg|_{\xi_1=1} = Bi(MC_s(1-\Gamma') - \Gamma') \quad (130)$$

Eq. (126) is itself a differential equation, and its boundary condition is:

$$\frac{\partial \Gamma'}{\partial \eta_1} = 0 \text{ (at the front and rear stagnant points)} \quad (131)$$

Beside the boundary conditions at the interface, the following initial and boundary conditions are specified for the governing equation (Eq. (51)):

$$C(\xi_1, \eta_1, \theta) = 1 \; (\theta = 0) \quad (132)$$

$$C(\xi_1, \eta_1, t) = 1 \text{ (at the far remote boundary, } \xi_1 = 0) \quad (133)$$

$$\frac{\partial C}{\partial \eta_1} = 0 \text{ (on the axis of symmetry, } \eta_1 = 0) \quad (134)$$

Validation of the present numerical procedure is focused on the part extended from that used in our previous work [16, 92], that is, the part of the computer code related to adsorption kinetics and interfacial transport of the surfactant. The underlying part of the program with respect to deformable, surfactant-free droplet dynamics has essentially been done previously by Li [92] and Mao *et al.* [16], who used a computer code based on the same principles and simulated a series of single drops with satisfactory comparison with the experimental data and the literature. To validate the mathematical procedure, the relevant numerical simulation [10] was run for drops at creeping flow for comparison with analytical results.

5.1.3. Effect of Surfactant

It is interesting to investigate the effects of various dimensionless groups on the flow field inside and outside the drop and to observe the surface concentration profile at the interface. A series of parametric studies were conducted with only one variable changed from the base set of parameters. The influencing nondimensional groups include $Re, We, \lambda, \zeta, Ma$, which arise in the Navier–Stokes equations and boundary conditions, and Bi, M, Pe, Pe_s, Pe^*, which arise in the convective–diffusion equation and boundary conditions. Among them, the density ratio and viscosity ratio were chosen to be somewhat arbitrarily fixed at representative $\lambda = 2.0$ and $= 2.0$, respectively. From existing simulation of drop flow in pure systems, it was observed that the flow structure was relatively insensitive to λ and ζ. Moreover, in drop flows, the bulk Peclet number, Pe, the sublayer Peclet number, Pe^*, and the surface Peclet number, Pe_s, are usually very large because of small diffusion coefficients of a surfactant in the liquid, and the distribution of a surfactant is dominated solely by surface convection and adsorption kinetics. Therefore, the stagnant-cap regime was thought the most common physically realizable regime [93], and only the effects of Biot number and dimensionless concentration on the motion of a drop are expected to be significant and worthy of numerical investigation. According to the above analysis, the reference state with $Re = 100$, $We = 0.5$, $\lambda = 2.0$, $\zeta = 2.0$, $Ma = 100$, $Bi = 0.1$, $M = 1 \times 10^{-2}$, $Pe = 10^3$, $Pe_s = 10^5$, $Pe^* = 10^3$ is selected. Since the Weber number is 0.5, the deformation of the drop is negligible and the emphasis is focused on the effect of the physicochemical properties of surfactants on the flow field of a nearly spherical drop. It should also be noted that the representative value of the Marangoni number, which characterizes the influence of surface concentration gradient on the motion of a droplet, was found to be about 100 at moderate Reynolds numbers.

The effect of M on the motion of the drop is first examined. M, the dimensionless concentration, characterizes the ratio of desorption to adsorption rates as indicated by Eq. (127). When the bulk concentration is large, the surface adsorption at the rear of the droplet should be accumulated to a high level due to adsorption flux in the front of the droplet, Γ reaches a very high value on the rear part of the droplet and significant $\nabla_s \sigma$ is built up. In this case, the surfactant has great effect on the motion of the droplet. In contrast, when $M \ll 1$, the surface adsorption over the whole interface is low and nearly constant at the value in equilibrium with the bulk phase. In this case, the surface tension gradient is small and exerts little effect on the motion of droplets.

Fig. **25** [10] illustrates the effect of variation in M on the flow field with M in the range from 0 to 0.1. In a pure system (M= 0), no recirculating wake is formed behind the drop as shown in Fig. **25**(a). However, when the drop is slightly contaminated (M= 0.005), a detached recirculating wake shown in Fig. **25**(b) appears. From the comparison among Figs. **25**(b) through **25**(e), it is noted that detached wakes become progressively closer to the drop rear surface when M increases. Eventually, when $M = 0.1$, it becomes difficult to tell whether the recirculating wake is marginally detached or simply attached. The recirculating wake behind a drop becomes progressively larger in size and shifted closer to the drop rear surface as M increases, which can be explained by the fluid mechanics and the theory of vorticity accumulation. The interfacial profiles of relevant variables (for example, adsorption, vorticity, velocity, interfacial tension) provide evidence quite contrary to the basic assumptions of the stagnant-cap model. The stagnant-cap model seems to lack the potential of becoming a predictive and quantitative tool for modeling the drop flow under the influence of surfactant and it saves little numerical work in the corresponding simulation [10].

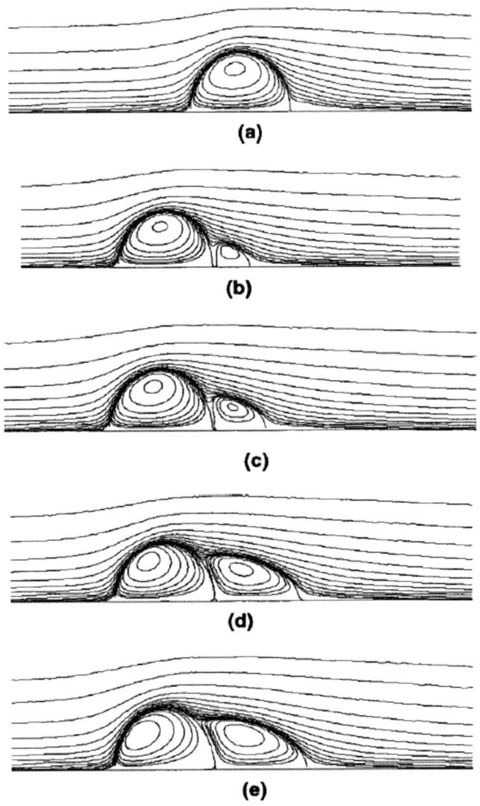

Figure 25: Influence of M on flow structure (contours of stream function) Conditions: (***a***) $M = 0$ (Pure system), $C_D = 0.796$; **(b)** $M = 0.005, C_D = 0.834$; **(c)** $M = 0.01, C_D = 0.855$; (***d***) $M = 0.04, C_D = 0.855$; (e) $M = 0.1$, $C_D = 0.956$.

Fig. **25** also demonstrates that the size of the recirculating wake is largely determined by M. According to Leal [94], the volume of recirculating wake depends on the relative rate of production of vorticity generated upstream on the drop surface and its transport downstream into the wake region by convection and lateral diffusion. From the interfacial vorticity distribution around the drop (Fig. **26**(b)), it is observed that the vorticity produced at the drop surface is increased, apparently due to reduced surface mobility as indicated by Fig. **26**(a). As M is increased the vorticity at the equator is increased significantly while it becomes negative at the rear surface, resulting in a higher driving force for lateral diffusion into the wake. It follows naturally that the accumulation of vorticity is promoted in favor of onset of recirculation in the wake. From the above, that the recirculating wake expands in volume as M increases seems a natural deduction from the theory of vorticity accumulation.

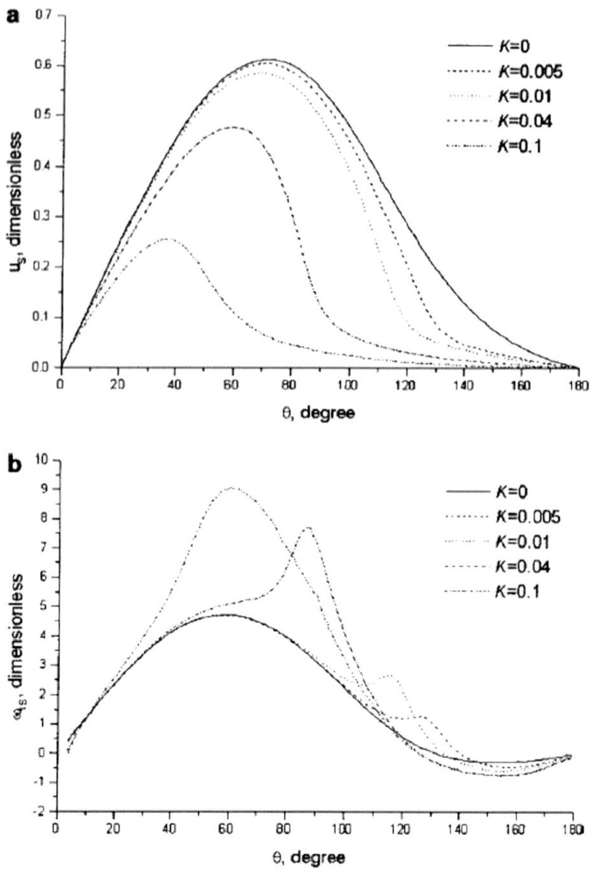

Figure 26: Distribution of tangential velocity and vorticity at the surface for the cases as in Fig. **25**.

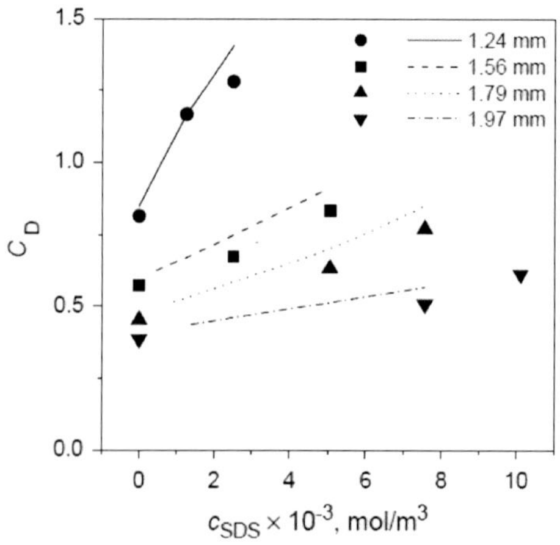

Figure 27: The effect of SDS concentration in the bulk phase on the drag coefficient of a MIBK drop with equivalent diameter varied from 1.24 to 1.97 mm (Solid symbols stand for the experimental data and lines for the simulated results).

For the experimental cases, the numerical drag coefficients of single MIBK drops whether or not contaminated by sodium dodecyl sulphate (SDS) are compared with the experimental data in Fig. **27** [95]. For the same diameter drop, it is obvious that the drag coefficient increases gradually with the increment of the SDS concentration as the drop progressively behaves more like a rigid sphere. It is also seen from Fig. **27** that a much higher SDS concentration is needed for a larger drop to reach such an asymptotic state.

5.2. Effect of Surfactant on Mass Transfer

The study of mass transfer of solutes to/from a single drop with adsorbed surfactants on the interface provides an essential base for better understanding of liquid extraction process and for scientific design of the equipments. A numerical approach on the effect of a surfactant on the motion of a buoyancy-driven drop at intermediate Reynolds numbers has been described in the above sections and explained in details in the previous work [10]. The transient convective diffusion equation is dealt with the same method shown in section 2 and adopts the same initial and boundary conditions with section 5.1. A large enough region for the exterior phase is chosen and the finite domain distortion function Eq. (32) is applied [15]. And for the interior phase, the distortion function is simply Eq. (33) [53].

If there is any interfacial resistance (or so-called energy barrier) caused by added surface active agents in a liquid-liquid extraction system, the solute concentration difference, $(mc_1^s - c_2^s)$, will prevail at the interface. The resistance can be evaluated from the concentrations in the vicinity of the interface during mass transfer, which is a direct indication of the barrier mechanism. On the other hand, when the drop surface is covered by the adsorbed surfactant, the mass transfer area is simply assumed to be $(1 - \Gamma')$ of interfacial area. Then the total instant diffusion flux over the interface can be formulated as:

$$N = k_s(1 - \Gamma')(mc_1^s - c_2^s) \tag{135}$$

In this equation, the interfacial mass transfer coefficient, k_s, is introduced, which can be evaluated later by the least square fitting. The following boundary conditions can be obtained on the local flux conservation of the solute through the surfactant film:

$$\frac{D_1}{h_{\xi_1}}\frac{\partial c_1}{\partial \xi_1} = -\frac{D_2}{h_{\xi_2}}\frac{\partial c_2}{\partial \xi_2} = k_s(1 - \Gamma')(mc_1^s - c_2^s) \tag{136}$$

Thus, Eq. (136) can be used as one of the boundary conditions to solve the convective diffusion equations, and the joint effect of hydrodynamics and energy barrier on mass transfer can be evaluated numerically. It can be shown by variable transformations that the characteristic value of Sh_{od} from the simulation is the same, no matter whether the direction of mass transfer is c → d or d → c.

In order to check the applicability of the developed numerical scheme, transfer of acetic acid from aqueous continuous phase to a rising MIBK drop is conducted in a droplet file column, and the surfactants used are sodium dodecyl sulphate (SDS), Triton X-100 and Tween 80. The effect of surfactants on terminal velocity, extraction fraction is obtained experimentally. Compared with the experimental results, the validation of the numerical procedure can be illustrated. The pertinent details are explained by Li *et al.* [95]. It is found that surface-active agents have great effects on the motion and mass transfer rate of MIBK drops, and that the rising velocity and mass fraction decrease gradually with the increase of surfactant concentration.

5.2.1. Hydrodynamic Effect on Transient Mass Transfer

In transient mass transfer to a MIBK drop contaminated by SDS is simulated. The comparison between the results predicted by numerical simulation and the experimental data is shown in Fig. **28** [95]. It is observed that the numerical simulation is quite successful, and both the extraction fraction and the overall mass transfer coefficient decrease with the increase of SDS bulk concentration. When SDS bulk concentration attains 5.04×10^{-3} mol/m^3, the value of k_{od} is only about one third of that in the pure system. This can be explained by the distribution of interfacial velocity and the local mass transfer coefficient at $t = 0$ along the drop surface, which are plotted against θ in Figs. **29** and **30** [95]. It is observed that the local Sherwood number at the nose ($\theta = 0°$) is, in general, much higher than that in the rear or wake region (θ approaching 180°). When the MIBK is contaminated heavily, it is found from the profiles of u_η that the interfacial surface becomes stagnant at the rear stagnant region and is much less than that shown in Fig. **29**. Correspondingly, the value of Sh_{loc} drops down abruptly along the stagnant surface (Fig. **30**) and assumes much lower values than that in the pure system (Fig. **28**).

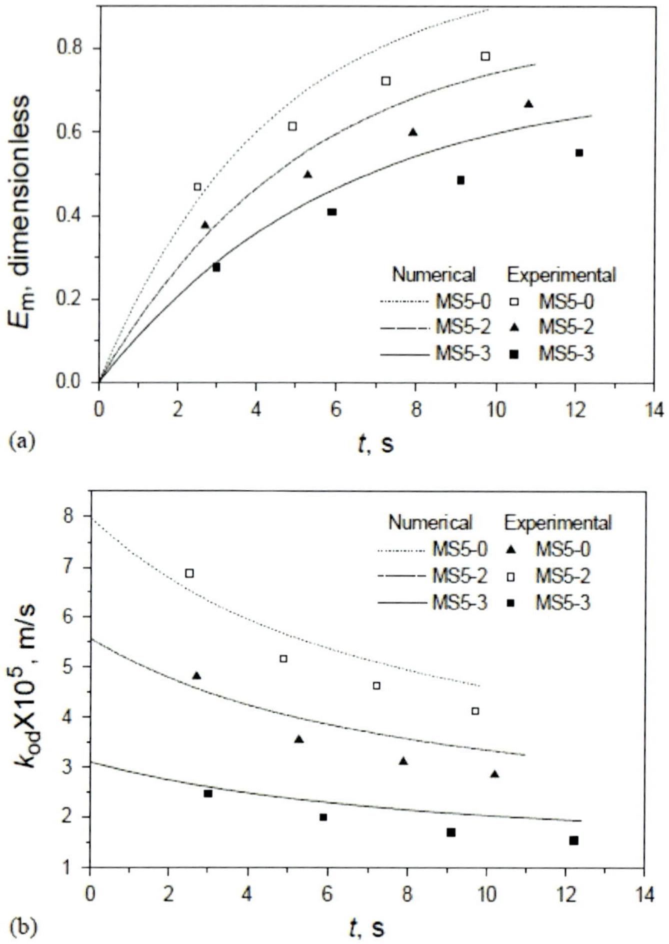

Figure 28: Contrast of simulation with experiment on the relationships of (a) extraction fraction E_{m}, and (b) overall mass transfer coefficient k_{od} with mass transfer time (d = 1.56 mm).

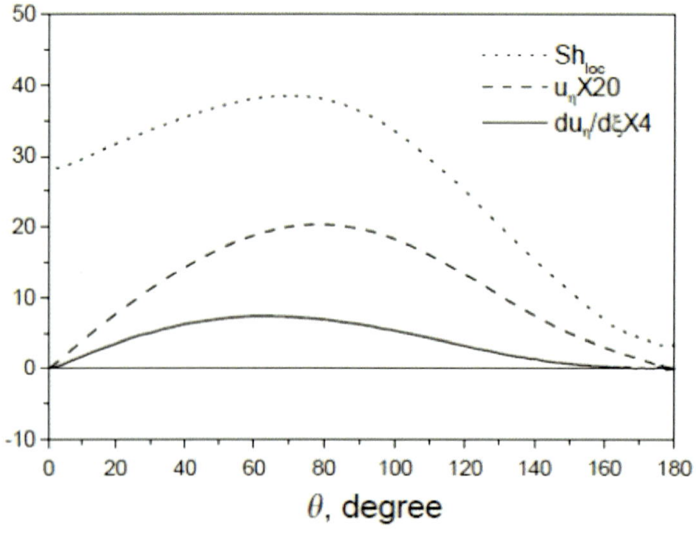

Figure 29: Profiles of interfacial shear, surface velocity and local Sherwood numbers (t = 0) along the clean drop surface (d = 1.56 mm).

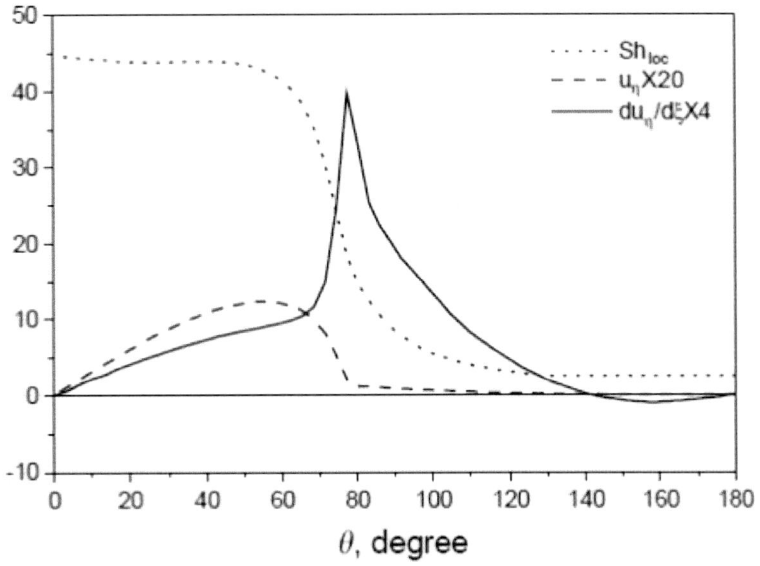

Figure 30: Profiles of interfacial shear, velocity and local Sherwood numbers ($t = 0$) along the heavily contaminated drop surface ($d = 1.56$ mm).

5.2.2. Effect of Interfacial Resistance on Transient Mass Transfer

To identify possible interfacial resistance to mass transfer by adsorbed surfactant layer, the mass transfer to a Tween 80 contaminated MIBK drop is simulated numerically, and the hydrodynamic and energy barrier contribution on mass transfer are taken into account altogether. It is found for the clean extraction system that the numerical results is about 12% higher than the experimental data (Fig. **31**) [95], whereas, about 30% higher for the Tween 80 contaminated system (Fig. **32**) [95]. Compared with the simulation results of the SDS contaminated system, it is deduced that Tween 80 molecules adsorbed on the interface has evident mass transfer resistance ($1/k_S$) on acetic acid, which can be estimated by the least square fitting to the experimental data. It can be noted that interfacial resistance in a surfactant monolayer film increases ten times higher than in a clean film, due to physic-chemical interaction between surface active agent and solute molecules.

To resolve the reliable prediction of motion and mass transfer in swarms of bubbles or drops, the study of hydrodynamic behavior of single bubbles and drops contaminated is of fundamental importance, as it provides a useful starting point toward the understanding of more complicated and practical multiphase systems. With the validated numerical procedure, the effect of surfactant on the motion and mass transfer of single drops in contaminated systems is examined in this section. The merits and defects of the stagnant cap model are discussed against the detailed results obtained from numerical simulation. It is also indicated that the numerical simulation can be used to predict the motion and mass transfer of a single drop, no matter whether it is contaminated by surfactant.

6. SUB-PARTICLE FLOW PATTERN AND MASS TRANSFER

6.1. Introduction

The well-known Marangoni effect on the sub-particle scale is an important factor in many practical cases for interpreting the inetrphase mass transfer and interfacial flow. With the induced local vortices at the interface, the convective currents normal to the interface would add to the interphase mass transfer otherwise dominated by molecular diffusion. The numerical approach is now available to simulate the surface tension gradient induced Marangoni convection at the interface and the thereby increased interphase mass transfer. The resultant information can shed light to the mechanisms based on which the interphase interaction is to be either promoted or suppressed.

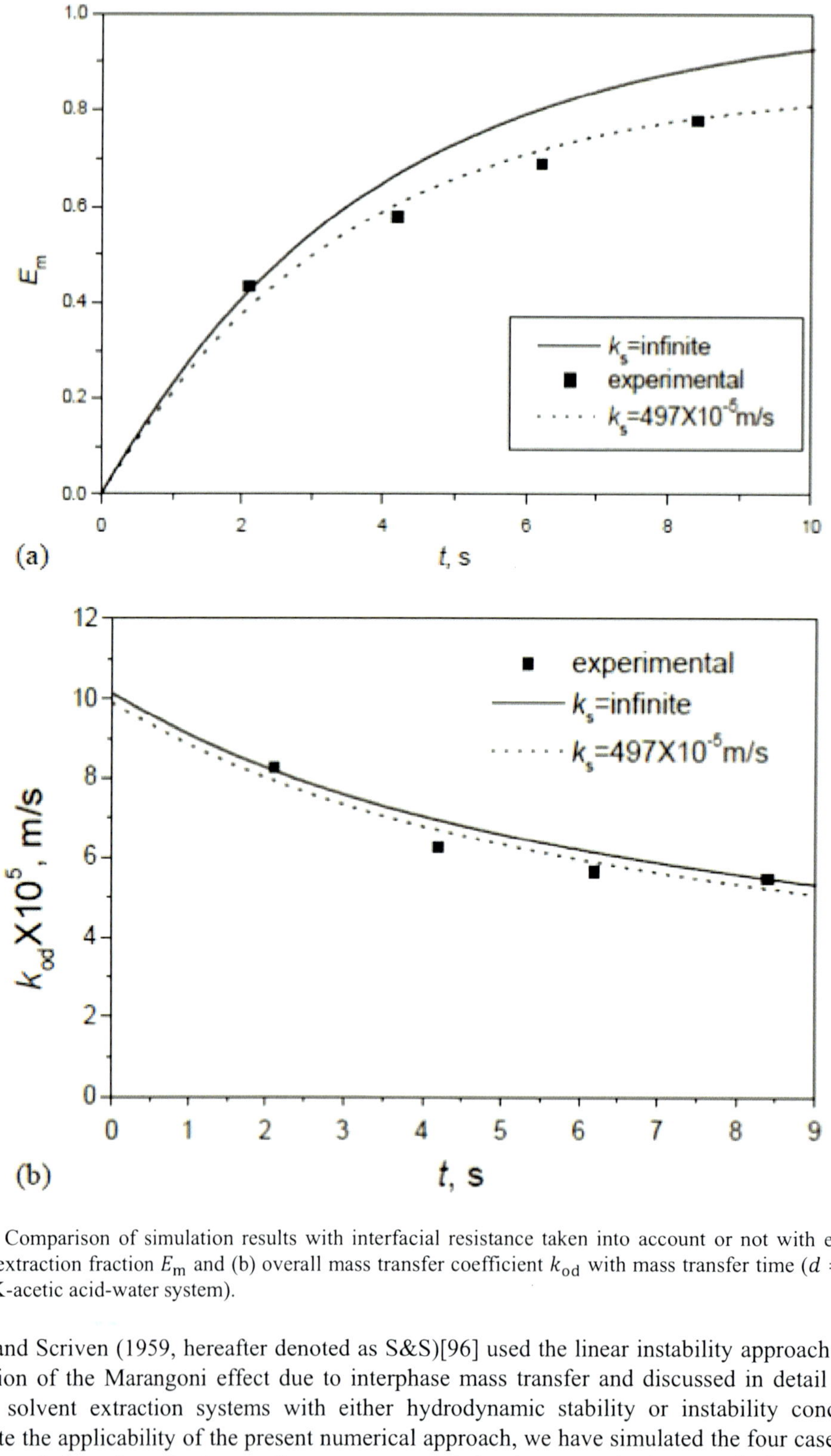

Figure 31: Comparison of simulation results with interfacial resistance taken into account or not with experimental data of (a) extraction fraction E_m and (b) overall mass transfer coefficient k_{od} with mass transfer time (d = 1.79 mm, clean MIBK-acetic acid-water system).

Sternling and Scriven (1959, hereafter denoted as S&S)[96] used the linear instability approach to analyze the induction of the Marangoni effect due to interphase mass transfer and discussed in detail four cases typical to solvent extraction systems with either hydrodynamic stability or instability concluded. To demonstrate the applicability of the present numerical approach, we have simulated the four cases adopting the orthogonal boundary-fitted coordinate system and level set method respectively. The original specifications of the cases studied are listed in Table **7**, and the corresponding non-dimensional parameters to start the numerical simulation are in Table **8**. In the four cases, label 1 denotes the mass transfer from the continuous phase to the drop phase and 2 denotes the reverse mass transfer direction.

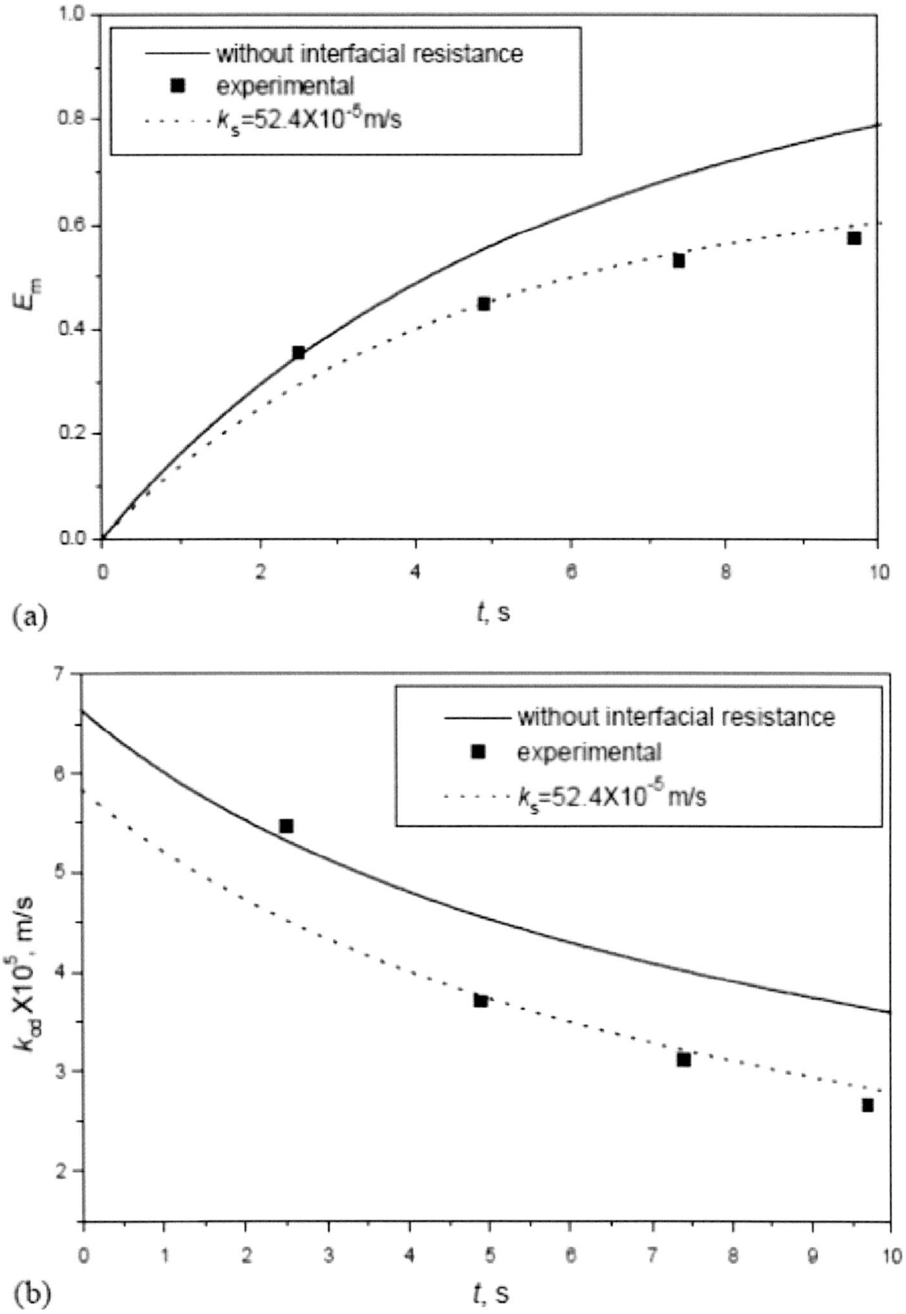

Figure 32: Comparison of simulation results with interfacial resistance taken into account or not with experimental data of (a) extraction fraction E_m and (b) overall mass transfer coefficient k_{od} with mass transfer time ($d = 1.79$ mm, $c_{Tween\ 80} = 0.005$ mol m^{-3}, heavily contaminated drop).

Table 7: Cases analyzed by Sternling and Scriven (1959).

Case	A	B	C	D
ν_2/ν_1	0.5	0.5	0.5	0.5
D_2/D_1	2	1	0.667	0.25

Note: $\mu_1 = 0.001$ Pa · s, $D_1 = 2 \times 10^{-9}$ m^2/s, $m = 1$, $\mu_2/\mu_1 = 2$, $d\sigma/dc = -10^{-6}$ N · m^2/kg.

Table 8: Parameters for current numerical simulation of Marangoni effect for a spherical drop corresponding to Table **7**.

Case	A	B	C	D
$\lambda = \mu_2/\mu_1$	2	2	2	2
$\zeta = \rho_2/\rho_1$	4	4	4	4
Re_1	0.1	0.1	0.1	0.1
Re_2	0.2	0.2	0.2	0.2
Pe_1	500	500	500	500
Pe_2	250	500	750	2000

Note: $D_1 = 2 \times 10^{-9}\ \mathrm{m^2/s}$, $m = 1$, $Ca = 5 \times 10^{-4}$, $\delta = -0.1$.

6.2. Simulation by Orthogonal Boundary-Fitted Coordinate System

The Marangoni effect is mathematically formulated and numerically simulated for slowly moving single spherical drops with the orthogonal boundary-fitted coordinate system. To resolve the Marangoni effect underlain by local interfacial hydrodynamic disturbance, the Navier-Stokes equation and continuity Eqs. (41)-(44) are to be solved in an axisymmetrical coordinate system. The possible local fluctuation of solute concentration at the drop surface would trigger surface tension gradient, which will in turn alter the local liquid flow near the interface in both phases. Thus, the convective diffusion equation of solute transport (Eq. (51)) must be solved at the same time in a coupled way. A necessary link between the two sets of governing equations is the constitutive equation describing the influence of the solute concentration on the surface tension of the liquid-liquid system.

Special attention is given to the balance of tangential force along the interface, where the variation of interfacial tension also contributes to the balance, as done by Li and Mao [10]:

$$\mu_1\left(\omega_1 - 2\kappa_{(\eta)}u_{\eta_1}\right) - \mu_2\left(\omega_2 - 2\kappa_{(\eta)}u_{\eta_2}\right) = -\frac{1}{h_{\eta_1}}\frac{\partial \sigma}{\partial \eta_1}\ \text{(shear stress balance)} \tag{137}$$

When non-dimensionalized, it is turned into the following form:

$$\Omega_1 - 2\kappa'_{(\eta)}U_{\eta_1} - \lambda\left(\Omega_2 - 2\kappa'_{(\eta)}U_{\eta_2}\right) = -\frac{\sigma_0}{\mu_1 U_T H_{\eta_1}}\frac{\partial \sigma'}{\partial \eta_1} = -\frac{1}{CaH_{\eta_1}}\frac{\partial \sigma'}{\partial \eta_1}\bigg|_{\xi_1=1} \tag{138}$$

with the surface tension variation due to variable solute concentration explicitly incorporated. Since surface tension is variable, the normal stress balance across the interface is also different from a system without solute mass transfer, but the normal balance becomes irrelevant due to the present assumption of spherical shape despite the local Weber number is variable along the interface. $\kappa'_{(\eta)}$ is the dimensionless surface curvature in the plane passing through the axis of symmetry, which is equal to 1 for a spherical drop as in our case. In this formulation, the solute at the surface is assumed to behave in the same manner as in the bulk phase (no adsorption of solute onto the interface). Therefore, no independent governing equation appears for the interface as a separate phase.

The link between the surface tension and the mass transfer of solute is necessary for providing the surface gradient appearing in Eq. (138). The relation of surface tension with the solute concentration is different from system to system and non-linear in general. The simplest model is the linear one:

$$\sigma' = 1 + \delta C_1^s \tag{139}$$

with the surface tension sensitivity coefficient δ as an index of the system to the effect of solute concentration. In most cases, δ is negative, reflecting the fact that a solute usually decreases the interfacial tension.

Numerical solution of Eqs. (41)-(44) under suitable boundary conditions provides the external and internal flow fields and the flow parameters relevant for numerical simulation of simultaneous interphase mass transfer. Limited by space, the details of numerical solution of fluid flow are referred to the literature [10, 13, 96].

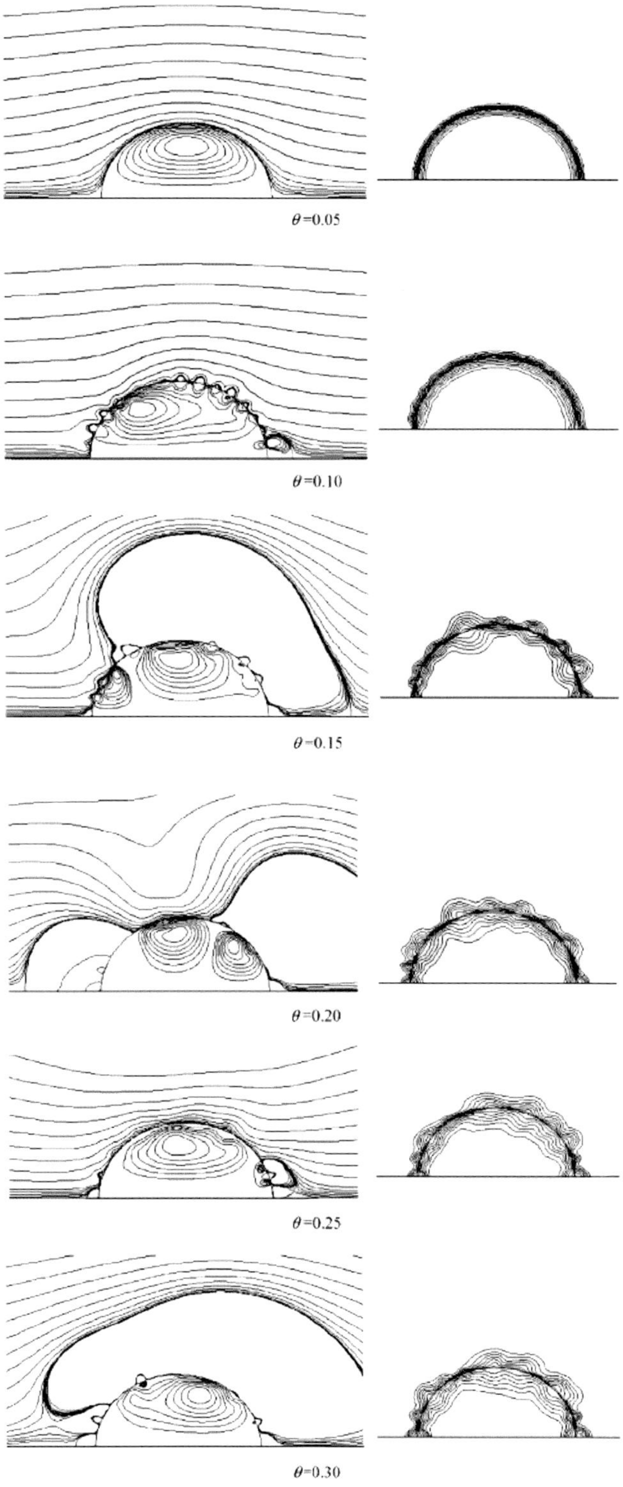

Figure 33: Marangoni effect induced evolution of flow patterns (left) and concentration contours (right) with time for a spherical drop (case A1) with $Re_1 = 0.1$, $\lambda = 0.5$, $\zeta = 0.91$, $Pe_1 = 100$, $Pe_2 = 100$, $Ca = 2 \times 10^{-4}$, $\delta = -1$, $m = 2$, concentration perturbation $\delta_p = 0.0001$.

To be closer to pendent drops in most experimental studies, the object of the simulation is a drop with $Re_1 = 0.1$. It is believed that the toroidal Marangoni roll cells on a slowly moving drop are more rational than that for a stationary drop, because the toroidal structure is more consistent with the situation of the flow and transport dominated by external axial flow in the continuous phase. In this case, the surface concentration is disturbed artificially in each time step:

$$\widetilde{C_1^s} = C_1^s(1 + \delta_p \Lambda) \tag{140}$$

where Λ is a pseudo-random number in the range from -1 to 1, and δ_p is a constant to designate the level of perturbation to the surface solute concentration.

Fig. **33** [97] is the illustration of development of surface instability with the solute transfer from the continuous phase into a blank drop. At $\theta = 0.05$, no surface turbulence appears, the flow field is similar to the steady flow without mass transfer and the concentration contour lines are concentrically distributed at two sides of the drop surface, because the interfacial turbulence needs certain time for itself to develop.

The surface Marangoni effect develops to a significant level only when the normal gradient of solute concentration at the surface decreases to a lower level, as indicated for $\theta = 0.10$. At this moment the Marangoni convection patterns much smaller than the drop scale appear in many isolated locations on the surface. When they develop larger in size so that they may bridge two ends of the concentration boundary layer, the convective transport in the normal direction to the interface becomes effective and displays its significant role in enhancing the interphase mass transfer. From the concentration contour plots at $\theta = 0.10$, the sub-drop scale Marangoni convection patterns are quite apparent. It is conjectured that these small patterns are more effective in mass transfer promotion than the large vortexes on the drop scale as occurring at $\theta = 0.15, 0.20$ and 0.30, since the large patterns cover mostly a wide area with C_1 very close to unity, where the convection contributes little to the overall mass transfer. The streamline maps also demonstrate that the surface turbulence is dynamic and transient in nature, the patterns change from time to time, and conceivably are swept downstream by the external bulk flow. As the mass transfer is concerned, up to $\theta = 0.30$ the mass transfer enhancement remains dominant since besides the large vortexes there exist sub-drop scale interfacial convection patterns at the drop surface.

Since the capillary number Ca is the key parameter to induction of the Marangoni effect, a parameter sensitivity test would be interesting. Fig. **34** [97] shows that the Sh_{oc} curve changes with the value of Ca, and therefore further numerical tests are performed. When Ca takes 2.5×10^{-3} or lower, the Marangoni effect occurs, but when it is assigned to be 3×10^{-3}, 5×10^{-3} or 1×10^{-2} with $\delta = 0.1$, no Marangoni effect on mass transfer is observed and the Sherwood number drops to the level of the base case with $\delta = 0$. It is evident that there is a critical value of $-\delta/Ca$ below which no Marangoni convection and enhancement of interphase mass transfer would be observed. It is also observed in coincidence with our intuitive guess that the larger the absolute value of $-\delta/Ca$, the earlier the Marangoni effect occurs.

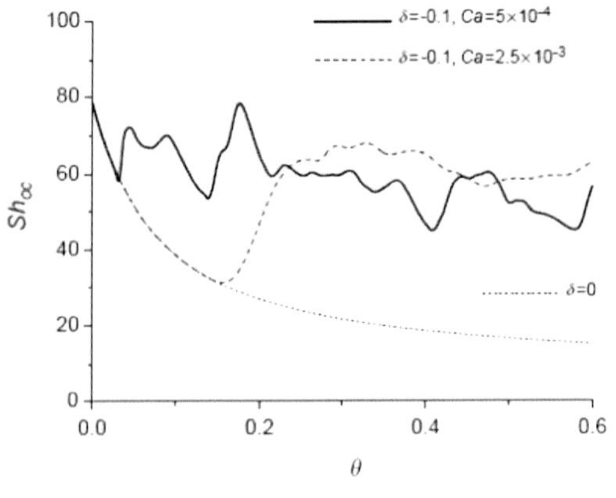

Figuer 34: Sherwood number for mass transfer outwards from a spherical drop (case A2) as influenced by the Marangoni effect ($Re_1 = 0.1, m = 1$).

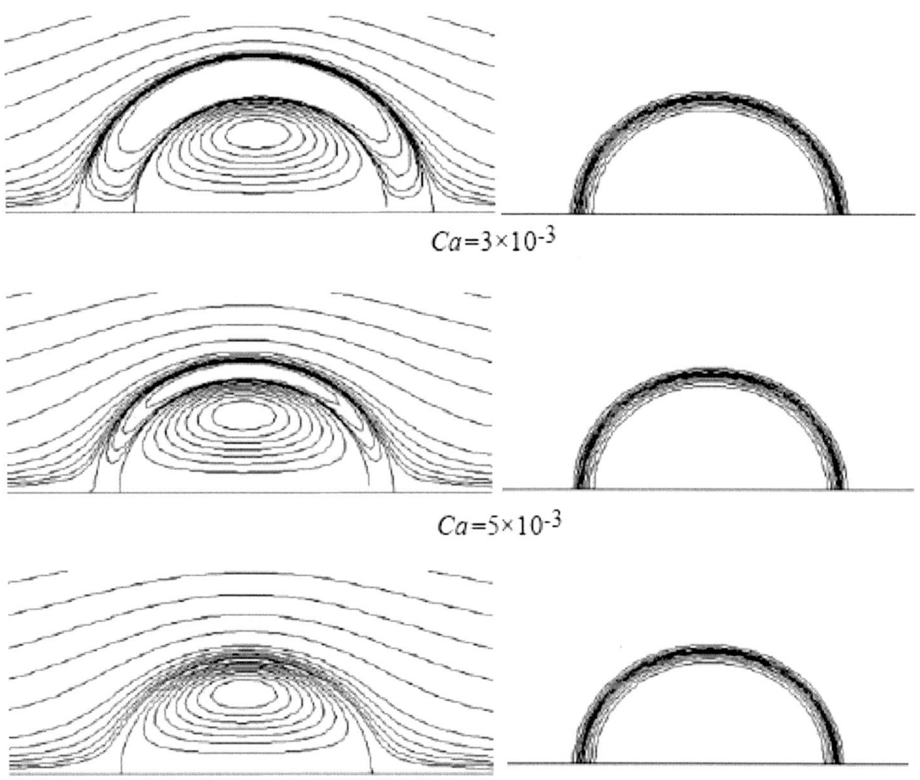

$Ca=3\times10^{-3}$

$Ca=5\times10^{-3}$

$Ca=1\times10^{-2}$

Figure 35: Variation of the Marangoni convection as influenced by Ca at $\theta = 0.30$ for case A2 when no mass transfer enhancement observed ($\delta = -0.1$).

Although no mass transfer enhancement is observed for the large Ca, the difference exists in the strength of the Marangoni convection. Fig. **35** [97] shows the different Marangoni convection patterns at $\theta = 0.30$ and the corresponding solute concentration contours. It is observed that the solute concentration contour maps consist of almost concentric circles, indicating the normal concentration gradient being little disturbed by possible surface concentration fluctuation. Although when $Ca = 3 \times 10^{-3}$, the Marangoni convection pattern is quite significant, but it wraps the drop and contributes little to interphase mass transfer rate. As Ca increases gradually, the surface tension stress becomes less sensitive to the solute concentration, leading to gradual vanish of the Marangoni convection pattern. When $Ca = 1 \times 10^{-2}$, the pattern is marginally observable, and the streamline map is now almost the same as that in the case without solute mass transfer.

6.3. Simulation by Level Set Method

Numerical simulation of the Marangoni effect induced by interphase mass transfer to/from deformable single drops in unsteady motion in liquid-liquid extraction systems is performed in a Eulerian axisymmetric reference frame. The drop surface evolution is captured by the level set method, in which the interface is represented by the embedded set of zero level of a scalar distance function defined in the whole computational domain.

The two-phase flow is governed by the continuity and Navier-Stokes Eqs. (1) and (2) and the source term **S** is the volume force induced by the surface tension, including the Marangoni effect. The volume force is based on the well-known Continuum Surface Force (CSF) model [98], so that the interfacial surface forces can be incorporated as body forces per unit volume in the momentum equations rather than as boundary conditions. **S** exerted on the interface [98] is expressed as:

$$\mathbf{S}(x) = \mathbf{f}_s(x)\delta(x - x_s) \tag{141}$$

where x_s is a point on the surface, δ the Dirac delta function, and $\mathbf{f}_s$ the surface tension force per unit interfacial area, is given by:

$$\mathbf{f}_s(x) = \sigma\kappa\mathbf{n} + \nabla_s\sigma \tag{142}$$

where σ is the surface tension coefficient, ∇_s the gradient operator along the interface, $\mathbf{n} = \nabla\phi/|\nabla\phi|$ the unit vector normal to the interface, and κ the mean curvature of the interface, $\kappa(\phi) = \nabla \cdot \mathbf{n}$. The first term in Eq. (140) is a kind of force acting normal to the interface, and the second term tangential to the interface in the direction of larger values of σ. The surface tension σ may be dependent on the temperature and concentration. If only the influence of concentration on the surface tension is considered, the following simple linear model is assumed:

$$\sigma = \sigma_0(1 + \delta C) \tag{143}$$

where δ is negative in most cases, reflecting that a solute usually decreases the interfacial tension. Eq. (140) is then expressed as:

$$\mathbf{f}_s(x) = \sigma\kappa\mathbf{n} - \frac{\partial\sigma}{\partial c}(\mathbf{I} - \mathbf{nn}) \cdot \nabla\mathbf{c} \tag{144}$$

where $\mathbf{I}$ is the identity matrix, and c the concentration. The source term $\mathbf{S}$ reads:

$$\mathbf{S}(x) = \left(\sigma\kappa\mathbf{n} - \frac{\partial\sigma}{\partial c}(\mathbf{I} - \mathbf{nn}) \cdot \nabla\mathbf{c}\right)\delta(\phi) \tag{145}$$

Thus, the continuity and Navier-Stokes Eqs. (1) and (2) can couple with the level set function to simulate the Marangoni effect induced by interphase mass transfer.

Fig. **36** [99] clearly shows that the Marangoni effect is a dynamic sub-drop scale phenomenon. Many local liquid circulation patterns much smaller than the drop itself appear at the both sides of drop surface, which promote the convective mass transfer across the concentration contour lines in the concentration boundary layer. These circulation patterns are swept downstream to the rear along the surface so that the drop surface is more frequently refreshed. These two factors promote in turn the interphase mass transfer on the drop scale. This mechanism suggests the significance of the spatial and temporal structures on a smaller scale when studying the physico-chemical phenomena on a larger scale.

Scrutiny of the streamline maps almost at the same instant $\theta = 0.2$, as shown in Fig. **37** [99], suggests that two figures show the similar strength of local circulation at the interface but the main vortices inside the drop are located at quite different positions, probably due to the different directions of mass transfer. From the positions of vortices, it may be conjectured that the surface tension gradient in these two cases plays the opposite effect on the drop surface mobility. In case A2, the surface flow would be retarded so that the interphase mass transfer is not promoted, while case A1 has promoted surface flow and hence the higher value of Sh number. More detailed simultaneous analysis of flow and concentration fields is needed before a thorough explanation can be provided.

Fig. **38** [99] compares the drop with the Marangoni effect with the one without the sensitivity of surface tension to the solute concentration at the same instant $\theta = 0.04$ after its release. From the velocity vector maps, we can see that near the interface there are a lot of small velocity vectors perpendicular to the interface for the case with the Marangoni effect; meanwhile, this kind of velocity vectors are missing for the case without Marangoni effect. We conclude that the small velocity vectors perpendicular to the interface cause the increase of the mass transfer rate and result in Marangoni instability.

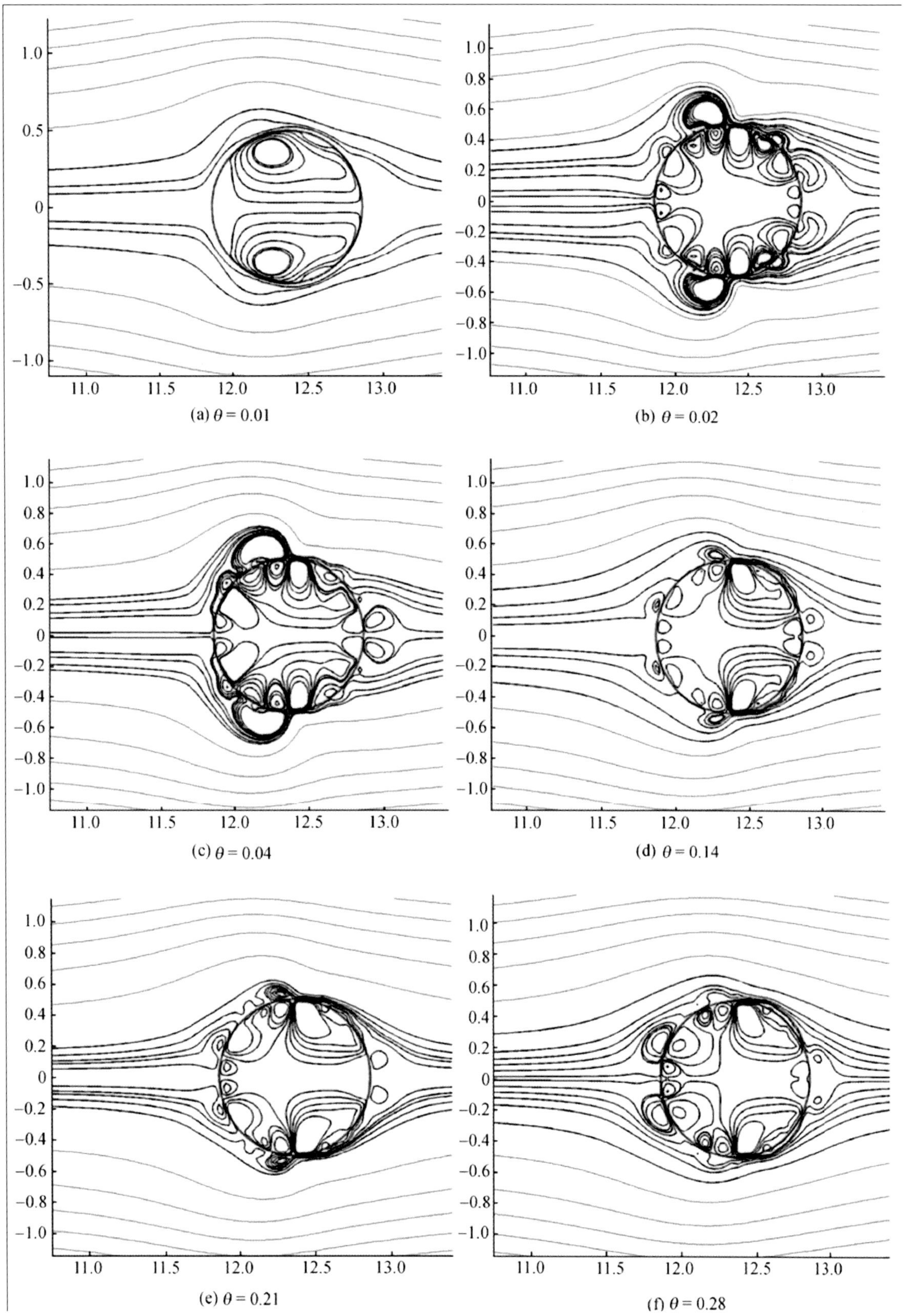

Figure 36. Marangoni effect induced evolution of flow patterns of Case A1 (c → d transfer, $\mu_1/\mu_2 = 2$, m = 1, $D_1 = 2 \times 10^{-9}\ \mathrm{m^2 \cdot s^{-1}}$, $\mu_1 = 0.001\mathrm{Pa \cdot s}$, $\delta = -0.1$, $\sigma_0 = 0.01\ \mathrm{N \cdot m^{-1}}$).

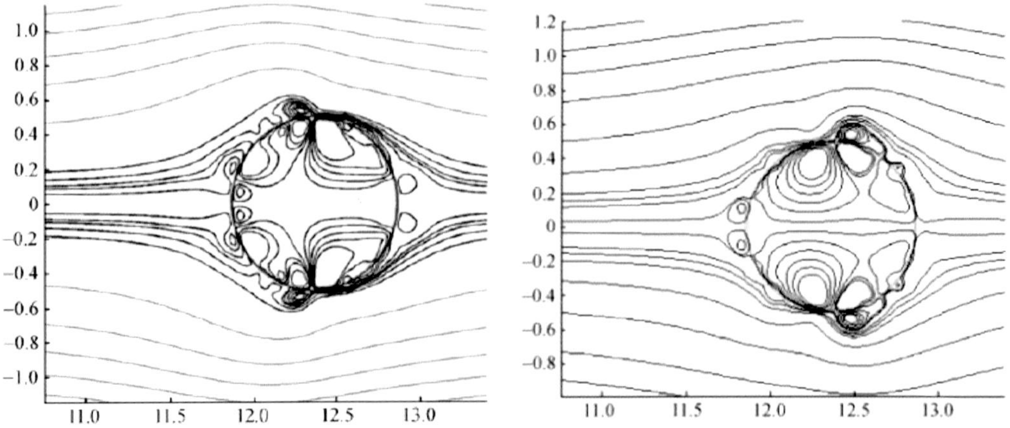

Figure 37: Marangoni effect induced evolution of flow patterns of Case A1 (left) and A2 (right) at dimensionless time $\theta = 0.2$.

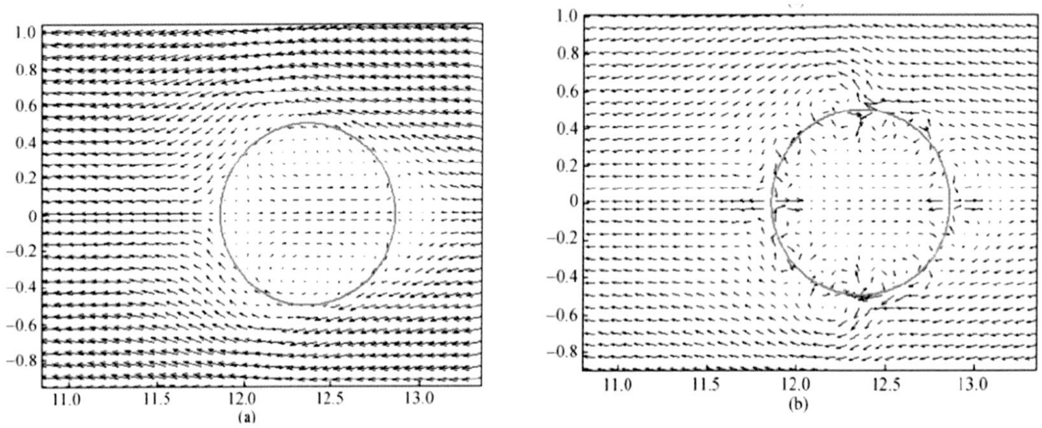

Figure 38: Velocity vector plots. (a) having no Marangoni effect (c $\rightarrow$ d transfer, the physical properties are the same as the case A1 in Table **8**); (b) having Marangoni effect (c $\rightarrow$ d transfer, the physical properties are the same as case A1 in Table **8**, $\theta = 0.04$.)

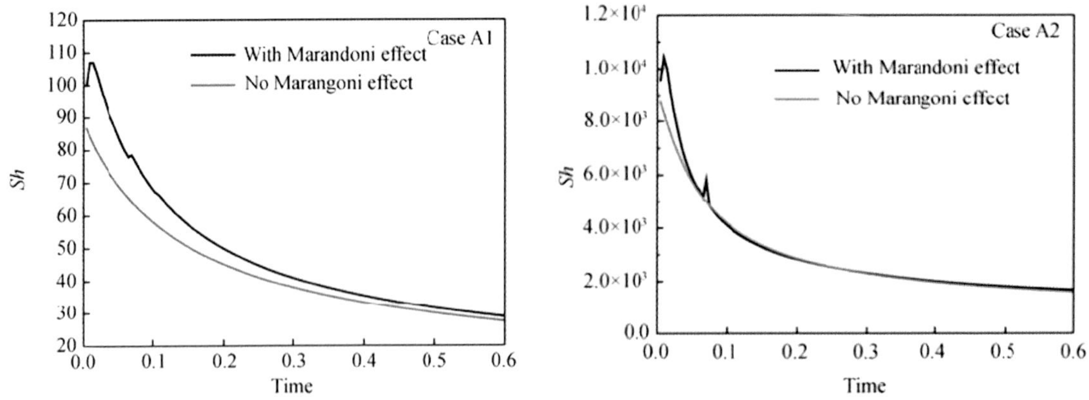

Figure 39: Influence of Marangoni effect on the Sherwood number for mass transfer to(left)/from(right) a deformable drop (case A).

Fig. **39** [99] presents the results of Sh for the continuous phase to the drop mass transfer (c-to-d transfer) and the counterpart of mass transfer from the drop to the continuous phase (d-to-c transfer). The physical reason of interfacial instabilities is the result of surface force along the interface arising from surface tension gradient, which depends on the local concentration variation at the interface. From Fig. **39**, the mass transfer coefficient of the case with Marangoni effect is above that of the case without the Marangoni effect, which is in accordance with ref. [96]. Cases A1 presents obvious Marangoni effect on mass transfer, which is also in agreement with Sternling and Scriven's analysis. For cases A2, the obvious improvement of overall coefficient is not found except for the initial stage that is different from [96]. The reason is probably that the simulated unsteady mass transfer in this present condition is different from the undisturbed steady state in [96].

Table 9: Comparison of the mass transfer instability predicted by numerical simulation with that by Sternling & Scriven [95], Mao & Chen [97] and Wang *et al.* [99].

Case	A	B	C	D
1.Transfer direction	c-to-d			
S&S	instability diffusion-limited	instability flow-limited	Instability flow-limited	instability flow-limited
Mao & Chen	MC: yes MT: enhanced	MC: yes MT: enhanced	MC: yes MT: enhanced	MC: yes MT: enhanced
Wang *et al.*	MC: yes MT: enhanced	MC: yes MT: enhanced	MC: yes MT: enhanced	MC: yes MT: enhanced
2.Transfer direction	d-to-c			
S&S	stable	stable	Instability diffusion-limited	instability diffusion-limited
Mao &Chen	MC: yes MT: enhanced	MC: yes MT: enhanced	MC: yes MT: enhanced	MC: yes MT: no
Wang *et al.*	MC: yes MT: enhanced	MC: yes MT: enhanced	MC: yes MT: enhanced	MC: yes MT: enhanced

Note: $D_1 = 2 \times 10^{-9}$ m^2/s, $\delta = -0.1$, MC: Maragoni convection, MT: mass transfer

6.4. Comparison with Literature

Table **9** indicates that our simulation [97, 99] is in fair agreement with the results from Sternling and Scriven [96], but there remains some disaccord. In [97], the predicted enhancement occurs some time after the start of interphase mass transfer, while the enhancement is predicted at the very early period of mass transfer in the work of Wang *et al.* [99]. This apparent difference may come from the steady state of drop motion that has been set up before the mass transfer starts, while both motion and mass transfer start at $t = 0$. It is believed that this study of unsteady motion of drops deals with a more realistic situation of drop mass transfer either in laboratory or in industry. Sternling and Scriven [96] analyzed the two liquid layer system with the linear concentration profiles in two liquids established as in steady interphase mass transfer, which is also quite different at least in case geometry.

A number of future topics deserve further study. As the Marangoni effect is three-dimensional phenomenon in nature, the present numerical methods should be extended to three-dimensional cases. More elements of mechanism should be incorporated into the mathematical model. Also, the numerical procedure should be tested by experiments in a quantitative sense.

7. BEHAVIOR OF PARTICLE SWARMS

7.1. Introduction

In practical multiphase processes, the dispersed phase is often present as small particles (including solid particles, bubbles and drops) so that the generated large specific surface area will result in high interphase mass and heat transfer rate and high process efficiency. The interaction between particles must be included when estimating the interphase forces between the dispersed and continuous phases. What are really needed are the correlations for interphase forces of particles in a swarm or assemblage for macroscopic numerical simulation of process equipments. It is necessary to establish the correlations in a swarm corresponding to those for single particles. Unfortunately, the work concerned with particle swarms is not sufficient as compared with the counterpart on single particles.

On one hand, we can conduct experiments on particle swarms and get the data and empirical correlations. On the other hand, theoretical and numerical work may help for this goal. For example, we may correct the correlations for single particles with introducing the phase fraction as an additional parameter, if we can find the functional relationship between the properties of single particles and those in a swarm.

Theoretical and numerical efforts to this task include roughly two approaches. The first one is direct numerical simulation of multi-particle systems, from two-particle system (tandem or side-by-side configurations) way up to thousands of particles. Due to the rapid progress in computer technique and numerical methods, the hydrodynamics of thousand plus solid particles [100], and 125 drops (boundary element method) [101] have been simulated satisfactorily. However, these numbers of particles are still too small than a large-scale commercial chemical reactor (up to millions of particles) to get reliable simulation results.

The second approach is simpler and computationally efficient. The so-called cell model takes the particle swarm as composed of the same average cells, and deals with a typical cell constituted by a central particle and the surrounding shell of the continuous phase in accordance to the average phase fraction [28, 29] as illustrated in Fig. **40**. Because of treating only an average cell, the cell model enjoys the advantages of simple formulations and low computational load. The weakness is that to accurately express the interparticle actions is difficult because the means available for such manipulation are the outer cell boundary conditions only.

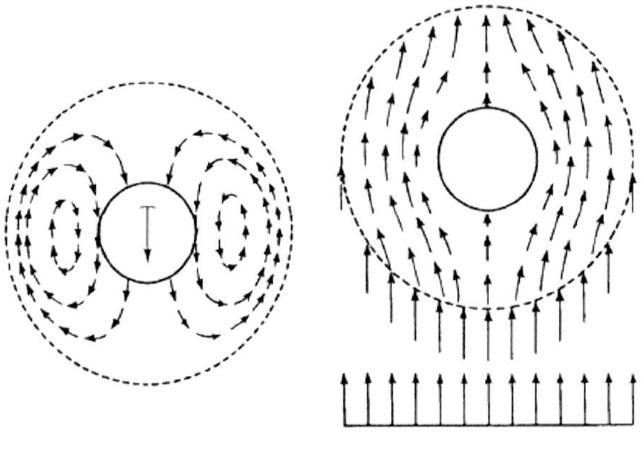

Figure 40: Spherical cell model in fixed and moving coordinate systems [28].

By the cell model, the only way to express the interparticle effect is the adoption of correct boundary conditions. In the literature, (1) the external uniform flow, (2) the free-shear condition or (3) the zero vorticity condition were adopted for the outer cell boundary. These conditions do not reflect the real flow conditions. Some improvements of cell models are suggested in Section 3.4 (models M1 and M2) [31]. The above three conditions are all biased from the real flow through particle swarms, and refinement is necessary. The improved cell models are considered as an effective and realistic tool for probing the motion and mass transfer in particle swarms and used for interpreting the increase of drag coefficient and the hindered mass transfer for solid particles, bubbles and drops.

7.2. Viscous Flow Through Spherical Particle Assemblage

Viscous flow through a spherical particle assemblage at $Re = 200$ with the voidage $\epsilon = 0.9$ is simulated with the cell model method using four different outer boundary conditions. The streamline contours of a typical single particle are presented in Fig. **41** [31], showing a recirculating vortex is formed in the wake, and the size of the eddy is much smaller than that for a single buoyant drop in an unbounded medium. Since Model K [30] imposes most stringent constraint on liquid flow by uniformity of liquid flow and zero vorticity at the outer boundary, it predicts the largest eddy length in the wake and the greater drag coefficient than other three models. The reason for Model M2 [31] gives the smallest drag and eddy length may be that the imposed fore-aft symmetry allows more liquid flows in the region closer to the outer cell boundary.

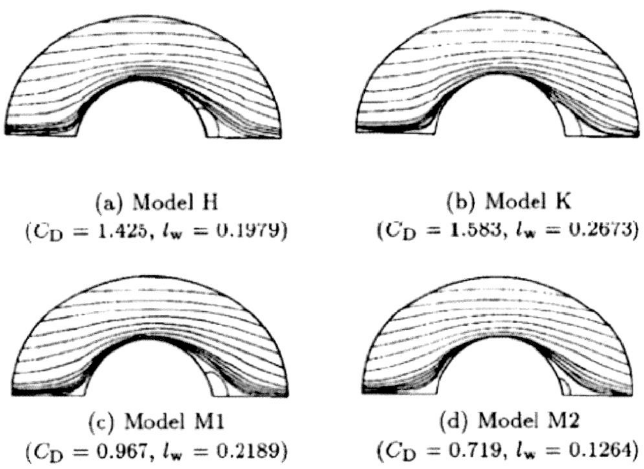

Figure 41: The comparison of the streamline maps and size of the recirculating zone for a cell with $Re = 200$ and $\epsilon = 0.9$.

El-Kaissey and Homsey [33] evaluated the free surface cell model (H) and free vorticity cell model (K) in terms of the familiar Carmen friction factor:

$$f = 150\frac{1-\epsilon}{Re} + 1.75 \quad \textbf{(146)}$$

The interrelation between C_D and f is:

$$f = \frac{4}{3}\epsilon^3 C_D \quad \textbf{(147)}$$

Fig. **42** [31] presents the simulated C_D or f as the function of Re and ϵ, showing the trend of monotonous dropping of f as $Re/(1-\epsilon)$ is increased. It is found from Fig. **42** that Model H coincided with the classical Ergun equation up to $Re/(1-\epsilon)$ about 20. The agreement between Model K and the Ergun equation is rather poor and presents more apparent sensitivity to voidage ϵ. It is observed that Model K fails to reflect the influence of ϵ on f, and the curves with different ϵ do not collapse into a single line correlation. The curves with $\epsilon = 0.9$ stand outside the curve bunch, suggesting that the Ergun equation does not suit to dilute suspension with the voidage above 0.7, since the current set of numerical parameters is evaluated based mainly on the data of fixed and fluidized beds. The agreement is quite satisfactory for Model M2 except for $\epsilon = 0.9$. Model M2 is evidently capable of clustering the curves of f in the range of ϵ from 0.4 to 0.7, suggesting it reflects more accurately the influence of ϵ and physical interaction between spherical cells.

7.3. Mass Transfer of Spherical Particle Assemblage

The cell model is applied to simulate the mass transfer in a swarm of spherical solid particles under intermediate Reynolds numbers based on the numerically resolved flow field in the cell. Simulations are conducted in the range of $1 \le Re \le 500$, $0.4 \le \epsilon \le 0.9$ and $1 \le Pe \le 3000$. Numerical simulation results suggest the significant effect of swarm voidage ϵ and the Peclet number Pe, and indicate that the conventional definition of the mass transfer factor as a simple function of $Re(1-\epsilon)$, as found in empirical correlations in the literature, fails to account for the effects of Pe and the voidage. Comparison of the simulation results for the mass transfer factor with two recommended correlations is reasonable and the deviation from the correlations seems to increase as $Re(1-\epsilon) > 100$.

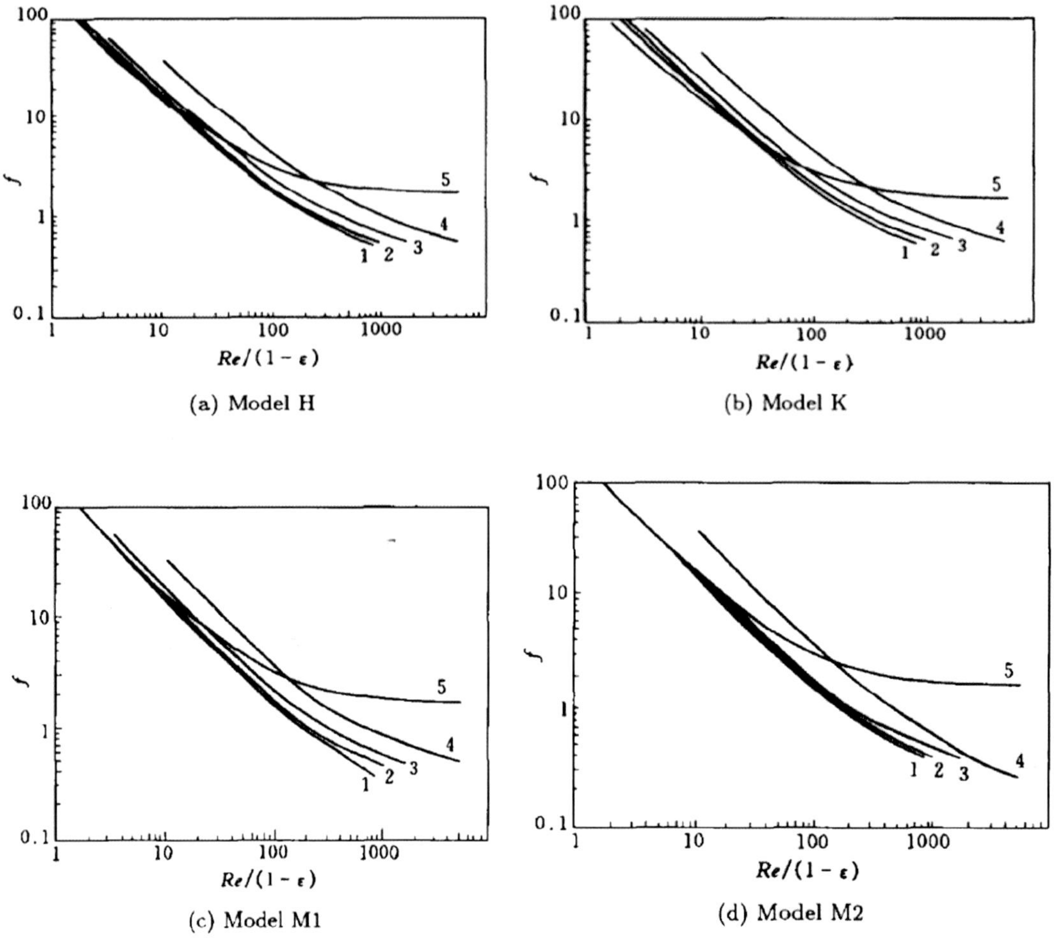

Figure 42: Dependence of friction coefficient f as a function of Re and ϵ for the original and modified cell models (ϵ: $1-0.4, 2-0.5, 3-0.7, 4-0.9, 5-$ Ergun's).

Due to the boundary conditions for Models H and K, the fluid flow in the cell is isolated from its neighbors, but Models M1 and M2 allow flow interchange among cells as viewed in the fixed reference frame. Fig. **43** [102] compares the streamlines around the center solid sphere in both the fixed and moving reference frames. From the figure, it is clear that Models H and K isolate the fluid flow between neighboring cells while the modified models, M1 and M2 subject to $\tau_{\xi\eta} = 0$, allow inter-cell fluid exchange. Therefore, the free surface (no friction) condition is not sufficient to exclude completely the interaction among cells.

7.3.1. Influence of Parameters on Sherwood Number

Several physical parameters have a strong effect on the mass transfer in a particle swarm. Numerical predictions shown in Fig. **44** [102] reveal that Pe and Re have significant influence on the mass transfer factor, particularly in the lower range of Pe. This indicates that the proposition of the mass transfer factor j_D fails to account for the overall effect of Pe, though the effect of Pe is expressed by the appearance of Sc in the definition of j_D ($j_D = Sh/ReSc^{1/3}$). This phenomenon is easily understood. Since mass transfer with small Pe means large diffusion coefficient and small concentration gradient as the driving force, the concentration boundary layer in this case would stretch far from the particle surface even into the outer cell boundary where the constant concentration is specified. Therefore, j_D will increase naturally due to the limited transfer distance to the center sphere. Small Re eases the concentration boundary layer stretching in all directions, particularly at the upstream direction, resulting in increased j_D, as expected. Effort is made to find an approximate functional form for representing the effect of Pe that is not covered by the previous definition of j_D. It is found that the modified form of j_D, Eq. (142), can transform the varying j_D into roughly constant j_D' as shown in Fig. **44**.

$$j_D' = j_D / \left(1 + \frac{1}{\epsilon Pe^{2/3}}\right) \tag{148}$$

Although the voidage is included as a variable in these correlations, it fails to account fully for the obvious influence of ϵ on j_D as indicated in Fig. **45** [102]. This influence, particularly obvious from the present results for lower Pe and ϵ, is due to the ease with which the concentration boundary layer reaches the cell boundary as a consequence of the shorter distance between the cell boundary and particle surface at low voidage. When j_D' is used to represent the effect of Pe and ϵ, the simulation results become easier to correlate (Fig. **45**). From these results, it seems that empirical correlations provide only a basic estimate of the mass transfer coefficient. A more accurate and reliable correlation may be obtained from straightforward numerical simulations.

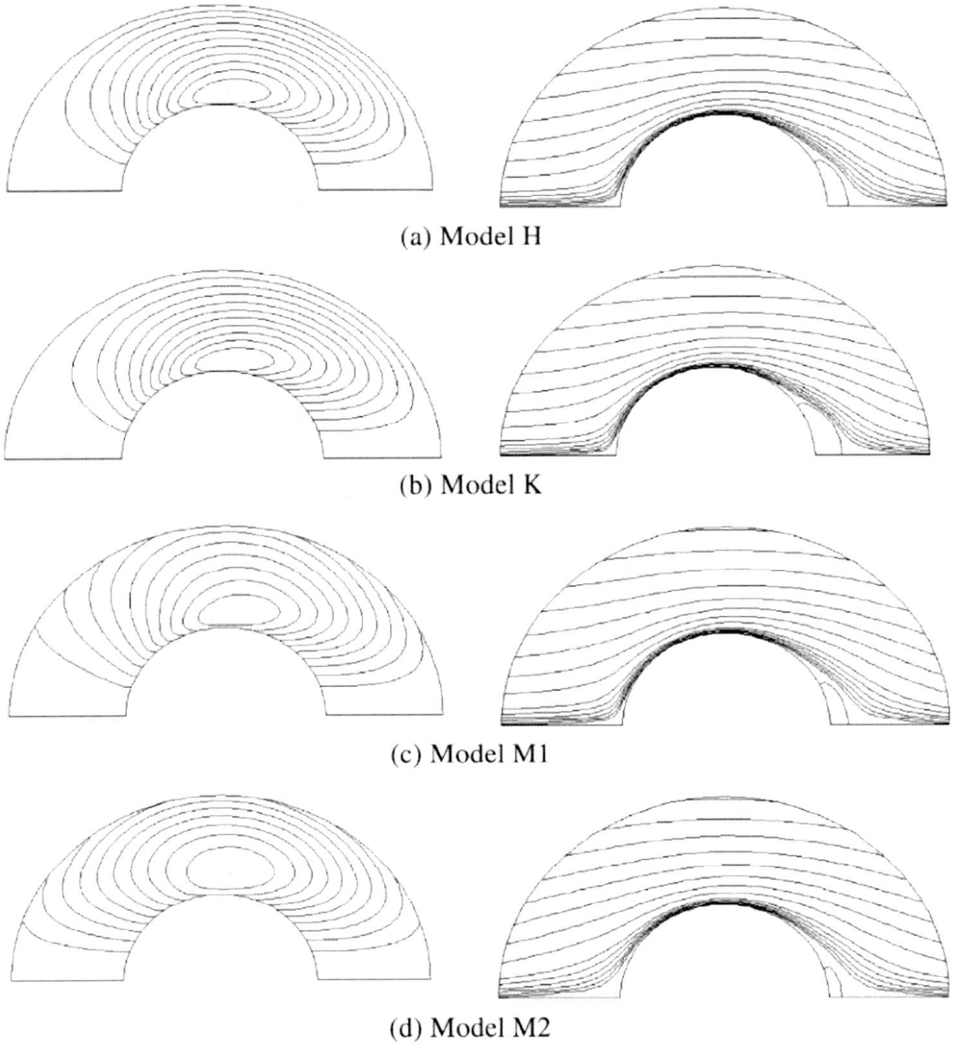

Figure 43: Streamline maps of fluid flow in a cell in the fixed reference frame (left: particle moving to the left) and in the moving reference frame (right: fluid flowing to the right) with $Re = 200$ and $\epsilon = 0.9$.

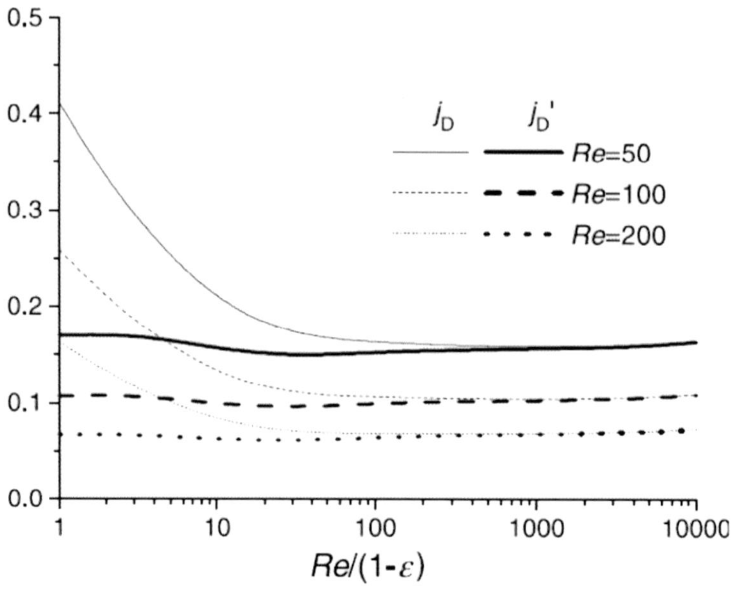

Figure 44: Dependence of mass transfer factor on the Peclet number at constant Re (model H, $\epsilon = 0.7$).

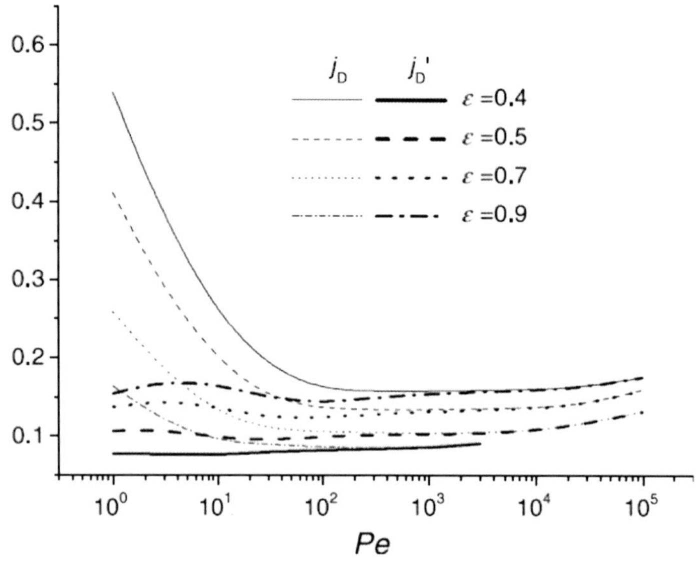

Figure 45: Dependence of mass transfer factor on the Peclet number at different ϵ (model H, $Re = 100$).

7.3.2. Comparison of Numerical Simulation with Correlations

Despite the previous observation, it is still desirable to compare the present simulation results with some available nondimensional correlations recommended in the literature. In Figs. **46a** and **47a** [102], comparison is made between the Model H simulation for Pe in the range from 1 to 3000 and two correlations mentioned in Introduction. The agreement is reasonable and it seems that the Sen Gupta and Thodos correlation [103] represents the effect of voidage a little better. Since their data are on the water vapor-air system with small Sc, the correlation can reflect the behavior of j_D more accurately in the lower range of Pe in Figs. **46** and **47**. When the modified mass transfer factor in Eq. (146) is used to represent the simulation results, the scatter of the simulated j_D is significantly reduced and now the simulation looks closer to the Chu correlation since after transformation by Eq. (146) j_D' is reduced to the level of j_D with medium Pe numbers.

7.4. Swarm of Fluid Particles

Cell models were applied to study solid sphere assemblages at the start [28, 29], including hydrodynamics [31] and mass transfer [102]. LeClair and Hamielec [35] extended it to swarms of spherical bubbles with $Re < 1000$ and $\epsilon = 0.4{\sim}1$, and observed good agreement between prediction and experiment for small Re. Sankaranarayanan and Sundaresan [104] used the lattice Boltzmann method in combination of VOF (volume-of-fluid) for tracking the interface to examine the motion and lift force of single bubbles in water in a 3D periodical cubic domain under the condition of low voidage, and this is essentially a cell model with a cubic shape. It is also used for studying the mass transfer in a drop swarm [105, 106]. In Fig. **48** [107], the predicted drag coefficients (symbols) are in close agreement with the Kumar-Hartland correlation [108] for $Re < 100$ only.

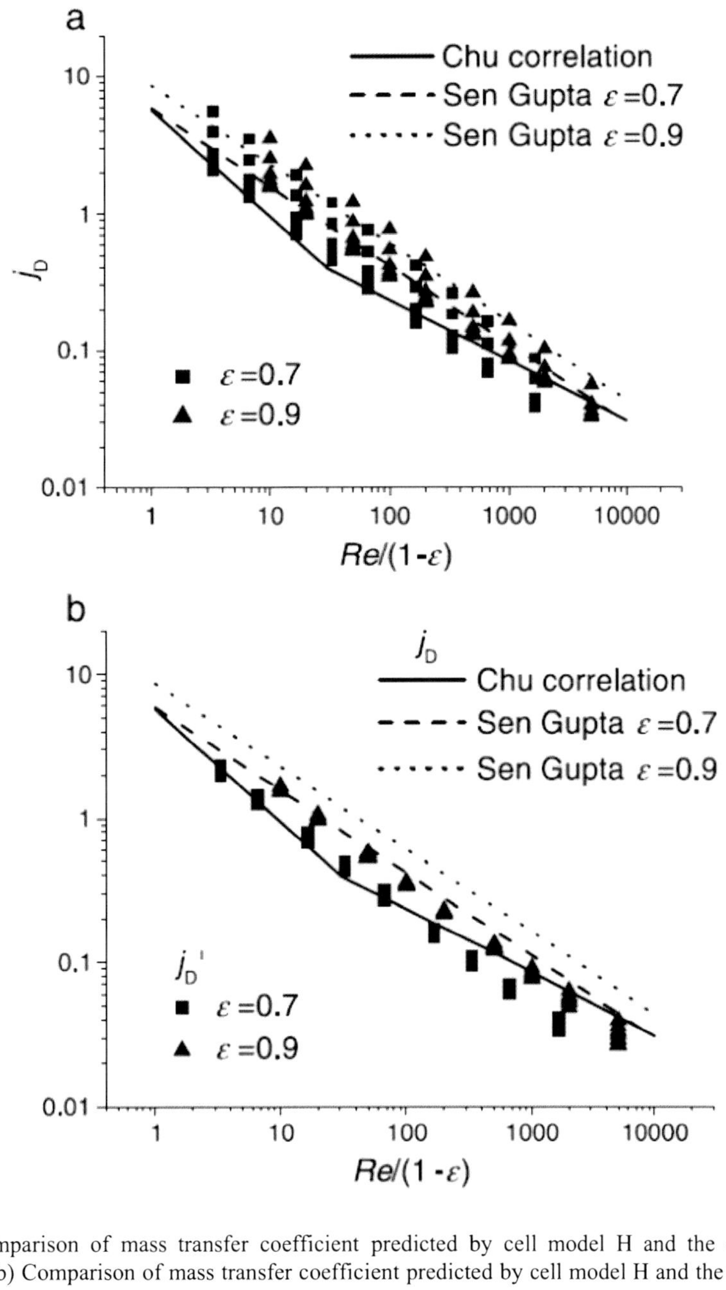

Figure 46: (a) Comparison of mass transfer coefficient predicted by cell model H and the empirical correlation ($\epsilon = 0.7$ and 0.9). (b) Comparison of mass transfer coefficient predicted by cell model H and the empirical correlation in terms of j_D' ($\epsilon = 0.7$ and 0.9).

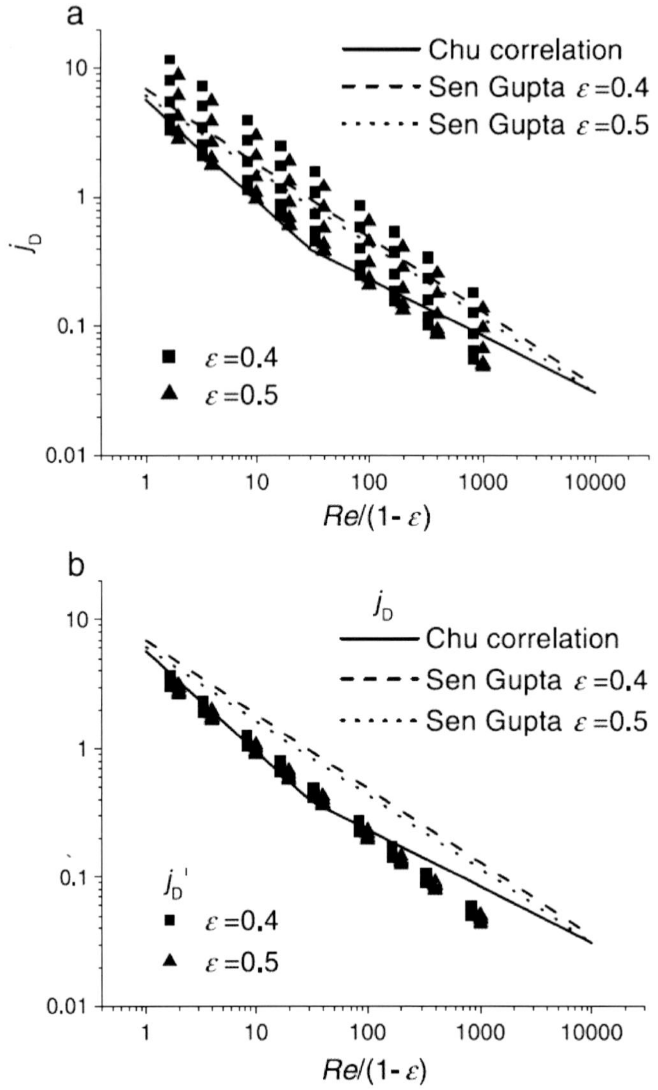

Figure 47: (a) Comparison of mass transfer coefficient predicted by cell model H and the empirical correlation ($\epsilon = 0.4$ and 0.5). (b) Comparison of mass transfer coefficient predicted by cell model H and the empirical correlation in terms of j_D' ($\epsilon = 0.4$ and 0.5).

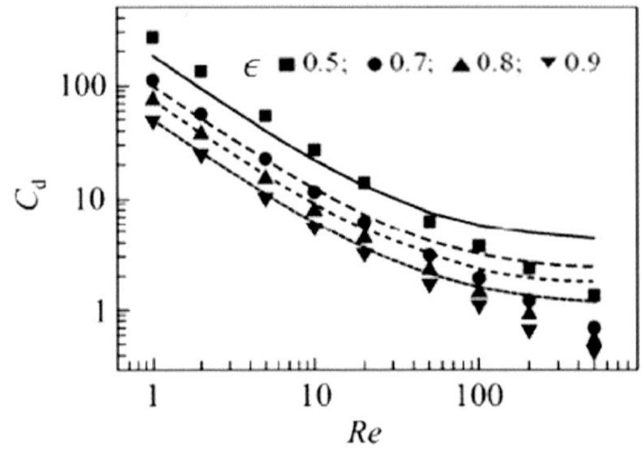

Figure 48: Drag coefficients of drops predicted by the cell model [106] with the Happel boundary condition versus the correlation [108].

Most of correlations for particle swarms are some kinds of extension with the phase fraction incorporated into ones for single particles. In general, the following form of correlation is so far preferred:

$$C_{\mathrm{D}} = C_{\mathrm{D0}}\, f(\alpha_{\mathrm{p}}) \tag{149}$$

in which C_{D0} is the drag coefficient of single particles. The function $f(\alpha_{\mathrm{p}})$ differs from case to case [107].

The case of bubbles is more complicated due to their deformation, breakage and coalescence, and empirical equations similar to Eq. (147) is used to correlate the drag coefficient in bubble swarms to that of a single bubble. However, the reported functional forms of $f(\alpha_{\mathrm{p}})$ are diversified, and the correlations from different sources are applicable only in their respective limited range. Pan *et al.* [109] redefined the gas Reynolds number in a swarm to express the effect of gas fraction and transformed the correlation for contaminated single bubbles by Tomiyama [4] to a bubble swarm based on a modified Reynolds number.

Less efforts are devoted to drop swarms. It is noticed that Kumar and Hartland [109] collected 998 data in 29 systems ($Re = 0.16 - 3169$, $\epsilon = 0.01 - 0.76$) and proposed a drag coefficient with average relative error of 8.8%.

It is thought that no sufficient and systematical studies exist on the constitutive equations for other forces, particularly for liquid drops, in the existing literature. The constitutive equations on drag, heat and mass transfer coefficients in particle swarms as simple function of Re, ϵ and the physical properties of the systems are the desired form to chemical engineers, though such functional dependence is in reality very complex. The data base for correlations can be compiled either from experimental measurements or from numerical simulation based on the cell models with suitable verification. Each way requires our laborious efforts.

7.5. Cell Configuration

When particles are associated with vortex in the wake, the simple geometric configuration with a spherical particle at the centre of a cell demands careful scrutiny. This part examines an important factor for the cell configuration: the location of a solid sphere in the spherical cell.

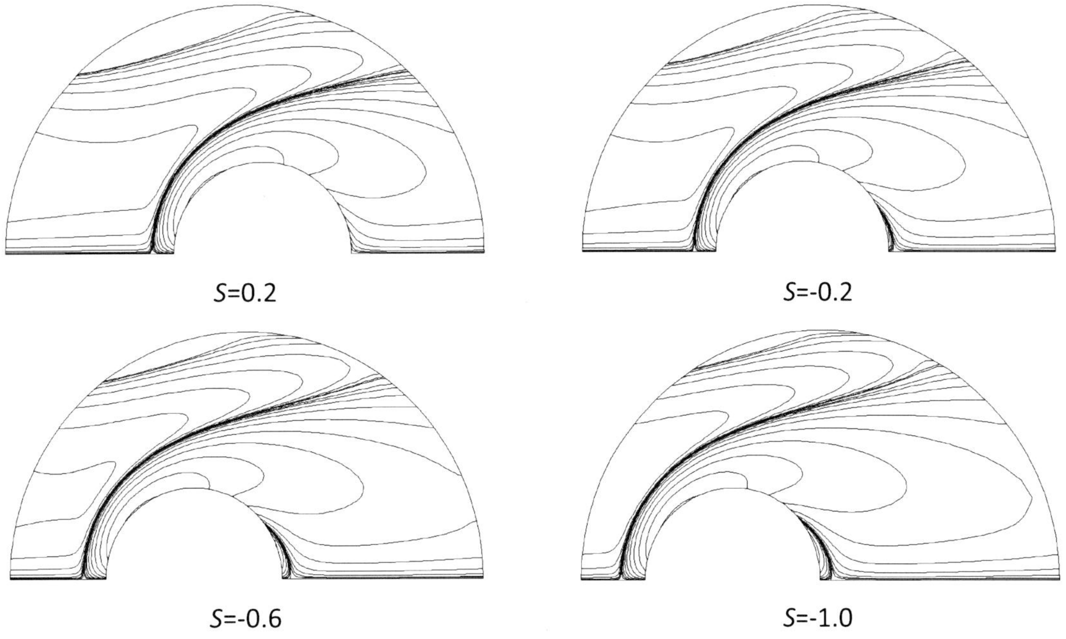

Figure 49: Contour plots of vorticity for $Re = 50$ and voidage of 0.95.

The flow in the wake behind the particle is obviously limited by the presence of downstream boundary. In the range of creeping flow, the fore-aft symmetry of the fluid flow is well kept and not much vorticity is swept to the downstream boundary. The symmetry is gradually destructed in extent as Re increases. To explore the effect of the centre location of a particle on the resulted drag coefficient, a series of simulation with different centre shift S are performed and the contour maps of vorticity are presented in Fig. **49** [110]. The streamline plots are not presented for being not very sensitive to the value of S. It is seen that the drag coefficient varies with S monotonically in Fig. **50** [110], but there is a minimum for the energy dissipation W roughly at $S = -0.6$. The minimum of W corresponds to the lowest rate of irreversible entropy production, which suggests a steady state of an irreversible non-equilibrium process. If there is a cell existing in the system, the centre particle may migrate to this location and be stabilized there. At the location, the simulation gives $C_D = 2.762$ and $W = 1.553$.

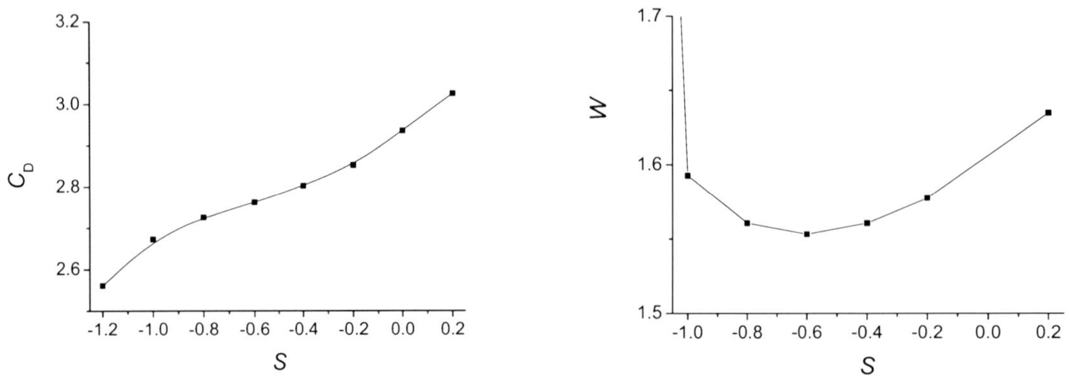

Figure 50: Effect of S on C_D and W ($Re = 50$, voidage=0.95).

It is quite surprising that the centre shift is so large as compared with the cell outer radius $R_o = 2.714$. This needs further examination. The effect of Reynolds number on the optimum shift is examined by numerical simulation on the particle assemblage at $Re = 0.1$ with well kept fore-aft symmetry. Similar results are presented in Fig. **51** [110]. The minimum W is roughly at $S = -0.05$, justifying centering a particle in the cell at low Re.

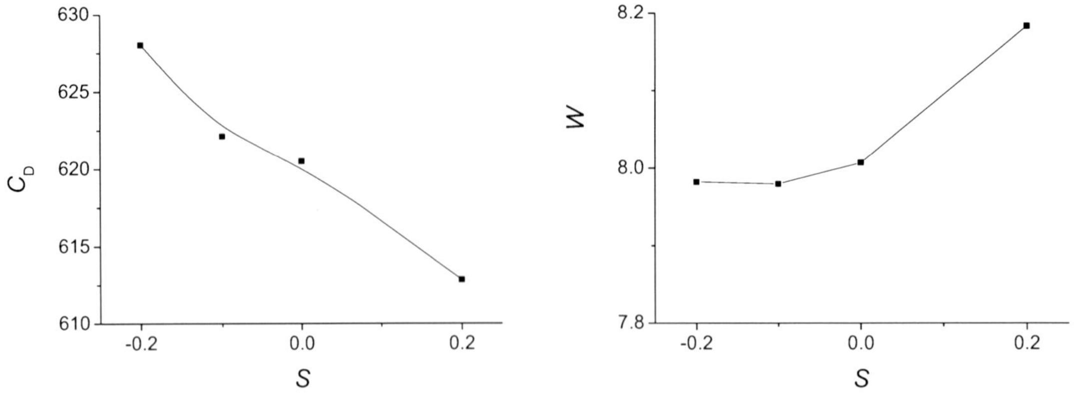

Figure 51: Effect of S on C_D and W (Re=0.1, voidage=0.95).

In practical multiphase processes, the dispersed phase is often present as small particles, bubbles and drops so that the generated large specific surface area will result in high interphase mass and heat transfer rate and high process efficiency. Therefore, the interactions between particles must be included when estimating the interphase forces between the dispersed and continuous phases. In this section the modified cell model is proposed to express the correlations for interphase forces of particles in a swarm or assemblage for macroscopic numerical simulation of process equipments. We also have made great progress in modifying the cell model to describe the real condition more accurately.

8. CONCLUSIONS AND PERSPECTIVE

From all of achievements we have gotten, numerical simulation has shown the powerful capacity to research multiphase systems in chemical engineering, including the numerical simulation of multiphase flow and interphase mass transfer on particle scale. On one hand, more detailed numerical information on the spatial distribution of flow field, species concentration, temperature, turbulent strength and chemical reaction rate and their temporal variations may be obtained than sole experimental measurements. On the other hand, increasing difficulty is frequently encountered to analyze theoretically the non-linear multiphase flow systems coupled with transport and chemical reaction. However, the nowadays computer and numerical simulation can solve the partial differential equation efficiently to probe into chemical engineering properties of heterogeneous systems. Therefore, we should spare no effort to develop high-efficiency numerical methods to meet the demands of scientific research and industrial application.

Both orthogonal boundary-fitted coordinate system and level set method for numerical simulation of the motion of and mass transfer from/to a fluid particle (drop or bubble), considering the initial unsteady accelerating stage of a falling/rising fluid particle, are developed. Although in the moving reference coordinates fixed on the fluid particle the CPU cost increases a little, it is worthy when dealing with the fluid particles in accelerating stage to get more accurate solution. The results are found in better agreement with the experimental data than that by the simulation based only on the stage of steady motion of fluid particles. With success on the multiphase flows with lower or moderate Reynolds numbers, it is desired to modify those methods to adapt higher Reynolds number, which corresponds with the practical commercial application conditions.

The mirror fluid method will be used to study the effects of irregular side walls, motion of other non-spherical (*e.g.*, slender) solid particles with rotation or tilting in a Newtonian or non-Newtonian fluid. Extension to simulate three-dimensional solid-fluid flow at high Reynolds numbers will also be tested for further validity and application of this novel numerical algorithm. The numerical simulation of multiphase flow in a stirred tank with complex geometrical shape impellers is under way.

The present study provides necessary basis for further work on three-dimensional numerical analysis of the Marangoni effect induced by surface tension gradient caused by interphase mass transfer. More elements of mechanism, such as other physico-chemical properties of interface, should be incorporated into the mathematical model. Also, the numerical procedure should be tested by experiments in a quantitative sense, which is currently under study. Meanwhile, the simulation of the Marangoni effect considering the initial stage of a falling/rising fluid particle based on the present algorithm is also currently under way. The effects of various surfactants on the motion, shape and Marangoni convection of drops/bubbles need to be numerically simulated and experimentally determined for manipulating multiphase transport and optimizing the design of multiphase reactors and separators.

Numerical simulation of mass transfer in a particle swarm can be accomplished in terms of cell models by first simulating the fluid flow in a typical cell followed by simulation of external mass transfer based on the fluid flow information in the same domain. It is believed that the modified cell model provides a sound basis for further investigation of the behavior of particle swarms and has to be further developed to improve the performance in the range of Reynolds numbers from 50 to 1000.

ACKNOWLEDGEMENTS

The authors acknowledge the financial supports from the National Natural Science Foundation of China (Nos. 20990224, 20976176), the National Science Fund for Distinguished Young Scholars (21025627), 973 Program (2010CB630904, 2009CB623406), the National Project of Scientific and Technical Supporting Program (2008BAF33B03), 863 key project (2007AA060904) and Jiangsu Province Projects (BY2009133, BE2008086). Authors wish to thank Prof. Jiayong Chen in our Institute, for his valuable advice and continuous encouragement. We are grateful to many other former Ph.D. and M.S. students of Prof. Mao and Prof. Yang for their contributions to the experimental and numerical studies of multiple phase system on particle scale.

NOMENCLATURE

a	acceleration, $\mathrm{m/s^2}$
A	adsorption rate constant, $\mathrm{m^3/(mol \cdot s)}$
A_0	initial total mass
Ac	acceleration number, $d_e a/u_b^2$
Ar	Archimedes number
$A(\tau)$	total mass
b	power-law index, dimensionless
B	desorption rate constant, $\mathrm{s^{-1}}$
c	concentration, $\mathrm{mol/m^3}$
C	dimensionless concentration
C^s	surface concentration, dimensionless
C_D	drag coefficient
C_{DA}	total drag coefficient
Ca	Capillary number
d	particle diameter, m
d_e	volume-equivalent particle diameter, m
D	molecular diffusivity of solute, $\mathrm{m^2/s}$
De	Deborah number, $\lambda U_T/d_e$
E_m	extraction fraction
Eo	Eotvos number, $d_e^2 \rho g/\sigma$
$f(\xi,\eta)$	Distortion function
f	Carmen friction factor
Fr	Froude number, $V^2/(Lg)$
F_A	added mass force, N
F_B	buoyancy force, N
F_D	drag force, N
F_H	history force, N
h_ξ, h_η	scaling factor, m
H_ξ, H_η	dimensionless scaling factor
$\mathbf{I}$	the identity matrix
j_D	Mass transfer factor, defined by $j_D = Sh/ReSc^{1/3}$
j_D'	modified mass transfer factor, defined by Eq.(148)
k	mass transfer coefficient
K	power-law consistency coefficient, $\mathrm{Pa \cdot s^n}$
L	characteristic length, m
L_w	wake length
m	distribution coefficient
M	dimensionless surface concentration
Mo	Morton number
$\mathbf{n}$	unit vector normal to a surface
n	Carreau model parameter, dimensionless

N	diffusive flux
Oh	Ohnesorg number, $[\mu_c^4/(\gamma\rho_b R)]^{1/2}$
p	pressure, Pa
Pe	Peclet number for mass transfer, dU_T/D
Pe^*	Peclet number at subsurface
R	volume-equivalent particle radius, m
R	universal gas constant,
R_o	cell outer radius
Re	Reynolds number, $\rho LV/\mu$
Re_M	Non-Newtonian number, defined by Eq. (98)
S	centre shift
S	source term
Sc	Schmidt number, $\mu/\rho D$
Sh	Sherwood number, $2Rk/D$
t	time, s
T	temperature, °C
u	velocity vector, m/s
u	axial velocity component, m/s
U_T	terminal velocity, m/s
v	radial or transverse velocity component, m/s
V	reference velocity, $\sqrt{dg}$
V_p	particle volume, m^3
W	energy dissipation
We	Weber number, $2R\rho_1 V^2/\sigma$
x, y	coordinate in physical plane, dimensionless
$\Delta x, \Delta y$	grid spacing in x and y directions, m
X, Y	coordinate in auxiliary plane, dimensionless

Greek Letters

α	dimension of system
β	empirical coefficient in Eq. (68)
$\dot{\gamma}$	shear rate, s^{-1}
γ	dimensional surface tension
δ	surface tension sensitivity coefficient
δ_p	perturbation level
$\delta_\varepsilon(\phi)$	delta function
Γ	surface adsorption, mol/m^2
Γ_∞	saturation surface adsorption, mol/m^2
ε	"half thickness" of the interface
ϵ	voidage
ζ	relative density, ρ_1/ρ_2
θ	dimensionless time
κ	Gaussian curvature
$\kappa_{(\xi)}$	surface curvature, m^{-1}
$\kappa_{(n)}$	normal curvature, m^{-1}
Λ	random number in [-1,1]
λ	relative viscosity, μ_2/μ_1
μ	viscosity, Pa · s
μ_0	zero shear viscosity, Pa · s
μ_a	apparent viscosity, Pa · s
ν	kinematic viscosity, m^2/s
ξ, η	coordinate in computational plane, $0 \le \xi, \eta \le 1$
ρ	density, kg/m^3

σ	surface tension, N/m
σ_0	surface tension of pure system, N/m
ς	characteristic time of fluid in Carreau model
τ	virtual time
$\boldsymbol{\tau}$	stress tensor component
ψ	stream function, m^3/s
Ψ	stream function, dimensionless
Ψ''	modified stream function defined in Eq. (114)
ω	vorticity, s^{-1}
Ω	vorticity, dimensionless
ϕ	level set function

Superscripts

∞	remote boundary
s	drop surface
$\wedge$	transformation form
*	in equilibrium with other phase
-	average
'	nondimensional variables

Subscripts

1	continuous phase
2	particle phase
c	continuous phase
cal	calculated value
d	drop or dispersed phase
exp	experimental value
in	first measurement location
g	gas phase
l	liquid phase
loc	local
o	overall
out	second measurement location
p	particle
s	Surfactant

REFERENCES

[1] Z.-S. Mao and C. Yang, "Challenges in study of single particles and particle swarms," *Chinese J. Chem. Eng.*, vol. 17 (4), pp. 535-545, 2009.

[2] R. Clift, J. R. Grace, and M. E. Weber, *Bubbles, Drops, and Particles*, Academic Press, New York, pp. 30-33, 1987.

[3] M. S. Kwauk and H. Z. Li, *Handbook of Fluidization*, Chemical Industry Press, Beijing, pp. 106, (in Chinese), 2008.

[4] A. Tomiyama, "Struggle with computational bubble dynamics," *Multiphase Sci. Technol.*, vol. 10 (4), pp. 369-405, 1998.

[5] S. Hu and R. C. Kintner, "The fall of single liquid drops through water," *AIChE J.*, vol. 1 (1), pp. 42-48, 1955.

[6] Z. Zapryanov and S. Tabakova, *Dynamics of Bubbles, Drops and Rigid Particles*, Kluwer Academic, Dordrecht, pp. 338-343, 1999.

[7] M. W. Reeks and S. McKee, "The dispersive effects of basset history forces on particle motion in a turbulent flow," *Phys. Fluids*, vol. 27, pp. 1573-1582, 1984.

[8] D. A. Drew and R. T. Jr. Lahey, "The virtual mass and lift force on a sphere in rotating and straining flow," *Int. J. Multiphase Flow*, vol. 13, pp. 113-121, 1987.

[9] L. Zhang, C. Yang, and Z.-S. Mao, "Unsteady motion of a single bubble in highly viscous liquid and empirical correlation of drag coefficient," *Chem. Eng. Sci.*, vol. 63, pp. 2099-2106, 2008.

[10] X. J. Li and Z.-S. Mao, "The effect of surfactant on the motion of a buoyancy-driven drop at intermediate Reynolds numbers: A numerical approach," *J. Colloid Interface Sci.*, vol. 240, pp. 307-322, 2001.

[11] G. Ryskin and L. G. Leal, "Orthogonal mapping," *J. Comput. Phys.*, vol. 50, pp. 71-100, 1983.

[12] G. Ryskin and L. G. Leal, "Numerical solution of free-boundary problems in fluid mechanics: Part 1. The finite-difference technique," *J. Fluid Mech.*, vol. 148, pp. 1-17, 1984.

[13] D. S. Dandy and L. G. Leal, "Buoyancy-driven motion of a deformable drop through a quiescent liquid at intermediate Reynolds number," J. *Fluid Mech.*, vol. 208, pp. 161-192, 1989.

[14] T. W. Li, C. G. Sun, Z.-S. Mao, and J. Y. Chen, "Influence of distortion function on the accuracy of numerical simulation of the motion of a single buoyancy-driven deformable drop," *Huagong Yejin*, vol. 20, pp. 29-37 (in Chinese), 1999, also in *Sel. Pap. Eng. Chem. Metall.* (China), 1999, Science Press, Beijing, pp. 92–102, (in English), 2000,

[15] Z.-S. Mao and J. Y. Chen, "Numerical solution of viscous flow past a solid sphere with the control volume formulation," *Chinese J. Chem. Eng.*, vol. 5 (2), pp. 105-116, 1999.

[16] Z.-S. Mao, T. W. Li, and J. Y. Chen, "Numerical simulation of steady and transient mass transfer to a single drop dominated by external resistance," *Int. J. Heat Mass Transfer*, vol. 44, pp. 1235-1247, 2001.

[17] T. W. Li, Z.-S. Mao, J. Y. Chen, and W. Y. Fei, "Terminal effect of drop coalescence on single drop mass transfer measurement and its minimization," *Chinese J. Chem. Eng.*, vol. 9 (3), pp. 204-207, 2001.

[18] C. Yang and Z.-S., Mao, "An improved level set approach to the simulation of drop and bubble motion," *Chinese J. Chem. Eng.*, vol. 10 (3), pp. 263-272, 2002.

[19] C. Yang and Z.-S. Mao, "Numerical simulation of interphase mass transfer with the level set approach," *Chem. Eng. Sci.*, vol. 60, pp. 2643-2660, 2005.

[20] X. Y. Li, Y. F. Wang, G. Z. Yu, C. Yang, and Z.-S. Mao, "A volume-amending method to improve mass conservation of level approach for incompressible two-phase flows," *Sci. China Ser. B-Chem.*, vol. 51, pp. 1132-1140, 2008.

[21] S. Osher and J. A. Sethian, "Fronts propagating with curvature-dependent speed: algorithms based on Hamilton-Jacobi formulations," *J. Comput. Phys.*, vol. 143, pp. 495-518, 1988.

[22] M. Sussman, P. Smereka, and S. Osher, "A level set approach for computing solutions to incompressible two-phase flow," *J. Comput. Phys.*, vol. 114, pp. 146-159, 1994.

[23] Y. C. Chang, T. Y. Hou, B. Merriman, and S. Osher, "A level set formulation of Eulerian interface capturing methods for incompressible fluid flows," J. *Comput. Phys.*, vol. 124, pp. 449-464, 1996.

[24] H. Zhang, L. L. Zheng, V. Prasad, and T. Y. Hou, "A curvilinear level set formulation for highly deformable free surface problems with application to solidification," *Num. Heat Transfer*, *Part* B, vol. 34, pp. 1-20, 1998.

[25] C. Yang and Z.-S. Mao, "Mirror fluid method for numerical simulation of sedimentation of a solid particle in a Newtonian fluid," *Phys. Rev. E*, 2005, vol. 71, 036704, 2005.

[26] R. Glowinski, T. W. Pan, T. I. Hesla, and D. D. Joseph, "Experimental investigation of transverse flow through aligned cylinders," *Int. J. Multiphase Flow*, vol. 25, pp. 755-794. 1999.

[27] R. Glowinski, T. W. Pan, T. I. Hesla, D. D. Joseph, and J. A. Périaux, "A fictitious domain approach to the direct numerical simulation of incompressible viscous flow past moving rigid bodies: Application to particulate flow," J. *Comput. Phys.*, vol. 169, pp. 363-426, 2001.

[28] J. Happel and H. Brenner, *Low Reynolds number hydrodynamics*, 2nd, Noordhoff, Leydon the Netherlands, 1973.

[29] J. Happel, "Viscous flow in multiparticle systems: slow motion of fluids relative to beds of spherical particles," *AIChE J.*, vol. 4, pp. 197-201, 1958.

[30] S. Kuwabara, "The forces experienced by randomly distributed parallel cylinders or spheres in a viscous flow at small Reynolds numbers," *J. Phys. Soc. Jpn.*, vol. 14, pp. 527-532, 1959.

[31] Z.-S. Ma and J. Y. Chen, "Numerical simulation of viscous flow through spherical particle assemblage with the modified cell model," *Chinese J. Chem. Eng.*, vol. 10 (2), pp. 149-162, 2002.

[32] V. Mohan and J. Raguraman, "A theoretical study of pressure drop for non-Newtonian creeping flow past an assemblage of spheres," *AIChE J.*, vol. 22, pp. 259-264, 1976.

[33] M. M. El-Kaissy and G. M. Homsy, "A theoretical study of pressure drop and transport in packed beds at intermediate Reynolds numbers," *Ind. Eng. Chem. Fundam.*, vol. 12, pp. 82-90, 1973.

[34] D. Z. Gu, Z. X. Li, and D. C. Wu, "Non-Newtonian flow through assemblage of uniform spheres," *J. Chengdu Univ. Sci. Technol.*, vol. 4 (1), pp. 1-12, 1992, (in Chinese).

[35] B. P. LeClair and A. E. Hamielec, "Viscous flow through particle assemblages at intermediate Reynolds numbers—A cell model for transport in bubbles swarms," *Can. J. Chem. Eng.*, vol. 49, pp. 713-720, 1971.

[36] B. P. LeClair and A. E. Hamielec, "Viscous flow through particle assemblage at intermediate Reynolds numbers—steady-state solutions for flow through assemblages of spheres," *Ind. Eng. Chem. Fundam.*, vol. 7, pp. 542-549, 1968.

[37] R. Tal(Thau) and W. A. Sirignano, "Cylindrical cell model for the hydrodynamics of particle assemblages at intermediate Reynolds numbers," *AIChE J.*, vol. 28, pp. 233-237, 1982.

[38] R. Tal(Thau), D. N. Lee, and W. A. Sirignano, "Hydrodynamics and heat transfer in sphere assemblage—cylindrical cell model," *Int. J. Heat Mass Transfer*, vol. 26, pp. 1265-1273, 1983.

[39] F. Losasso, R. Fedkiw, and S. Osher, "Spatially adaptive techniques for level set methods and incompressible flow," *Comput. Fluids*, vol. 35, pp. 995-1010, 2006.

[40] M. Sussman and P. Smereka, "Axi-symmetric free boundary problems," *J. Comput. Phys.*, vol. 341, pp. 269-294, 1997.

[41] M. Sussman and E. Fatemi, "An efficient, interface-preserving level set redistancing algorithm and its application to interfacial incompressible fluid flow," *SIAM J. Sci. Comput.*, vol. 20, pp. 1165-1191, 1999.

[42] M. Sussman, A. S. Almgren, J. B. Bell, P. Colella, L. H. Howell, and M. L. Welcome, "An adaptive level-set approach for incompressible two-phase flows," *J. Comput. Phys.*, vol. 148, pp. 81-124, 1999.

[43] Z. L. Wang and Z. W. Zhou, "An improved level-set re-initialization solver," *Appl. Math. Mech.*, vol. 25, pp. 1083-1088, 2004.

[44] M. Sussman, "A second order coupled level set and volume-of-fluid method for computing growth and collapse of vapor bubbles," *J. Comput. Phys.*, vol. 187, pp. 110-136, 2003.

[45] M. Sussman and E. G. Puckett, "A coupled level set and volume-of-fluid method for computing 3D and axisymmetric incompressible two-phase flows," J. *Comput. Phys.*, vol. 162, pp. 301-337, 2000.

[46] M. Sussman, K. M. Smith, M. Y. Hussaini, M. Ohta, and R. Zhi-Wei, "A sharp interface method for incompressible two-phase flows," *J. Comput. Phys.*, vol. 221, pp. 469-505, 2007.

[47] D. Enright, R. Fedkiw, J. Ferziger, and I. Mitchell, "A hybrid particle level set method for interface capturing," *J. Comput. Phys.*, vol. 183, pp. 83-116, 2002.

[48] D. Enright, F. Losasso, and R. Fedkiw, "A fast and accurate semi-Lagrangian particle level set method," *Comput. Struct.*, vol. 83, pp. 479-490, 2005.

[49] J. Strain, "Tree methods for moving interfaces," *J. Comput. Phys.*, vol. 151, pp. 616-648, 1999.

[50] R. R. Nourgaliev, S. Wiri, N. T. Dinh, and T. G. Heofanous, "On improving mass conservation of level set by reducing spatial discretization errors," *Int. J. Multiphase Flow*, vol. 31, pp. 1329-1336, 2005.

[51] H. Oka and K. Ishii, "Numerical analysis on the motion of gas bubbles using level set method," *J. Phys. Soc. Japan*, vol. 68, pp. 823-832, 1999.

[52] D. Bhage and M. E. Webber, "Bubbles in viscous liquids: Shapes, wakes and velocities," *J. Fluid Mech.*, vol. 105, pp. 61-85, 1981.

[53] G. Ryskin and L. G. Leal, "Numerical solution of free boundary problems in fluid mechanics, Part 2. Buoyancy-driven motion of a gas bubble,through a quiescent liquid," *J. Fluid Mech.*, vol. 148, pp. 19-35, 1984.

[54] G. Ryskin and L. G. Leal, "Numerical solution of free boundary problems in fluid mechanics, Part 3. Bubble deformation in an axisymmetric straining flow," *J. Fluid Mech.*, vol. 148, pp. 37-43, 1984.

[55] A. Tomiyama, I. Zun, and A. Sou, "Numerical analysis of bubble motion with the VOF method," *Nucl. Eng. Des.*, vol. 141, 69-82, 1993.

[56] T. J. Lin, J. Reese, T. Hong, and L. S. Fan, "Quantitative analysis and computation of two-dimensional bubble columns," *AIChE J.*, vol. 42, pp. 301-318, 1996.

[57] K. Sankaranarayanan, X. Shan, I. G. Kevrekidis, and S. Sundaresan, "Bubble flow simulation with the lattice Boltzmann method," *Chem. Eng. Sci.*, vol. 54, pp. 4817-4823, 1999.

[58] Y. F. Wang, C. Yang, Z.-S. Mao, and J. Y. Chen, "Application of the level set approach for numerical simulations of two-phase flow." (in Chinese), *Prog. Nat. Sci.*, vol. 14 (2), 220-222, 2002.

[59] M. Manga and H. A. Stone, "Buoyancy-driven interactions between two deformable viscous drops," *J. Fluid Mech.*, vol. 256, pp. 647-683, 1993.

[60] J. F. Wang, C. Yang, and Z.-S. Mao, "A simple weighted integration method for calculating surface tension force to suppress parasitic flow in the level set approach," *Chinese J. Chem. Eng.*, vol. 14 (6), pp. 740-746, 2006.

[61] F. Odar and W. S. Hamilton, "Forces on a sphere accelerating in a viscous liquid," *J. Fluid Mech.*, vol. 18, pp. 302-314, 1964.

[62] F. Odar, "Verification of the proposed equation for calculation of the forces on a sphere accelerating in a viscous fluid," *J. Fluid Mech.*, vol. 25, pp. 591-592, 1966.

[63] P. M. Lovalenti and J. F. Brandy, "The force on a bubble, drop, or particle in arbitrary time-dependent motion at small Reynolds number," *Phys. Fluids A*, vol. 5, pp. 2104-2116, 1993.

[64] R. Mei and J. F. Klausner, "Unsteady forces on spherical bubbles at finite Reynolds number with small fluctuations in the free-stream velocity," *Phys. Fluids A*, vol. 4, pp. 63-70, 1992.

[65] R. Mei, J. F. Klausner, and C. J. Lawrence, "A note on the history force on a spherical at finite Reynolds number," *Phys. Fluids*, vol. 6, pp. 418-420, 1994,

[66] J. Magnaudet and I. Eames, "The motion of high-Reynolds-number bubbles in inhomogeneous flows," *Annu. Rev. Fluid Mech.*, vol. 32, pp. 659-708, 2000.

[67] L. Zhang, C. Yang, and Z.-S. Mao, "Unsteady motion of a single bubble in highly viscous liquid and empirical correlation of drag coefficient," *Chem. Eng. Sci.*, vol. 63, pp. 2099-2106, 2008.

[68] R. H. Thomas and K Walters, "The unsteady motion of a sphere in an elastico-viscous liquid," *Rheol. Acta*, vol. 5, pp. 23-27, 1966.

[69] K. R. Frater, "Drag on a sphere oscillating in an elastico-viscous fluid," *Z. Angew. Math. Phys.*, vol. 18, 798-803, 1967.

[70] M. J. King and N. D. Walters, "The unsteady motion of a sphere in an elastico-viscous liquid," *J. Phys. D: Appl. Phys.*, vol. 5, pp. 141-150, 1972.

[71] R. Y. S. Lai, "Drag on a sphere accelerating rectilinearly in a Maxwell fluid," *Int. J. Eng. Sci.*, vol. 12, pp. 645-655, 1974.

[72] R. Y. S. Lai, "Accelerating motion of a sphere in a Maxwell fluid," *Appl. Sci. Res.*, vol. 30, pp. 208-220, 1975.

[73] R. K. Gupta, "Unsteady motion of a spheroid in an elastic-viscous liquid," Z. *Angew. Math. Phys.*, vol. 27, pp. 273-279, 1976.

[74] A. Bagchi and R. P. Chhabra, "Accelerating motion of spherical particles in power law type non-Newtonian liquids," *Powder Technol.*, vol. 68, pp. 85-90, 1991.

[75] R. P. Chhabra, A. A. Soares, and J, M. Ferreira, "A numerical study of the accelerating motion of a dense rigid sphere in non-Newtonian powerlaw fluids," *Can. J. Chem. Eng.*, vol. 76, pp. 1051-1055, 1998.

[76] W. C. Park, J. F. Klausner, and R. Mei, "Unsteady forces on spherical bubbles," *Exp. Fluids*, vol. 19 (3), pp. 167-172, 1995.

[77] E. I. V. Van den Hengel, N. G. Deen, and J. A. M. Kuipers, "Application of coalescence and breakup models in a discrete bubble model for bubble columns," *Ind. Eng. Chem. Res.*, vol. 44, pp. 5233-5245, 2005.

[78] V. V. Buwa, D. S. Deo, and V. V. Ranade, "Eulerian-Lagrangian simulations of unsteady gas-liquid flows in bubble columns," *Int. J. Multiphase Flow*, vol. 32, pp. 864–885, 2006.

[79] L. Zhang, C. Yang, and Z.-S. Mao, "An empirical correlation of drag coefficient for a single bubble rising in non-Newtonian liquids," *Ind. Eng. Chem. Res.*, vol. 47, pp. 9767-9772, 2008.

[80] V. G. Levich, P*hysicochemical Hydrodynamics*, NJ, Prentice-Hall: Englewood Cliffs, 1962.

[81] R. P. Chhabra, Bubbles, *Drops, and Particles in Non-Newtonian Fluids*, Boca Raton, FL, CRC Press, 1993.

[82] R. P. Chhabra, *Bubbles, Drops, and Particles in Non-Newtonian Fluids* (second ed.), Boca Raton, FL, CRC Press, 2006.

[83] A. A. Kulkarni and J. B. Joshi, "Bubble formation and bubble rise velocity in gas-liquid system: A review," *Ind. Eng. Chem. Res.*, vol. 44, pp. 5873-5931, 2005.

[84] L. Zhang, C. Yang, and Z.-S. Mao, "Numerical simulation of a bubble rising in shear-thinning fluids," *J. Non-Newtonian Fluid Mech.*, vol. 165, pp. 555-567, 2010.

[85] J. F. Harper and D. W. Moore, "The motion of a spherical liquid drop at high Reynolds number," *J. Fluid Mech.*, vol. 32, pp. 367-391, 1968.

[86] T. W. Li, Z.-S. Mao, J. Y., Chen, and W. Y. Fei, "Experimental and numerical investigation of single drop mass transfer in solvent extraction systems with resistance in both phases," *Chinese J. Chem. Eng.*, vol. 10, pp. 1–14, 2002.

[87] J. Petera and L. R. Weatherley, "Modeling of mass transfer from falling drops," *Chem. Eng. Sci.*, vol. 56, pp. 4929-4947, 2001.

[88] J. F. Wang, P. Lu, Z. H. Wang, C. Yang, and Z.-S Mao, "Numerical simulation of unsteady mass transfer by the level set method," *Chem. Eng. Sci.*, vol. 63, pp. 3141-3151, 2008.

[89] P. Y. Huang, H. H. Hu, and D. D. Joseph, "Direct simulation of the sedimentation of elliptic particles in Oldroyd-B fluids," *J. Fluid Mech.*, vol. 362, pp. 297-326, 1998.

[90] D. Qi, "Lattice-Boltzmann simulations of particles in non-zero-Reynolds-number flows," *J. Fluid Mech.*, vol. 385, pp. 41-62, 1999.

[91] R. Bel Fdhila and P. C. Duineveld, "The effect of surfactant on the rise of a spherical bubble at high Reynolds and Peclet numbers," *Phys. Fluids*, vol. 8, pp. 310-321, 1996.

[92] T. W. Li, "Numerical simulation of mass transfer of single drops with low or intermediate Reynolds numbers in steady-state motion and its experimental verification," Ph.D. thesis, Institute of Chemical Metallurgy, Chinese Academy of Sciences, Beijing, China, 1998.

[93] Z. He, C. Maldarelli, and Z. Dagan, "The size of stagnant caps of bulk soluble surfactant on the interfaces of translating fluid droplets," *J. Colloid Interface Sci.*, vol. 146, pp. 442-451, 1991.

[94] L. G. Leal, "The stability of drop shapes for translation at zero Reynolds number through a quiescent fluid," *Phys. Fluids A*, vol. 1 (8), pp. 1309-1313, 1989.

[95] X. J. Li, Z.-S. Mao, and W. Y. Fei, "Effect of surface-active agents on mass transfer of a solute into single buoyancy driven drops in solvent extraction systems," *Chem. Eng. Sci.*, vol. 58, pp. 3793-3806, 2003.

[96] C. V. Sternling and L. E. Scriven, "Interfacial turbulence: Hydrodynamic instability and the Marangoni effect," *AIChE J.*, vol. 5, pp. 514-523, 1959.

[97] Z.-S. Ma and J. Y., Chen, "Numerical simulation of the Marangoni effect on mass transfer to single slowly moving drops in the liquid-liquid system," *Chem. Eng. Sci.*, vol. 59, pp. 1815-1828, 2004.

[98] J. U. Brackbill, D. B. Kothe, and C. Zemach, "A continuum method for modeling surface tension," *J. Comput. Phys.*, vol. 100, pp. 335-354, 1992.

[99] J. F. Wang, C. Yang, and Z.-S. Mao, "Numerical simulation of Marangoni effects of single drops induced by interphase mass transfer in liquid-liquid extraction systems by the level set method," *Sci. China Ser. B-Chem.*, vol. 51 (7), pp. 684-694, 2008.

[100] T. W. Pan, D. D. Joseph, R. Bai, R, Glowinski, and V. Sarin, "Fluidization of 1204 spheres: Simulation and experiment," *J. Fluid Mech.*, vol. 451, pp. 169-191, 2002.

[101] A. Z. Zinchenko and R. H. Davis, "An efficient algorithm for hydrodynamical interaction of many deformable drops," *J. Comput. Phys.*, vol. 157, pp. 539-587, 2000.

[102] Z.-S. Mao and Y. F. Wang, "Numerical simulation of mass transfer in a spherical particle assemblage with the cell model," *Powder Technol.*, vol. 134, pp. 145-155, 2003.

[103] A. Sen Gupta and G. Thodos, "Direct analogy between mass and heat transfer to beds of spheres," *AIChE J.*, vol. 9, pp. 751–754, 1963.

[104] K. Sankaranarayanan and S. Sundaresan, "Lift force in bubbly suspensions," *Chem. Eng. Sci.*, vol. 57, pp. 3521-3542, 2002.

[105] Z.-S. Mao and J. Y., Chen, "Numerical approach to the motion and external mass transfer of a drop swarm by the cell model," In: International Solvent Extraction Conference (ISEC'2002), Sole, K.C., Cole, P.M., Preston, J.S., Robinson, D.J., eds., South African Institute of Mining and Metallurgy, Johannesburg, South Africa, 2002, pp. 227-232.

[106] Z.-S. Mao and J. Y. Chen, "An attempt to improve the cell model for motion and external mass transfer of a drop in swarms at intermediate Reynolds numbers," In: International Solvent Extraction Conference (ISEC'2005), Fei, W.Y., Liu, H.Z., Yan, C.H., eds., Beijing, A417, 2005.

[107] Z.-S. Mao and C., Yang, "Challenges in study of single particles and particle swarms," *Chinese J. Chem. Eng.*, vol. 17 (4), pp. 535-545, 2009.

[108] A. Kumar and S., Hartland, "Gravity settling in liquid/liquid dispersions," *Can. J. Chem. Eng.*, vol. 63, pp. 368-376, 1985.

[109] Y. Pan, M. P. Dudukovic, and M. Chang, "Numerical Investigation of gas-driven flow in 2-D bubble columns," *AIChE J.*, vol. 46, pp. 434-449, 2000.

[110] Z.-S.Mao and C. Yang, "The cell model approach to solid particle assemblages: further justifications," *Lecture Series on Computer and Computational Sciences*, vol. 7(A), pp. 352-357 (*Int. Conf. Comput. Sci. Eng.*, Oct. 27-Nov.1, 2006, Chania, Greece), 2006.

CHAPTER 4

The Use of the Ultrasonic Technique for the Study and Online Monitoring of Multiphase Flows

Ricardo D. M. Carvalho[1,*] and Antonio C. Bannwart[2]

[1]Universidade Federal de Itajubá (UNIFEI), Itajubá, Brazil and [2]Universidade Estadual de Campinas (UNICAMP), Campinas, Brazil

Abstract: Multiphase flows are very common in the petroleum, chemical, and nuclear industries, oftentimes involving harsh media, strict safety regulations, access difficulties, long distances, and aggressive surroundings; accordingly, there is a need to determine the dispersed phase holdup using noninvasive fast responding techniques. Moreover, knowledge of the flow structure is essential for the assessment of the transport processes involved. The ultrasonic technique fulfills these requirements and could have the capability to provide the information required. In this chapter, the current status of the ultrasonic technique in the context of multiphase flow metering (MFM) is thoroughly reviewed; the focus is on multiphase flows of interest in the oil industry. Specific aspects involving the application of the technique to gas-liquid and gas-liquid-solid flows as well as dispersions and slurries are addressed. It is expected that this chapter will provide the reader with a clear view of what is already known about the application of ultrasonics to multiphase flows, its potential, and the research needs so that the technique could be effectively brought to the field.

Keywords: Multiphase flows, ultrasonic technique, monitoring, flow metering, gas liquid, gas liquid solid, dispersion, slurries.

1. INTRODUCTION

Multiphase flows are very common in the petroleum, chemical, and nuclear industries, oftentimes involving harsh media, strict safety regulations, access difficulties, long distances, and aggressive surroundings. In the oil industry, the types of multiphase flow of interest are [1]:

- Gas-liquid (oil/natural gas);
- Liquid-liquid (oil/water);
- Gas-liquid-liquid (natural gas/oil/water);
- Solid-liquid-liquid-gas (sand/oil/water/natural gas).

As discussed in [2], the output of an oil/gas reservoir can vary widely, depending on the location and age of the well; in addition to the components pointed out above, wax may also be present in the produced well stream. Multiphase flow meters are meeting with increased use for well testing and allocation measurement. In general, the use of multiphase flow meters should result in increased production, increased recovery, and lower investment cost. Ideally, a three-phase (oil-water-gas) flow meter needs to be reasonably accurate (typically ±5% of rate for each phase), nonintrusive, reliable, flow regime independent, and suitable for use over the full component fraction range. While this ideal three-phase flow meter has yet to be developed, the multiphase metering needs of major oil companies have been identified as [3] a gas volume fraction range of 0-99% and a water cut range of range 0-90%; it was also suggested that the total liquid and gas flow rates should be measured with a relative error of less than 5-10% while the absolute error in the water cut measurements be less than 2%.

***Address correspondence to Ricardo D. M. Carvalho:** Universidade Federal de Itajubá (UNIFEI), Itajubá, Brazil; E-mail: ricardo.2008@oi.com.br.

Lixin Cheng and Dieter Mewes (Eds)

Significant sand content is unlikely in a normal well stream because sand producing wells are normally treated or managed to minimize the amount of sand entering the production stream; however, even small quantities of particles in the well flow can cause significant damage. Once the prediction and sand handling issues have been carefully considered, a reliable method for measuring and monitoring the amount of sand in the flow would ensure that the design criteria are not exceeded; this would in turn increase the well productivity and profitability.

Regarding gas-liquid flows, the horizontal semi-slug flow, which is characterized by large (often frothy) surface waves, causes large fluctuations in liquid delivery along the pipe [1]. The horizontal slug flow, in which frothy slugs completely fill the cross-section of the pipe, is a source of serious difficulties in operation of horizontal pipelines. The semi-slug and slug flows as well as the plug flow can be considered sub-classes of a main class of intermittent flows. There are several factors that dictate the flow pattern of a multiphase flow in a conduit (phase properties, fractions and velocities; operating pressure and temperature; conduit diameter; *etc.*); yet, the transition to slug flow is particularly strongly influenced by pipe inclination. Even very small variations in pipe inclination can cause significant variations in the flow pattern of a given mixture, all the other variables staying the same [1]. It is therefore of great practical interest to have a measurement system able to detect the flow regime present in a given situation.

In connection with the production and transportation of heavy oils, water can be injected into the oil flow line to reduce friction and increase the flow rate for a given inlet pressure. The large difference in viscosity between oil and water leads to two clearly distinct liquid phases, with the possible presence of gas. In these water-assisted flow patterns, a continuous water phase is in contact with the pipe wall (annular configuration or core-annular flow), thus reducing friction and increasing the flow rate. Because highly viscous oils are involved, the use of noninvasive measurement techniques is mandatory.

Accordingly, there is a growing interest in the use of noninvasive techniques for measurement of the dispersed phases concentration, flow pattern detection, monitoring of void wave propagation, and other basic variables in multiphase flows; the ultimate goal, nonetheless, would be to obtain the flow rates of the different phases. Falcone *et al.* [1] provide a comprehensive discussion on Multiphase Flow Metering (MFM) techniques; the various potential devices fall into categories according to the parameter they measure (density, velocity, elemental analysis of individual atomic species, *etc.*). The flow rates of each phase are then inferred from these measured parameters in different ways depending on the specific metering technology. Ultrasonic sensors belong to the category of acoustic instrument, *i.e.* instruments that somehow depend upon sound waves.

Powell [4] presents a review of noninvasive methods used for measurement of several parameters of multiphase systems, especially the concentration of the dispersed phase and the velocity field. The focus is on techniques that can be applied to multiphase fluids of immediate practical interest. Magnetic Resonance Imaging (MRI) is said to be the most versatile technique among those studied as it can provide spatially resolved measurements of concentration, velocity, chemical constituents, and diffusivity. However, the equipment is expensive and requires the multiphase flow experimental apparatus to be brought to the MRI equipment rather than having the MRI unit readily available in a fluid mechanics laboratory. In addition, it requires highly trained personnel to operate and long times to acquire data. X-ray radiography is also quite specialized and has been used in a rather limited number of studies of multiphase systems; the same holds true for neutron radiography, which is mainly used for the study of multiphase systems involving liquid metals. Both techniques are affected by the fact that the images obtained represent the projection of a three-dimensional object onto a plane. Electrical Impedance Tomography (EIT) is a fairly new technique that has been used for measurement of concentration distributions of pipe flow suspensions. As yet, this technique has not met with widespread use.

Regarding impedance measurement techniques, Falcone *et al.* [1] point out that the measured impedance is likely to be a function of both phase fraction and phase configuration. If the configuration is not known beforehand, then the phase fraction cannot be determined unambiguously. Besides, if the mixture switches from oil-continuous to water-continuous the meter should also change from capacitance to conductivity measurements. Jaworek *et al.* [5] present a simple noninvasive capacitance system for void fraction measurement, but this technique is still at the laboratory stage.

The ultrasonic technique is already well established in many fields of application such as medicine and flaw detection in solid materials. The transducers and electronics are readily available commercially at relatively low cost; these systems are also compact and rugged. Besides, the use of ultrasound for application with multiphase systems has been the object of a large number of research papers, *e.g.* [6-17], which confirms the potential and adds to the reliability of the technique. Ultrasonic signals are information rich and can penetrate pipe, vessel and process chamber walls to successfully interrogate optically opaque fluids and dense suspensions. In addition, the signals are not significantly degraded by a wide range of process conditions and can provide real-time in-situ measurements and visualizations [6]. The drawback of current ultrasonic techniques is the need for prior signal attenuation calibration. The main ultrasonic parameters normally used in multiphase processes monitoring, measurement, and control are the sound pressure intensity, energy attenuation and the wave transit time; these parameters can be measured across the pipe diameter or along another acoustic path through the multiphase flow in case a lateral or backward reflection – or pulse-echo mode is used. As the concentration and size distribution of the dispersed media change (bubbles, oil droplets or solid particles), these signal properties vary due to the combined effect of acoustic attenuation and transmission phenomena-reflection, refraction, and scattering of the sound beam by the dispersed phases and acoustic attenuation by the continuous phase. Thus, the resulting multiphase flow signal is usually referenced to the signal for the continuous phase as a single phase fluid at the same temperature.

Tomographic techniques can be used either to produce a stand-alone tomographic flow meter or to enhance the performance of existing instruments. In the first case, unlike traditional industrial and medical imaging applications, the process being imaged moves quickly (typically up to 10 m/sec) and can change structure continuously. As a result, large amounts of data need to be processed quickly if a real-time system is to be feasible [2]. Ultrasound Computerized Tomography (UCT) has been studied by many researchers since the early 1980s as a means to reconstruct the distribution of a gas/liquid two-phase flow over the cross-section of a pipe. Warsito *et al.* [14] present a review of noninvasive techniques for monitoring gas-liquid bubbly flows. It is speculated that UCT is more appropriate for application in media poorly penetrated by light or other electromagnetic radiation; this would be the case of a relatively homogeneous, high density media. Nonetheless, two main limitations of the UCT technique pose a challenge to its use in applications where real-time performance is vital [14]: the complex sound field sensed by the transducers leads to errors due to overlapped or multiple reflected pulses; and the inherent slow propagation speed of ultrasound lowers the scanning speed. Another difficulty pointed out by Xu *et al.* [18] is the strong mismatch between the acoustic impedance of the phases as, for instance, in air-water flows. Despite these difficulties, Warsito *et al.* [14] could use the UCT technique to measure the time-averaged cross-sectional distributions of gas bubbles and solid particles in a slurry bubble column. Xu and Xu [19] and Xu *et al.* [18] were able to successfully reconstruct the distribution of simulated and actual gas-liquid two-phase flows over the cross-section of a pipe using the UCT technique. The experimental setup was fairly complex and the diameter of the smallest bubble in the reconstructed image corresponded to 0.6% of the cross-sectional area, but the spatial resolution was not uniform over the cross-section. In general, the image reconstruction time was much longer than the data acquisition time and could only be done off-line.

Simpler instrumentation setups involving the ultrasonic technique for the characterization of flow patterns in horizontal gas-liquid flows and for the measurement of the liquid film thickness and interface characteristics in inclined pipes are discussed in [20] and [21]. The ultrasonic technique has also been used to measure the dispersed phase concentration in three-phase and two-phase flows – air-oil; glass beads-oil; and air-glass beads-oil [12,16]. Zheng and Zhang [16] observed the signal attenuation to increase exponentially with increasing interfacial area of both the solid and gas phases. The air bubbles were said to have caused an increase in the acoustic wave transit time. Besides confirming the observations by Zheng and Zhang (16), Vatanakul *et al.* [12] also verified that the scatter in the transit time and in the signal amplitude increased as the gas volumetric concentration increased. Carvalho *et al.* [17], on the other hand, observed the acoustic wave initial transit time (time when the wave first hits the acoustic receiver) to remain essentially constant as the void fraction increased. Wada *et al.* [13] used the ultrasonic pulse-echo mode to identify the air-water flow pattern – bubbly, slug, or annular – in vertical Plexiglas and carbon steel pipes; flow pattern recognition was based on the delay (transit) time and the relative energy intensity of the mechanical wave in pulse echo mode. The authors were able to successfully determine the gas phase profiles in the cross-section and the flow pattern; however, no attempt was made to correlate the ultrasonic data with the void fraction.

Tsouris and Tavlarides [22] presented a method for measuring the volume fraction of toluene-water and water-kerosene dispersions based on the velocity differences of sound in the two phases. Later on, Tsouris *et al.* [23] successfully used a pulse-echo ultrasonic probe to measure the dispersed phase volume fraction (holdup) in liquid-liquid (toluene-water) dispersions; the researches claimed that the probe was small enough for the measurements to be considered local. Investigators in [24] were able to determine droplet sizes and size distribution in two-phase fat emulsions from broad-band ultrasonic attenuation measurements and velocity spectra. From the single particle scattering model, the authors could establish the relationship between particle size distribution and ultrasonic spectrum. Melnikov *et al.* [25] developed an ultrasonic apparatus and a method to measure the velocity and flow rates of the individual phases of a gas-oil-water multiphase flow. The device operated on the transit time variations as a function of the volume concentration of the phases in the stream. Despite the many devices and results reported above, the ultrasonic technique for flow pattern recognition and determination of the dispersed phases concentration in real applications has not yet been brought to the commercial level.

Existing commercial ultrasonic instruments currently available are used to measure single-phase fluid flow rates. There are two main types - instruments based on the acoustic wave transit time (pulse and return) and instruments based on the Doppler effect. When operating with liquid hydrocarbons, pulse-and-return instruments allow for a maximum solid particle content of less than 5% by volume and maximum gas content of less than 2% by volume. Doppler effect instruments, on the other hand, require a minimum of 25 ppm solid particles or gas bubbles 30 μm in diameter or greater; nevertheless, these fluids contain only a very small fraction of inhomogeneities.

Commercial ultrasonic systems exist that make use of the ultrasound generated by solid particles impacting on the inside of a pipe wall to detect and estimate the amount of sand in a multiphase flow [26]. Special filtering techniques are used to suppress unrelated noise (flow noise, electrical interference, *etc.*) from the ultrasonic signals. The ability to monitor the amount of sand in the flow in real time makes it possible to decide on safe production levels of an oil or gas well. Similarly, ultrasonic leak detectors are available which detect the high-frequency noise created by a leak of oil or gas. Despite its use in the field, these systems fall short of multiphase flow rate measurements.

In broad terms, the use of acoustic techniques in multiphase flows is made difficult by the complex attenuation, scattering, and reflecting processes that occur at the interfaces. The use of Doppler effect instruments for multiphase flow rate measurement assumes the phases to be at the same velocity and, therefore, requires homogenization. Acoustic cross-correlation is a technique for determining the velocity of a multiphase flow in a pipe by measuring the temporal fluctuations in the transmission of two axially spaced ultrasonic beams. The cross-correlation mathematical procedure then provides the time interval for the flow to move between the two ultrasonic stations. The auto-correlation technique assumes the flow to be homogeneous and, hence, for improved accuracy an effective upstream homogenizer is necessary. If the flow is not homogeneous, significant errors can be produced which are dependent on velocity and voidage.

In summary, the ultrasonic technique does seem to have the potential for application in multiphase flows. However, research is still needed for a clear understanding of the interactions between the ultrasonic wave and the multiphase structures. As this basic knowledge is developed, the need for calibration diminishes and the technique can be more confidently taken to the field. In what follows, aspects of the application of the ultrasonic technique to specific multiphase flow patterns or mixtures are discussed.

2. THE STUDY OF GAS-LIQUID FLOWS

2.1. Bubbly Flows

Carvalho *et al.* [17] obtained acoustic data and simultaneous high-speed images of vertical, upward, air-water two-phase flows in the bubbly flow regime (0 to 15% void fraction) in 54-mm ID Plexiglas pipes. A schematic overall view of the ultrasonic instrumentation is shown in (Fig. **1**); Panametrics Videoscan 2.25 MHz, 13 mm diameter ultrasonic transducers were mounted on Plexiglas adapters as shown. The direct transmission receiver (θ = 180°) was expected to get most of the energy transmitted through the bubbly flow while the dual element, emitter/receiver transducer (θ = 0°) was intended to provide the main complementary information about reflections off the same gas structures in the ultrasonic path. The side sensors (θ = 45° and 135°) were expected to receive at least part of the energy scattered off the acoustic beam and thus provide additional information about the interaction between the ultrasonic wave and the two-phase flow. A filming station located just downstream of the test section allows for high speed motion pictures (1000 fps) of the gas-liquid flow.

The fast pictures were analyzed for different flow patterns as the void fraction increased and it was found that very different gas structures – dispersed bubbles, bubble clusters, coalesced bubbles, and incipient Taylor bubbles – formed the air-water mixture depending on the void fraction and flow operating conditions. In order to check these results from the visual data against the acoustic attenuation data, average energy ratios for each operating condition were calculated as discussed by Carvalho *et al.* [17] Briefly, for the 0° and 180° sensors the energy ratios are the energy carried by the ultrasonic pulses as they intercept the two-phase flow referenced to the corresponding single-phase value. For the 45° and 135° sensors, the reference value was the energy carried by the ultrasonic pulses at 11% void fraction; in this condition, the energy scattered to the side sensors is close to its maximum and inception of the first Taylor bubbles is about to occur.

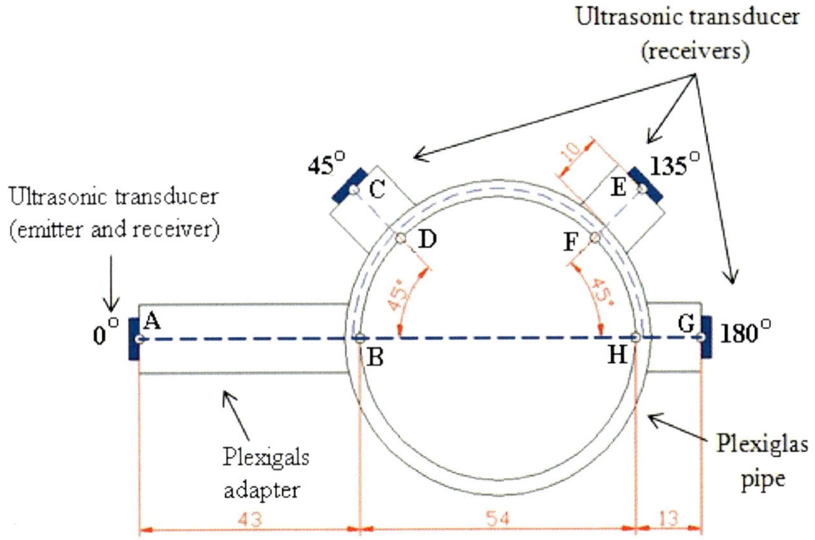

Figure 1: Schematic view of the ultrasonic instrumentation used by Carvalho *et al.* [17].

Fig. **2** shows the results for all receivers in the bubbly flow regime on a single plot both in terms of average energy ratios and average energy values. A clear interrelated behavior can be identified in (Fig. **2a**) - the energy received by the 180° sensor decreases continually, demonstrating the increasing blockage of the ultrasonic path; the pulse-echo signals (0° sensor) do not decrease to the same levels due to the contribution of the reflected energy. The 45° and 135° sensors get approximately the same amount of acoustic energy, which indicates that energy is uniformly diverted from the acoustic beam to the pipe circumference. In Fig. (**2b**), it is seen that for 6% void fraction onwards the 45°, 135°, and 180° receivers get approximately the same amount of acoustic energy. This shows that transmission through the two-phase flow to the 180° sensor no longer prevails upon energy scattering to other directions. The main conclusion is that all transducers provide a clear correlation with the void fraction in air-water bubbly flows and thus could be used to determine this flow parameter. In addition, the 180° and 0° (pulse-echo) sensors can be more readily used to measure the void fraction because single-phase water reference signals can be established for them, the received signals are located within a well bounded time window, and they are more sensitive to void fraction variations than the 45° and 135° sensors.

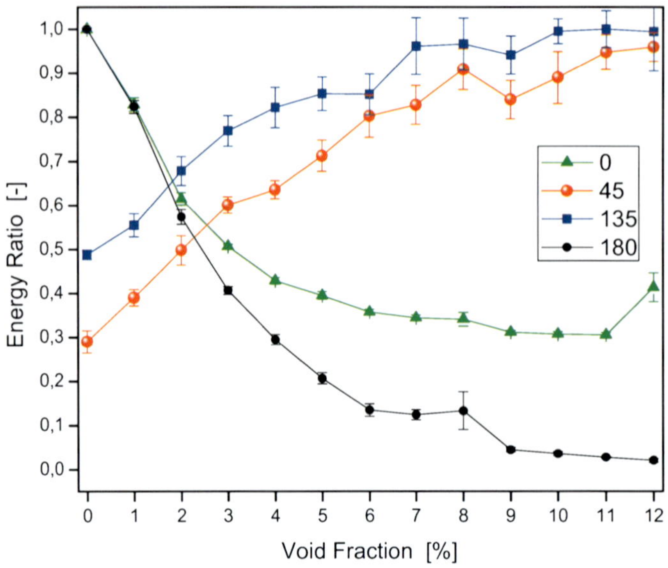

Figure 2a: Average energy ratios as a function of void fraction [17]

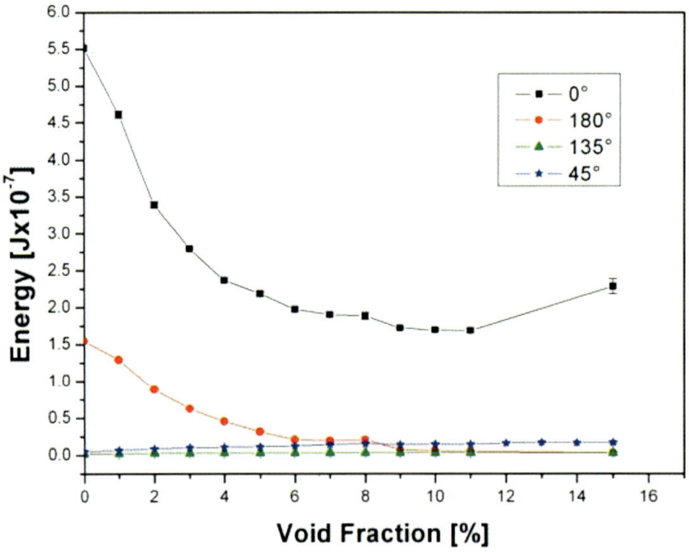

Figure 2b: Average energy values as a function of void fraction [17].

Despite the clear correlation between the acoustic attenuation and the void fraction, in the absence of a model for the interaction between the ultrasonic wave and the two-phase flow the determination of the void fraction from ultrasonic data requires calibration. Jones *et al.* [27] developed such a model, but it requires empirical data and is restricted to very low void fractions. In an attempt to gain insight into the acoustic dissipation phenomena as the ultrasonic wave traverses the two-phase flow and to more precisely relate the acoustic data to two-phase flow structures, plots were made of individual pulses energy ratios as a function of time alongside selected corresponding images. (Fig. **3**) shows one example of such plots and Table **1** represents an attempt at relating specific two-phase flow structures to combinations of instantaneous energy ratios. In this connection, a neural network is being developed by Gonçalves *et al.* [28] to detect the presence of flow structures such as those illustrated in Table **1** having as input the energy ratio combinations from the several sensors. If successful, this could lead to flow pattern recognition from simple ultrasonic measurements as it may not be necessary to fully reconstruct the flow image in order to identify the flow regime.

Table 1: Energy ratio ranges and typical flow patterns [17].

α	Energy Ratio Threshold				Flow Pattern	Image
	180°	0°	135°	45°		
2%	< 0.2	-	< 0.8	> 1.2	Bubbles near the emitter transducer	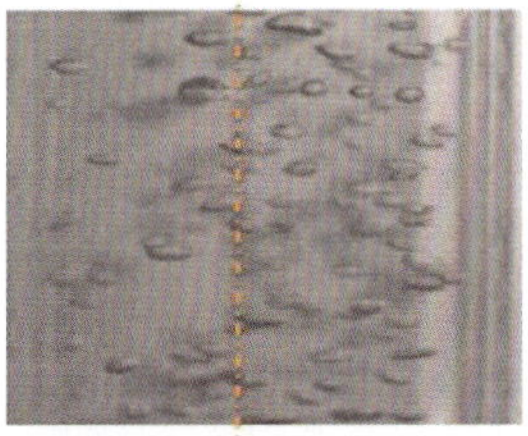
	< 0.2	-	> 0.8	< 0.6	Bubbles near the 180° transducer	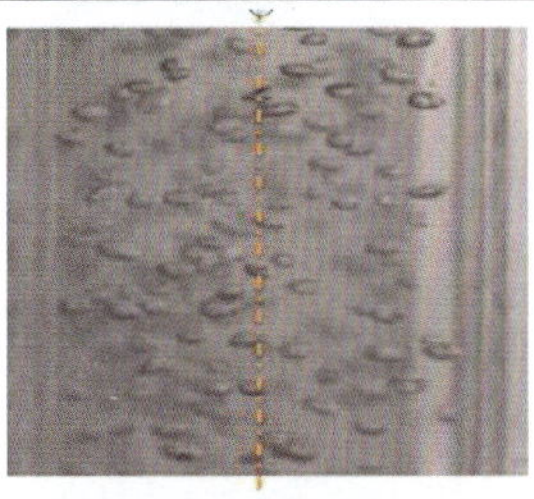
	> 0.9	-	< 0.6	< 0.6	Very scattered bubbles	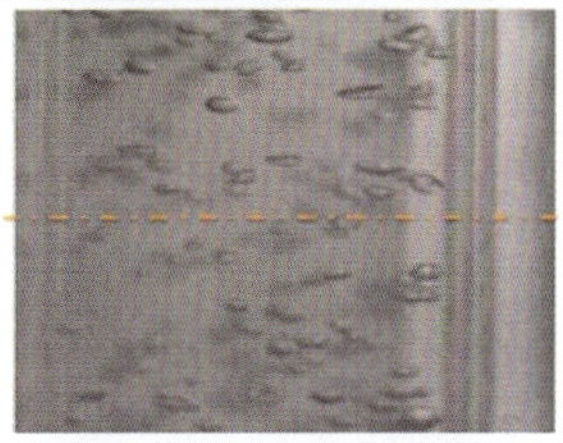
4%	> 0.6	-	< 0.6	< 0.6	Bubbles evenly distributed	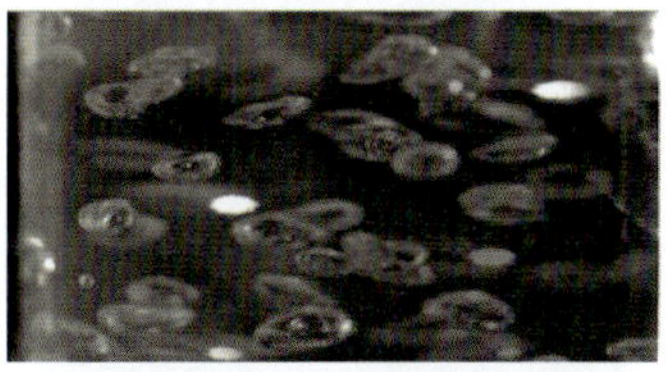
8%	> 0.6	-	< 0.6	< 0.6	Bubbles evenly distributed (denser population)	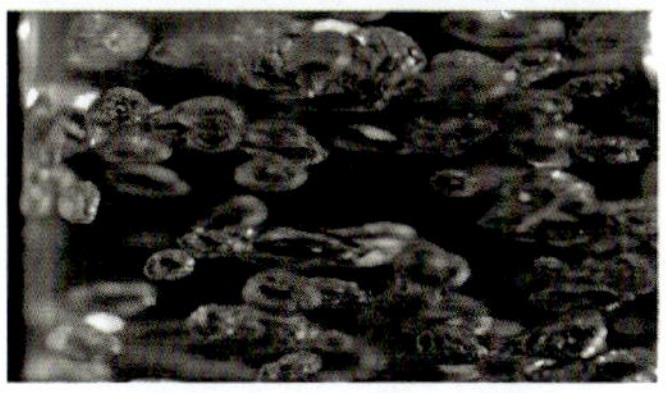
	< 0.1	-	-	> 1.2	Bubble cluster in the vicinity of the emitter sensor	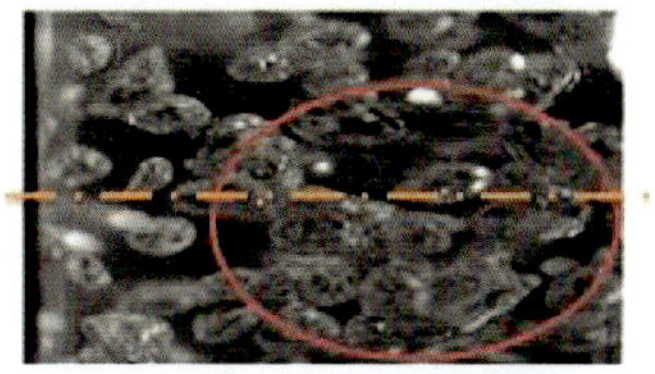
	< 0.1	-	> 0.8	-	Bubble cluster in the vicinity of the 180° sensor	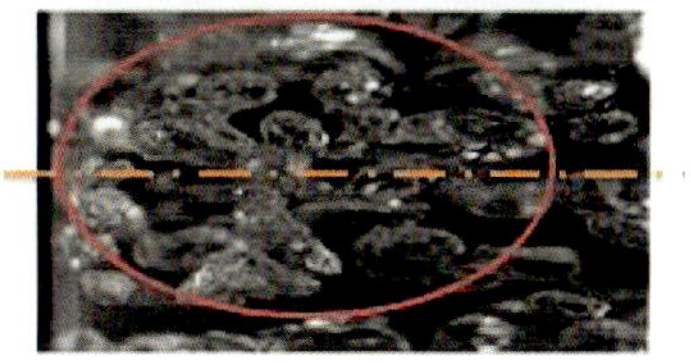

Table 1: Energy ratio ranges and typical flow patterns (continued) [17].

α	Energy Ratio Threshold				Flow Pattern	Image
	180°	0°	135°	45°		
10%	> 0.4	-	> 0.8	-	Bubbles evenly distributed (denser population)	
	< 0.1	-	< 0.6	-	Gas clot	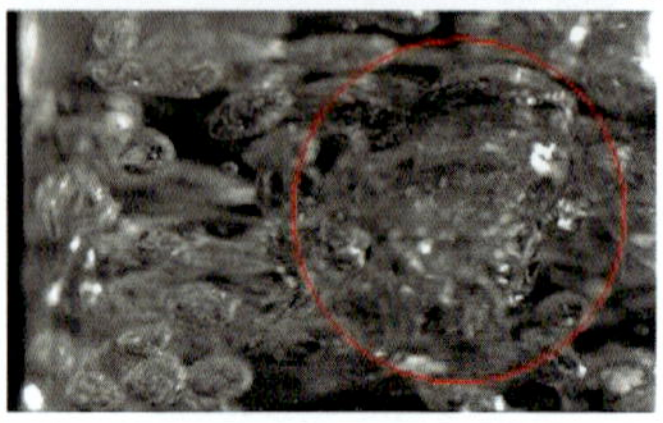
> 13%	< 0.1	-	-	-	Taylor bubble	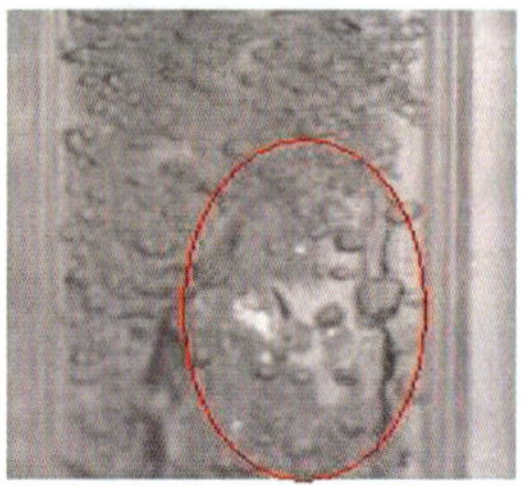

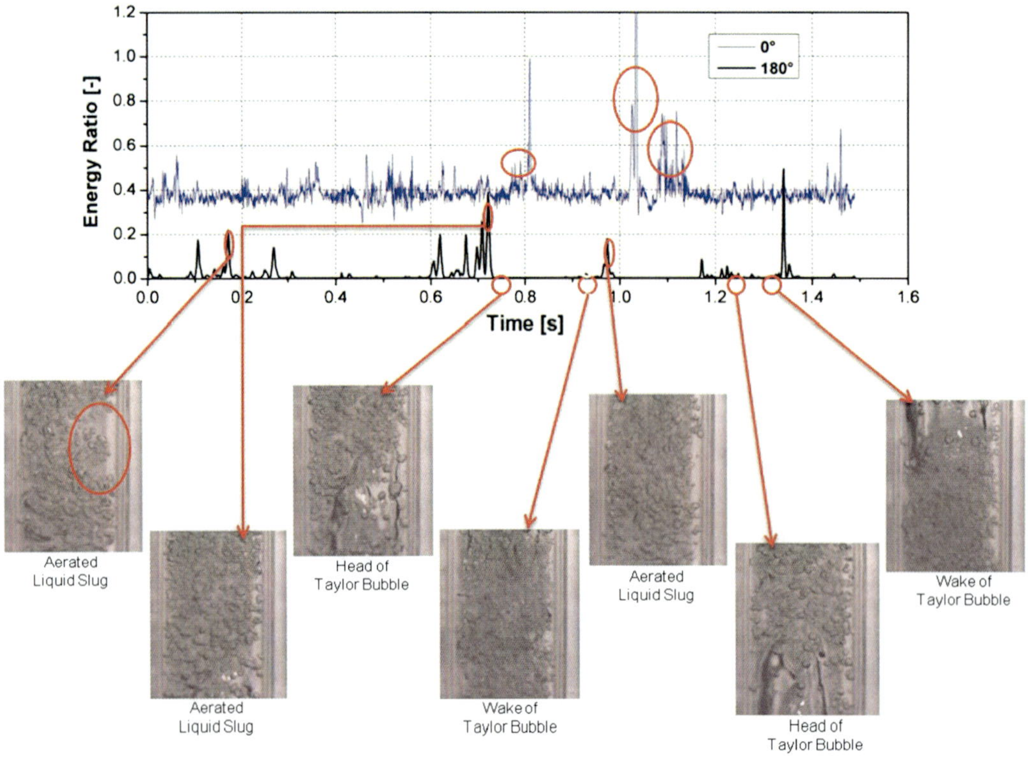

Figure 3: Instantaneous energy ratios and selected corresponding images for α = 18% [17].

Carvalho *et al.* [17] also verified only a very slight increase (less than 1 μsec) in the acoustic wave initial transit time – the time the wave first hits the 180º receiver – throughout the entire bubbly flow regime. The behavior is consistent with the claim by Vatanakul *et al.* [12] that the speed of sound is independent of gas holdup due to the great distortion of ultrasound around bubbles; in other words, the initial transit time would be associated with the portion of the acoustic energy that gets to the receiver transducer with minimum or no disturbance by the bubbles, thus representing mainly transmission through the continuous phase. In this way, despite the ease with which it can be measured, the initial transit time would not lend itself to reliable measurements of the void fraction in liquid-gas systems.

In an attempt to bring the ultrasonic technique closer to applications in the oil industry, Gonçalves *et al.* [29] obtained preliminary ultrasonic data for bubbly air-oil mixtures in 54-mm steel pipes using 1 and 2.25 MHz transducers, *e.g.* Fig. (**4**). The trends observed were very similar to those verified by Carvalho *et al.* [17] for air-water bubbly flows. The 45° and 135° sensors exhibited a gradual increase in the energy ratio as the void fraction increased while the 0° and 180° sensors exhibited a gradual decrease in this parameter. From this work, Gonçalves *et al.* [29] expect to clearly verify the correlation between the acoustic parameters and the gas fraction in oil as well as to systematically investigate the influence of transducer frequency on the quality of the acoustic data. In doing so, the authors will more confidently apply the technique to experimentations in the oil field.

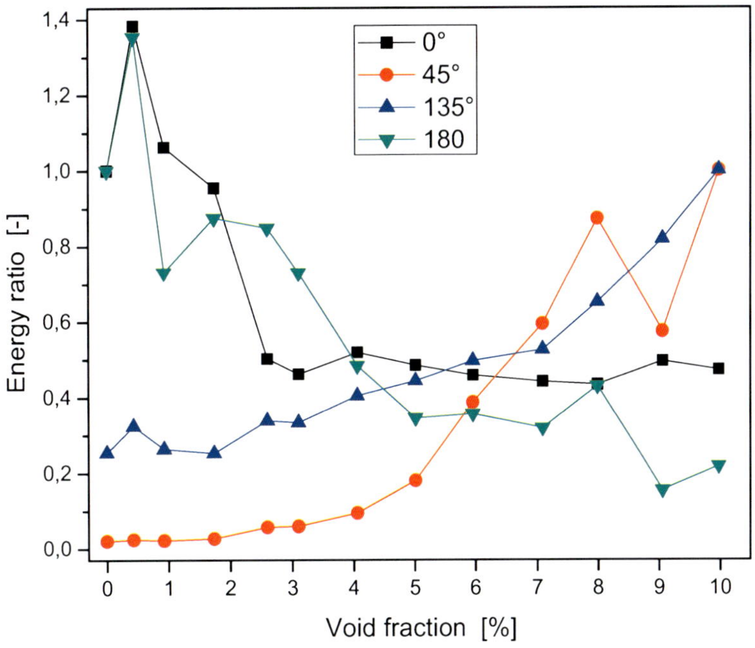

Figure 4: Energy ratio as a function of void fraction (steel pipe and 1 MHz transducers).

2.2. Intermittent Slug Flows

An extensive study of gas-liquid flows in pipelines has been carried out by Hanratty [30]. He and his co-workers focused mainly on problems related to the petroleum industry, namely, the prediction of flow patterns, the fraction of liquid entrained in the gas in annular flow, the slug frequency in slug flows, pressure drop and holdup. Woods and Hanratty [31] studied the effect of the liquid Froude number on the slug formation for co-current air-water flow in a horizontal pipeline. As pointed out by the authors, a critical issue in developing mechanistic models for the slug flow is the prediction of the slug frequency or the slug length. Taitel and Barnea [32] presented a thorough review of the steady-state and transient two-phase slug flow. Referring to (Fig. **5**), the slug body or slug unit is subdivided into two main sections, the liquid aerated slug zone of length L_s and the film zone of length L_f. The film zone consists of a liquid film and an elongated gas bubble; in vertical and off-vertical pipes the bubble is in the center of the pipe and is termed Taylor bubble. The translational velocity of the elongated bubble, U_t, is the velocity at which it propagates downstream. Unlike the liquid slug region, which is considered axially homogeneous, the liquid and gas velocities in the film zone vary along the pipe due to the variation of the film thickness behind the liquid slug. As a slug unit propagates downstream at velocity U_t, part of the liquid in the film moves upstream (backwards) relative to the gas liquid interface and is captured by the following slug. For aerated liquid slugs, the velocity of the dispersed bubbles is usually smaller than the velocity of the gas in the elongated bubbles; in addition, the liquid film seems to be essentially free of small bubbles. It is then supposed that the dispersed bubbles in the liquid slug coalesce at the nose of the elongated bubble, while gas bubbles are re-entrained from the back of the elongated bubble into the liquid slug.

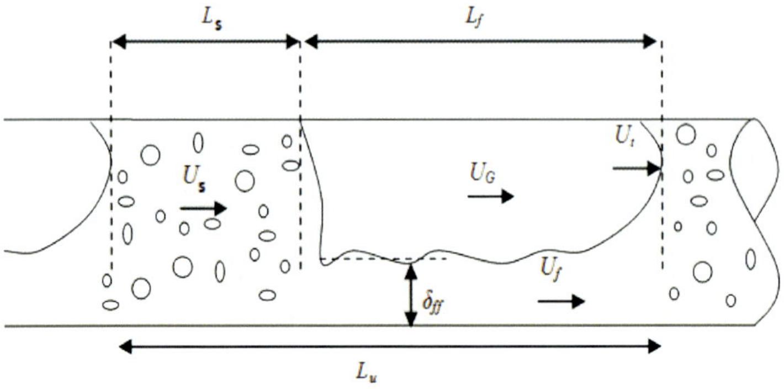

Figure 5: Schematic view of the slug flow (adapted from [33]).

Taitel and Barnea [32] demonstrated that the average void fraction of a slug unit is independent of the bubble shape, the bubble length, the liquid slug length, and the film thickness; as a consequence, the gravitational pressure drop can be calculated independently of the detailed slug structure. On the other hand, the length of the liquid film, its shape, $h_f(z)$, and, especially, the film thickness and its velocity just before pickup, h_{fe} and U_{fe}, are important parameters for calculating the pressure drop and heat and mass transfer in slug flow. However, the shape of the liquid film is a very complex structure, especially near the tail of the liquid slug; it is a three-dimensional problem and only approximate solutions are possible.

In connection with the liquid slugs, the studies by Woods and Hanratty [31] revealed that the mechanisms responsible for the formation of a slug, and thus the slug frequency and length, depend on the liquid Froude number of the wavy stratified flow, the gas velocity, and the location in the pipeline at which the slugs form. The important parameters in determining the transition from stratified to slug flow are the film height, h_s, and the necessary height for the stability of a slug, h_o. If h_s is greater than h_o, an instability at the gas-liquid interface has the possibility of creating a stable slug; on the other hand, for situations in which h_o is greater than h_s the stability of the slug, rather than the stratified flow, determines the initiation of the slug flow pattern.

From experimental observations for air-water systems in vertical upward and horizontal slug flows, Taitel and Barnea [32] suggested that the stable liquid slug length is relatively insensitive to the gas and liquid flow rates and is fairly constant for a given tube diameter. From simulations of the mixing process between the film and the slug by a wall jet entering a large reservoir, a value of 16D was obtained for the minimum stable liquid slug length in vertical upflow and 32D for the case of horizontal flow. Shemer and Barnea [34] used the hydrogen-bubble technique to record the velocity profiles behind the elongated bubbles in gas-liquid slug flow. The observed velocity field was then used to estimate the minimum stable slug length; a value of the order 20D was found.

Besides the slug length or frequency, other additional information necessary for a solution of slug flow models are the translational velocity, U_t, the dispersed bubbles velocity, U_b, and the liquid holdup in the liquid slug zone, R_s [32]. These parameters can be obtained from semi-empirical models or directly measured. Most slug flow models assume constant lengths and shapes of liquid slugs and elongated bubbles as well as constant

Taylor bubble propagation velocity; however, due to the intrinsically unsteady and irregular nature of slug flows, statistical means are required for its proper description [35]. These authors used local optical fiber probes in vertical gas-liquid slug flows to measure the instantaneous local velocities of nose and tail of elongated (Taylor) bubbles simultaneously with the slug length ahead of each bubble. Masala *et al.* [36] used four pulse-echo transducers to detect the gas-liquid interface and to identify separated flow patterns such as stratified smooth, stratified wavy, and annular horizontal flows of air-water mixtures in a horizontal pipe. The transducers frequency was 10 MHz and they were mounted onto the pipe outer wall in a staggered arrangement between the bottom and top surfaces. With this setup, oscillations in the liquid-gas interface as small as 2 mm could be detected and, from these oscillations, it was possible to tell the transitions from stratified smooth to stratified wavy flows (SS/SW) and from stratified wavy to slug flow (SS/SL). The accurate detection of the instantaneous gas-liquid interfaces made it possible to calculate the instantaneous and cross-sectional averaged void fraction for the different flow regimes on the basis of simple geometrical considerations.

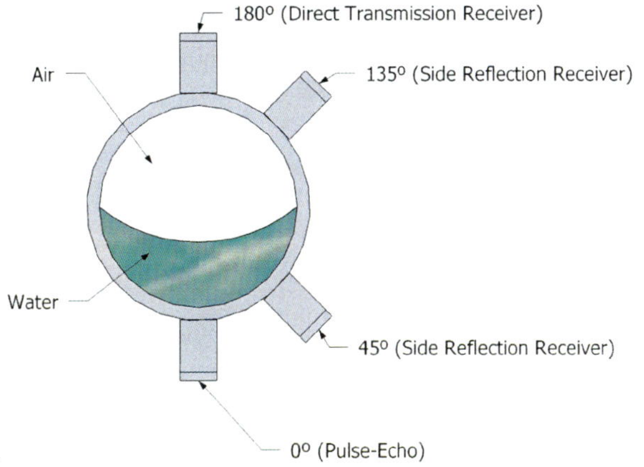

Figure 6: Schematic view of the ultrasonic measuring station used in [38].

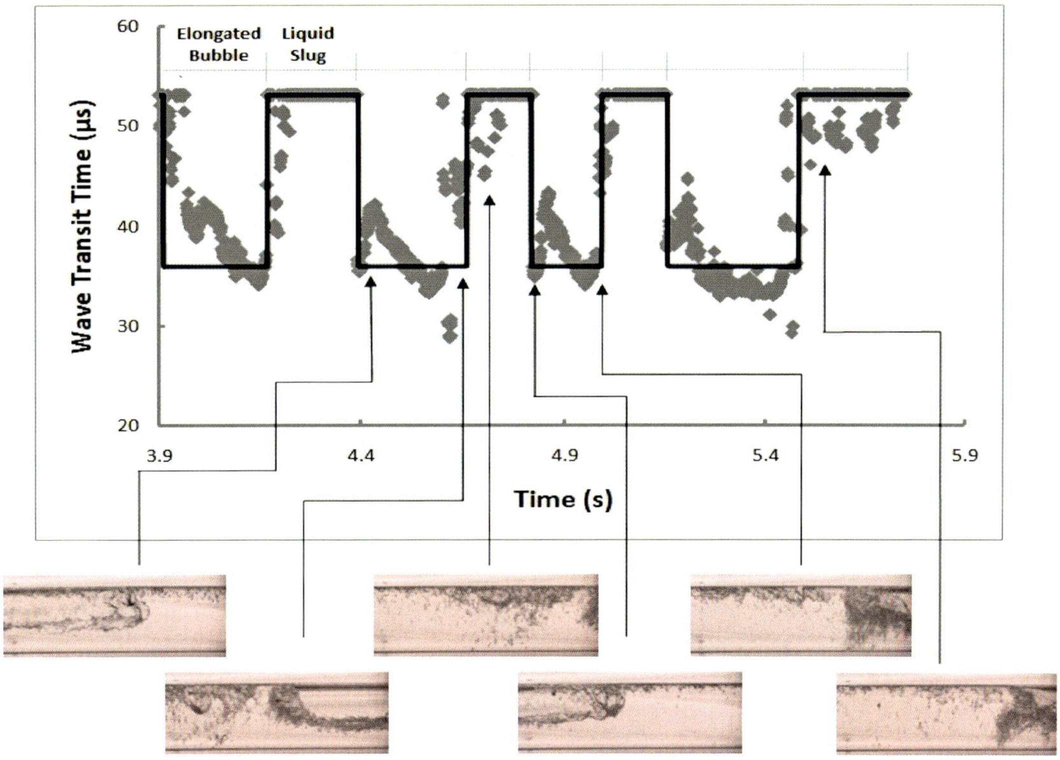

Figure 7: Ultrasonic transit time and selected corresponding images [38].

Grangeiro [37] and Tanahashi *et al.* [38] investigated the applicability of the ultrasonic technique for the study and monitoring of intermittent horizontal slug flows. In broad terms, the main goal was to use the technique to detect the liquid-gas interfaces and to use this information to calculate parameters related to the flow structure. The ultrasonic measuring apparatus used by the authors is similar to that used by Carvalho *et al.* [17] and is shown schematically in (Fig. **6**). Two such ultrasonic stations were positioned 0.645-m apart from each other on the outside of the Plexiglas pipe; L/D = 444 at the test location. Synchronized high-speed filming was used to corroborate the observations from the ultrasonic signals. For the emitter/receiver transducer, the initial transit time — the instant the sound wave first hits this transducer after traversing the two-phase flow and back — is a direct measurement of the interface location and, consequently, the liquid film thickness. As shown in (Fig. **7**), a computer routine was then written that would return appropriate values for data points identified as part of the liquid slug or the elongated bubble. It can be seen that the elongated bubble nose is generally associated with a rather gradual decrease in the signals whereas its tail causes a rather sharp increase with persistent fluctuations. This was an expected behavior as the elongated bubble wake is very turbulent. The agreement between results from the ultrasonic technique and high-speed filming was very good for lower gas fractions but not so good for higher gas fractions [17]. Yet, results for the lengths of the elongated bubble (liquid film), L_F, and liquid slug, L_S, for these two conditions had a fairly low uncertainty interval.

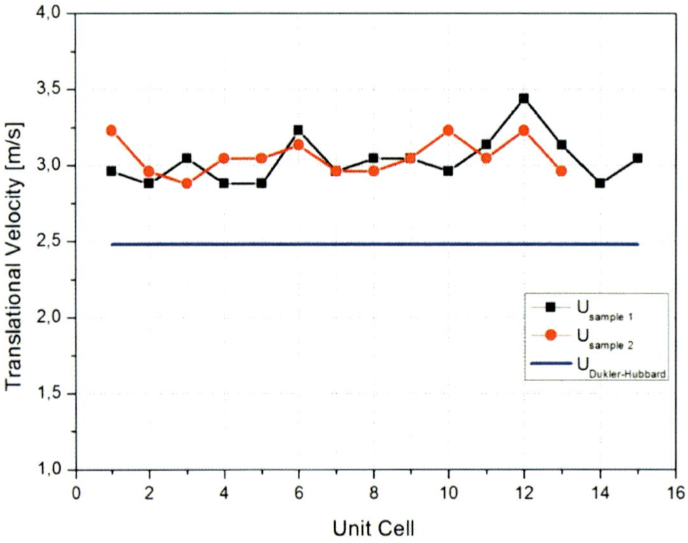

Figure 8: Elongated bubbles translational velocity for j_l = 1.49 m/s and j_g = 0.11 m/s [38].

From the data from the two ultrasonic stations, it was possible to calculate the elongated bubbles translational velocities (Fig. **8**). The x-axis refers to the intermittent flow unit cell number (elongated bubble/liquid slug) as they flowed past the ultrasonic test section during the sampling time. The experimental values were compared with the result from the Dukler and Hubbard model [33] with the closure relationship by Benjamin [39]; it can be seen that the model underestimated the elongated bubbles velocity. The average experimental velocity was 3.04 ± 0.15 m/s at the 95% confidence level. The fluctuations in the translational velocity were thus rather small and exhibited a wavelike behavior. (Fig. **9**) illustrates the elongated bubbles and liquid slugs lengths for the unit cells in sample # 2 in (Fig. **8**). The observed values for the liquid slugs spanned the range from 0.4 to 1.2 m ($7 < L_S/D < 22$), below the $L_S/D = 32$ for the minimum stable liquid slug length proposed in [32] but close to the $L_S/D = 20$ proposed by Shemer and Barnea [34]. The lengths were also observed to follow a wavelike behavior; for the other flow conditions tested, the same general behavior was observed. As discussed below, from the results for the transit time and length and frequency of passage of the flow structures, an attempt was made by Grangeiro [37] to estimate the liquid and gas flow rates for the intermittent horizontal flow and thus further extend the potential of the ultrasonic technique. It should be emphasized that with proper computational facilities these flow parameters can be calculated in real time.

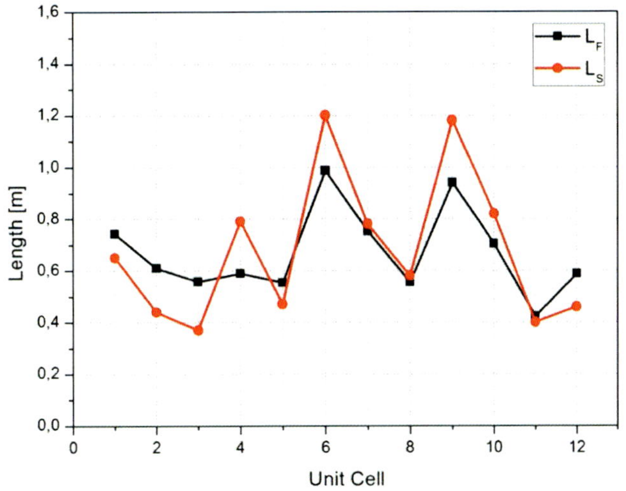

Figure 9: Elongated bubbles and liquid slugs lengths for j_l = 1.49 m/s and j_g = 0.11 m/s [38].

(Fig. **10**) shows a very simple model devised for the intermittent horizontal flow; the elongated bubbles were assumed to be sections of cylinders with flat bottom interfaces and the liquid film and the liquid slug were assumed to have no gas content. From geometrical considerations, estimates were then made for the liquid and gas flow rates for three operating conditions; condition #1 refers to j_l = 2.07 m/s and j_g = 0.11 m/s, condition #2 to j_l = 1.49 m/s and j_g = 0.09 m/s, and condition #3 to j_l = 0,90 m/s and j_g = 0.14 m/s. The results from the model were then compared with those measured by a coriolis meter and a laminar flow element, respectively. (Figs. **11** and **12**) show that the liquid flow rate is overestimated while the gas flow rate is underestimated, which is in agreement with the simple physical model depicted in (Fig. **10**). However, once the model is refined the ultrasonic technique can be used to measure the flow rates for this particular flow pattern.

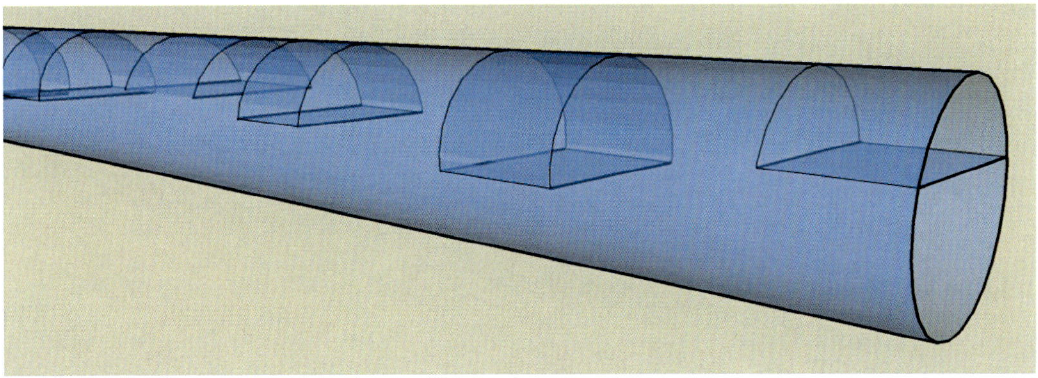

Figure 10: Mechanistic model for the estimation of the liquid and gas flow rates in the intermittent horizontal flow [37].

The improvement to the model that is being worked on presently refers to estimates of the axial void fraction distribution in the liquid slugs. Gonçalves *et al.* [28] are using the ultrasonic data and high-speed filming of Carvalho *et al.* [17] for bubbly air-water flows to develop a neural network capable of providing this distribution based on combinations of the instantaneous values of the energy ratios for the 0°, 45°, 135°, and 180° sensors.

The works discussed above show that it is possible to use the ultrasonic technique for reliable detection of gas-liquid interfaces and calculation of important flow parameters in air-water flows. However, to extend the technique to oil-gas flows it will be necessary to accurately establish the acoustic properties of the particular oils of interest.

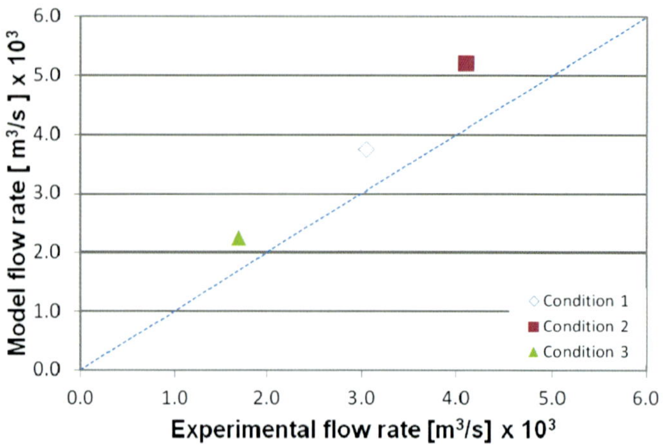

Figure 11: Comparison of the liquid flow rates estimates for the intermittent horizontal flow based on the ultrasonic technique and measurements by the Coriolis meter [37].

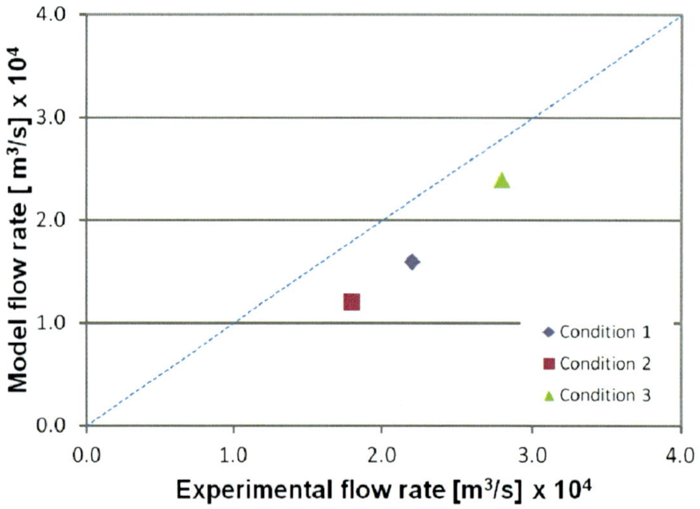

Figure 12: Comparison of the gas flow rates estimates for the intermittent horizontal flow based on the ultrasonic technique and measurements by the laminar flow element [37].

3. THE STUDY OF GAS-LIQUID-SOLID FLOWS

Kytömaa [10] presents a discussion of the mechanisms of acoustic attenuation in suspensions of solids and liquids. He identifies three different regimes of compressional wave propagation on the basis of the dominant length scales of the process; two main parameters are used to determine which regime a specific situation belongs to. The multiple-scattering regime (short wavelength regime) occurs for $ka >> 1$, which means that the particles are large compared to the wavelength and sound scatters uniformly in all directions. In a suspension of many such randomly arranged particles, sound will scatter randomly and will penetrate poorly. On the other hand, the long wavelength regime occurs when the particles are much smaller than the wavelength ($ka << 1$); the pertinent length then is the thickness of the boundary layer, δ, surrounding the particles, often called the Stokes layer thickness. A corresponding Reynolds number is then defined as $Re = a/\delta$. When Re << 1, the boundary layer is thick compared to the particle radius and the viscous relaxation time is shorter than the period of excitation. This particular situation is referred to as the *viscous regime* in which the drag is well represented by the Stokes law. When Re >> 1, the particle radius is much larger than the boundary layer thickness and the fluid outside it is governed by inviscid inertial effects; the drag force then becomes dominated by inertia through the added mass effect. However, the added mass effect is conservative and does not contribute to losses; the dominant loss term is the in-phase component of the Basset force. Additional effects such as unsteady thermal conduction through a thermal boundary layer surrounding the particle and resonant behavior are small in liquid-solid systems and can be neglected. Regarding the prediction of solids concentration from acoustic attenuation measurements, Kytömaa [10] pointed out that the common assumption that these are linearly related is reasonable only up to concentrations of around 10%. At higher concentrations particle interactions become significant in the viscous regime as well as in the inertial regime. From theoretical considerations, this author showed the attenuation is grossly nonlinear for both dissipation regimes and exhibits a maximum at an intermediate concentration. Regarding the sound speed, Kytömaa [10] presented expressions for the low- and high-frequency limits (the viscous and the inertial regimes) as a function of solids concentration. The viscous curve exhibits a minimum while the inertial one does not; the transition from the viscous regime to the inertial one was said to occur at $1 < Re < 10$. Data for 24 μm radius lead glass spheres in water at 450 kHz (inertial regime) displayed the expected monotonic asymptotic rise.

Carlson and Martinsson [40] present a simple theoretical model for ultrasound attenuation in suspensions; the model assumes there is no multiple scattering and is thus valid only for low particle mass fractions. In addition, the model assumes most of the acoustic energy to be backscattered by the solid particles, a scattering pattern that is valid for the long wavelength regime ($ka \ll 1$). Furthermore, the acoustic wave is modeled as a plane wave and the solid particles as incompressible rigid spheres; the suspension is taken as homogeneous with no conglomeration of the solid particles. The expression for the attenuation coefficient so derived will be proportional to the particle mass fraction alone as long as the distance between the transducers is constant. The theoretical predictions were checked against experimental data for 0 to 10% wt. suspensions of Dolomite particles ($(Ca,Mg)CO_3$) in water; the agreement was very good in the range from 1% to 7% mass fraction. Because the model results in a simple linear relationship between the attenuation coefficient and the mass fraction, the authors claim that it can be easily implemented in a measuring system with very limited computational power. However, calibration would be needed to determine the received ultrasound energy in a single-phase flow of the continuous phase alone; in real applications of the oil industry where different types of oil are encountered this could pose some difficulty.

In his derivation of a model for the sound velocity in finely divided suspensions, Urick [41] assumed that the particles were infinitely small compared to the wavelength of the sound (long wavelength limit) and that accordingly the effects of scattering could be neglected. The ratio of the sound velocity in the suspension to that in the suspending liquid alone was shown to be a parabolic function of the solids volume concentration and therefore to have a maximum or a minimum at some particular concentration; data for suspensions of kaolin in water clearly exhibited a minimum in the sound velocity (increase in transit time) for 20% vol. of kaolin. It is interesting to note, though, that in experiments with emulsions of xylene in water and water in xylene, Urick [41] measured the same sound velocity when the xylene concentration was the same; this indicated that the sound velocity was determined mostly by the emulsion composition. In all these experiments by Urick [41], particle size was much smaller than the wavelength, thus corroborating the assumption of long wavelength limit. When this criterion is not met, as in the present case, reflections and refractions of acoustic waves at the interfaces between the suspended particles and the continuous phase need to be taken into account.

Soong *et al.* [42] carried out acoustic attenuation measurements in a three-phase reactor containing water, glass beads, and nitrogen bubbles. The ultrasonic pulses nominal frequency was 2.5 MHz; the glass beads were 80 ± 5 μm in diameter and the concentration was varied from 1 wt.% to 35 wt.% for each nitrogen flow rate tested. The bubbles varied in size from 1 to 5 mm, the specific size distribution being dependent on the nitrogen flow rate. A decrease in the transit time with increasing solids concentration was observed; the transit time was defined as the time interval corresponding to the first distinct zero crossing of the acoustic wave at the receiver transducer. The nitrogen flow rate had no effect on the transit time so defined. It was argued that the portion of the acoustic beam that is transmitted through the solid particles is speeded up by the higher sound velocity in the new medium, which would explain the decrease in the transit time. A simple model was derived on these premises that indicated a cube root dependence of the percentage change in the transit time with concentration. Stolojanu and Prakash [43] carried out acoustic measurements in two-phase and three-phase systems. The transducers had a 4 MHz nominal frequency; the solid loadings were up to 90 wt.% (about 40% vol.) and the gas holdup up to 40% vol. Glass beads 35 μm in diameter were used as the solid phase while tap water and compressed air constituted the liquid and gas phases, respectively. Apparently, even for high gas holdups the flow was homogeneous with bubbles up to 5 mm in diameter. In two-phase systems, the transit time was observed to decrease monotonically with increasing solids concentration, the corresponding plot exhibiting a linear relationship throughout the entire concentration range tested. The measured transit times, however, were quite different from the values predicted by the phenomenological approach proposed by Urick [41]. Regarding the amplitude ratio, an exponential decay was observed with increasing solids concentration. In three-phase systems, the variations in the transit time for different solids concentration were significant only for high gas velocities. The amplitude ratios retained the exponential decay format with a relatively minor influence of varying solids concentration. Zheng [44] also presented ultrasonic data for two-phase and three-phase systems. The liquid, gas, and solid phases consisted respectively of water, air, and 500-μm glass beads with 2,500 kg/m^3 density. The discussion was focused on the standard deviations of the normalized amplitude and transit times referenced to single-phase liquid water. In air-water bubbly flows, the standard deviations of both parameters were seen to increase with increasing gas holdups. In liquid-particles flows, the standard deviations of the transit time increased with increasing solids concentration at the same time the standard deviations of the normalized amplitude decreased. Moreover, the standard deviations in liquid-solid flows were substantially smaller than those in liquid-gas flows, especially for the normalized amplitude. Regarding three-phase systems, the presence of solid particles was observed to have no influence on the fluctuations of the normalized amplitude whereas, for the transit time, the solids concentration greatly increased the fluctuations. The authors then suggested that the concentrations of gas bubbles and solid particles could be detected through careful examination of the fluctuations in the ultrasonic signals.

A description of an ultrasonic particulate sensing device can be found in [45]. A pulse is sent across the fluid and the amplitude of the energy scattered by the particulates is measured at a preselected angle; this amplitude is, in turn, related to the density and elasticity of the particulates. The particulates are then identified by comparing these values with those for known particulates; in addition, a direct comparison of the amplitude measurements provides the particulate size. Apparently, the device does not allow for measurements of the particulate concentration.

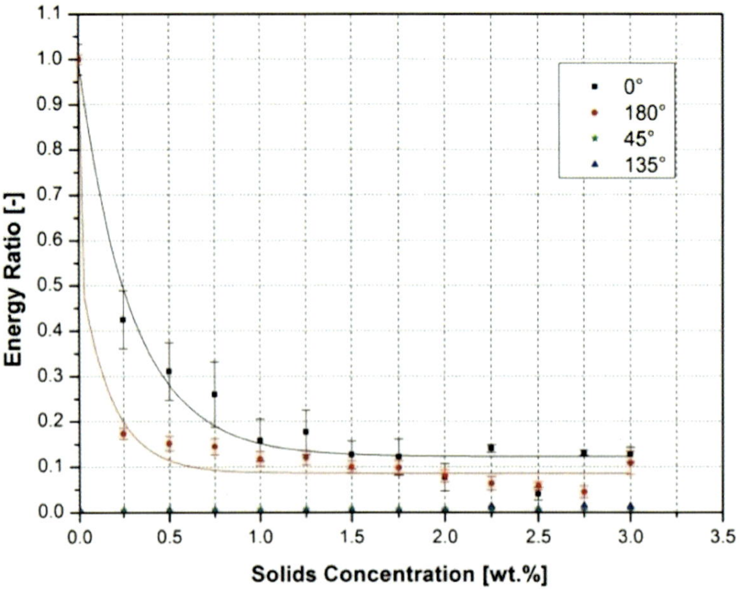

Figure 13: Average energy ratios as a function of solids concentration [38].

Tanahashi *et al.* [38] obtained ultrasonic data for suspensions of glass beads in water from 0 to 3% mass concentration in 0.25% steps. The acoustic instrumentation used was the same as that shown in (Fig. **1**). The glass beads size distribution ranged from 150 to 595 μm in diameter with $0.72 < ka < 2.84$ and $200 < Re < 800$; however, for 96% of the particles $ka > 1$ and $Re >> 1$ and the attenuation mechanisms were thought to be the combined effect of multiple scattering and inertial losses (in-phase component of the Basset force); as ka is not much larger than unity the multiple scattering process was not expected to be very intense. For the remainder 7% of the particles, $ka < 1$ and $Re >> 1$ so that acoustic attenuation was expected to be mostly due to inertial effects. In Table **2** a comparison is made with acoustic data in the literature for liquid-solid suspensions [10, 38, 41-43, 46]. The ultrasonic signals in [38] were observed to be fast attenuated even for very small sand concentrations while the signals duration remained essentially unchanged. (Fig. **13**) exhibits the average energy ratios as a function of solids concentration for all receivers. At the 0° and 180° sensors the energy received decreased exponentially with concentration, the correlation coefficients being 0.97 and 0.98 respectively. It is noteworthy that even for concentrations as low as 0.25% the energy ratio was already less than 0.5 and decreased to much lower values within the narrow concentration range tested. A similar behavior was observed by Soong *et al.* [42] for 40 μm radius particles in water, which also correspond to the inertial regime. Data from Stolojanu and Prakash [43], on the other hand, begin at amplitude ratios close to unity and display a much more gradual decrease despite being in the inertial regime also and with a Reynolds number somewhat close to that of Soong *et al.* [42]. Also shown in (Fig. **13**), at the 45° and 135° sensors no measurable energy was received throughout the entire concentration range tested; this behavior confirmed the expectations that multiple scattering should not be very intense for the particles sizes in this experiment.

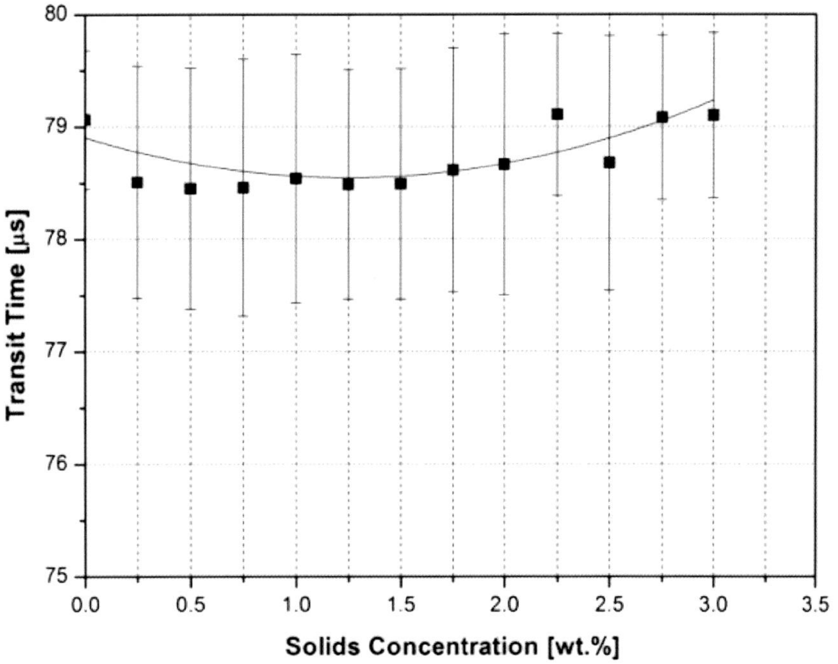

Figure 14: Initial transit time as a function of solids concentration [38].

(Fig. **14**) shows the initial transit time as a function of solids concentration [38]. A parabolic function was fitted to the averages with a correlation coefficient of 0.52; a minimum occurs at approximately 1.5%. The scatter in the averages could be due to concentration variations in the ultrasonic test section as compared to the nominal value in the suspension reservoir where it was prepared. The uncertainties in the averages were calculated from the standard deviation in 10,000 pulses in each sample and six samples for each concentration (60,000 in the ensemble). It is clear that the uncertainties far exceed the 0.5 μm overall variation in the averages; nevertheless, the parabolic trend in the averages was compared with the trends in other data sets in the literature (Table **2**). The uncertainty in the averages in these data sets averages was not reported. The parabolic trend with a minimum observed herein does not agree with the other authors who, for a much broader concentration range, observed only a monotonic decrease in the initial transit time. However, in (38) most particles had ka = 2.84 whereas for the other authors $ka < 1$; these trends were also the opposite of those verified by Urick [41] for the viscous dissipation regime and $ka \ll 1$. Kytömaa [10] points out that sound speed and attenuation each contain complementary information and measurements that can make use of both are likely to be more robust. Nevertheless, in view of the apparent strong dependence of the transit time variation on the *ka* parameter, the use of the initial transit time to measure solids concentration would be very much dependent upon calibration for the particles sizes expected in the real application. In case this distribution could not be known beforehand, the transit time would not yield reliable information.

As a conclusion, the selection of transducers frequency for measurement of solids concentration depends on the particle size distribution, the continuous phase, the concentration range desired, and the resolution required. The initial transit time variation with solids concentration seems to be very much dependent on the particles size relative to the wavelength and, thus, prior knowledge of size distribution is a requirement for reliable measurements. For the energy or amplitude ratio, on the other hand, there is agreement on the trends among the various authors, but the use of these parameters for solids concentration measurements also depends on calibration. Finally, poor ultrasonic penetration due to intense multiple scattering would cause the sound beam as received by the opposite sensor to be fast attenuated. In these cases, the side sensors could provide useful information that, together with the 0° and 180° sensors, would make the measurement more robust as a whole. Gonçalves *et al.* [29] obtained preliminary acoustic data for oil-sand suspensions in steel pipes and Paiva *et al.* [47] are presently investigating the estimation of the gas fraction and solids concentration in liquid-gas-solid mixtures based on ultrasonic signals; besides, the influence of the *ka* parameter on the mechanisms of acoustic attenuation is also being systematically established. In doing so, it is hoped that advances will be made towards setting up the requirements for reliable identification and measurement of dispersed phases concentration using ultrasonics.

Table 2: Comparison of acoustic data in liquid-solid systems from several authors in the literature.

Author (s)	Liquid-Solid System	Concentration Range [wt.%]	Particles Radius [μm]	ka	δ [μm]	Re	Acoustic Dissipation Regime	Amplitude or Energy Ratio with Increasing wt.%	Initial Transit Time with Increasing wt.%
Tanahashi *et al.* (38)	Glass beads in water	0 – 3%	75 - 300	0.72 - 2.84	0.37	203 – 806	Inertial	Exponential decrease	Parabolic, minimum at 1.5%
Kytömma (10)	Glass beads in water	0 - 60%	120	0.023	0.83	14.6	Inertial	Nonlinear with a minimum	Exponential decrease
Shukla and Prakash (46)	Glass beads in water	0 – 40%	0.17 – 0.55	0.23 - 0.75	0.31	55 – 178	Inertial	-	Exponential decrease
Soong *et al.* (42)	Glass beads in water	0 – 35%	40	0.42	0.35	114	Inertial	Exponential decrease	Linear decrease
Stolojanu and Prakash (43)	Glass beads in water	0 – 100%	17.5	0.30	0.26	63.3	Inertial	Exponential decrease	Linear decrease
Urick (41)	Kaolin in water	0 – 63.4%	0.2 – 2.5	0.0009 – 0.011	0.55	0.36 – 4.52	Viscous	-	Parabolic, maximum at 39.4%

4. THE STUDY OF SLURRIES AND LIQUID-LIQUID FLOWS

Simultaneous flow of two immiscible liquid phases has important applications in industry such as process industries and, particularly, in petroleum engineering. Very often during oil production, oil from a deep high-pressure reservoir comes up to surface facilities mixed with a water phase, which may itself already be present in the reservoir or thereto injected to keep the pressure constant. Furthermore, a free gas phase may appear early in the production line as soon as the pressure falls below the bubble point.

As pointed out by Tsouris and Tavlarides [22], an accurate, reliable, and continuous measurement of the volume fraction of one phase dispersed in another fluid is of great importance in many practical situations. One example from the oil industry is the pumping of heavy (viscous) crude oil in the form of an oil-water emulsion with water content as high as 58 vol.% [48]. Another example refers to the transportation of oil through pipelines without emulsification; even though the amounts of water involved are very small, because of the huge volumes involved, the water content must be accurately known. Capacitance probes have been available commercially for measurement of the volume fraction of dispersed phases in these cases; however, they are subject to coating by paraffins, which render them inaccurate after a short period of time [22]. On the other hand, laboratory analysis of samples is labor intensive, time consuming, and does not allow the continuous monitoring of the fluid-fluid system.

Tsouris and Tavlarides [22] then performed ultrasonic measurements of the volume fractions of toluene-water and water-kerosene dispersion systems. The measurements were based on a comparison of the acoustic wave transit time through the dispersions with the transit times through the pure phases. Improved accuracy in the calculation of the volume fractions was obtained by considering refraction and reflection phenomena of sound at the droplets interface; the effect of temperature on the transit times was also accounted for in the results. The experimentation demonstrated the ability of the ultrasonic technique to measure volume fractions as low as 0.005 with a relative error of 8%; yet, in order to reduce the error for volume fractions lower than 0.05, an empirical correction was introduced in the calculations.

Ribeiro *et al.* [49] used the ultrasonic technique to measure the holdup fraction of dispersed toluene in water. Starting with the expression for the propagation velocity of a longitudinal acoustic wave in liquid-liquid systems, the authors arrived at a simple expression for the holdup fraction as a function of the acoustic wave transit time through the dispersion; this expression does not take into account the drop size distribution or the mixture pattern. In this respect, Ribeiro *et al.* [49] performed their acoustic measurements through confined stirred dispersions and made reference to the work of Havilicek and Sovová [50], who showed the dependence of the ultrasound velocity on the holdup and its *independence* on the size of the droplets forming the dispersion. The transit time definition used to correlate the data was the time for the received signal to reach 25% of the maximum amplitude of the rectified signal envelope; this definition was preliminarily verified by the authors to be the least affected by signal spreading over the holdup range tested (0 to 40% toluene). An average 5.4% variation in the ultrasound velocity, from 1,480 m/sec for saturated water down to approximately 1,400 m/sec for 40% toluene, was observed by Ribeiro *et al.* [49]. As continual development of electronics allows for increasingly more precise transit time measurements, this variation in the ultrasound velocity could lead to precise holdup measurements in many situations of practical interest.

Bamberger and Greenwood [7] used the ultrasonic technique to noninvasively measure the density and solids concentration of liquids and slurries flowing through a pipeline. The density measurements were based on multiple reflections of the ultrasonic signal at the pipe wall/fluid interface, which increased the instrument sensitivity. In addition, the data reduction method used to obtain the reflection coefficient and the attenuation in the liquid or slurry made the measurement independent of any particular pulser voltage. However, in order to obtain the fluid density a separate measurement of the acoustic wave transit time through the liquid or slurry was required. Regarding the determination of the solids concentration, it was demonstrated that, upon calibration, measurements of the slurry density and acoustic attenuation could provide information about the particulate weight percentage; this requires previous knowledge of the particulate density.

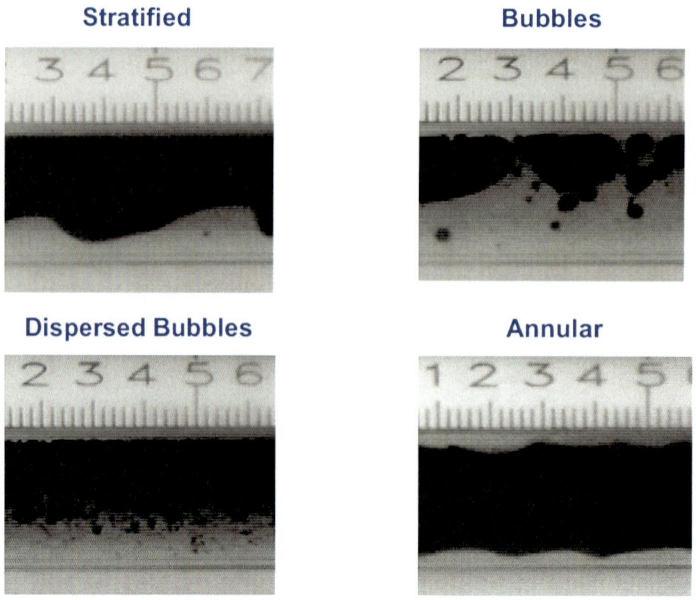

Figure 15: Flow patterns of water-assisted horizontal flows of heavy oil [51].

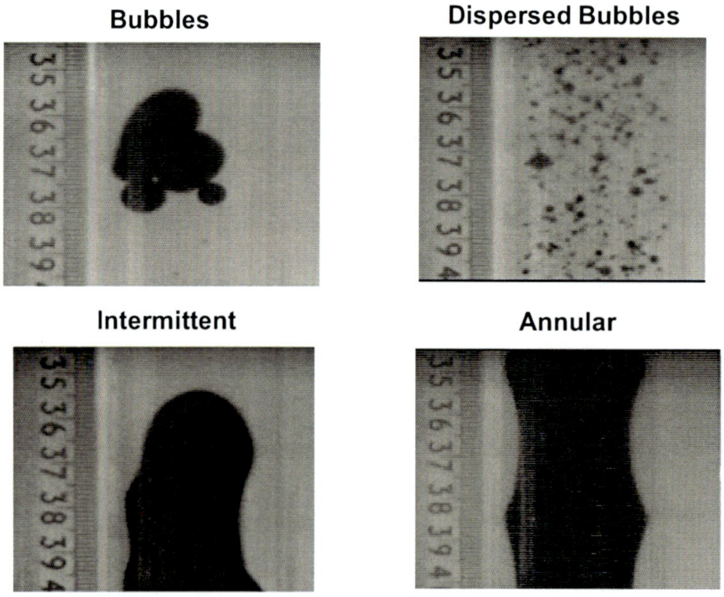

Figure 16: Flow patterns of water-assisted vertical upward flows of heavy oil [51].

Bearing in mind the need for calibration placed by the ultrasonic technique and, at the same time, its potential for nonintrusive measurements, attention was focused on the use of high-speed filming as a means to calibrate ultrasonic probes in the future, especially in the case of heavy oils. (Figs. **15** and **16**) show basic flow patterns of heavy oil/water flows in horizontal and vertical 2.84 cm i.d. tubes [51]. From the high-speed motion pictures (500 to 1000 fps), flow parameters can be obtained to calibrate the ultrasound. For instance, the axial velocities at the oil-water interfaces can be measured from sequential frames. If the interface is that of a constant shape oil bubble, then the interface velocity will correspond to the bubble velocity itself, V_o, and the corresponding oil volume fraction, ε_o, can be determined from the oil superficial velocity, V_{so}, as $\varepsilon_o = V_{so}/V_o$. However, for core-annular flows the viscous oil core is a deformable body subjected to turbulent shear at the interface and, thus, the interface velocity is not the same as the fluid velocity. In these cases, Bannwart [52] and Rodriguez and Bannwart [53] proposed to interpret the interfacial waves observed in stable core-annular flow as kinematic waves moving at the wave speed determined by the oil superficial velocity, V_{so}, the oil volume fraction, ε_o, and the oil-water mixture velocity, V_M [54]. From the high-speed filming, it was possible to determine the wave speed, which in turn allowed for the determination of the oil-water velocity ratio, s, and the corresponding oil volume fraction. The technique was successfully validated using data for horizontal and vertical core-annular flows from different sources; (Fig. **17**) illustrates this point for vertical oil-water flows inside a 2.84-cm id pipe. It can then be concluded that high-speed filming of core-annular flows can be satisfactorily used to calibrate ultrasonic probes.

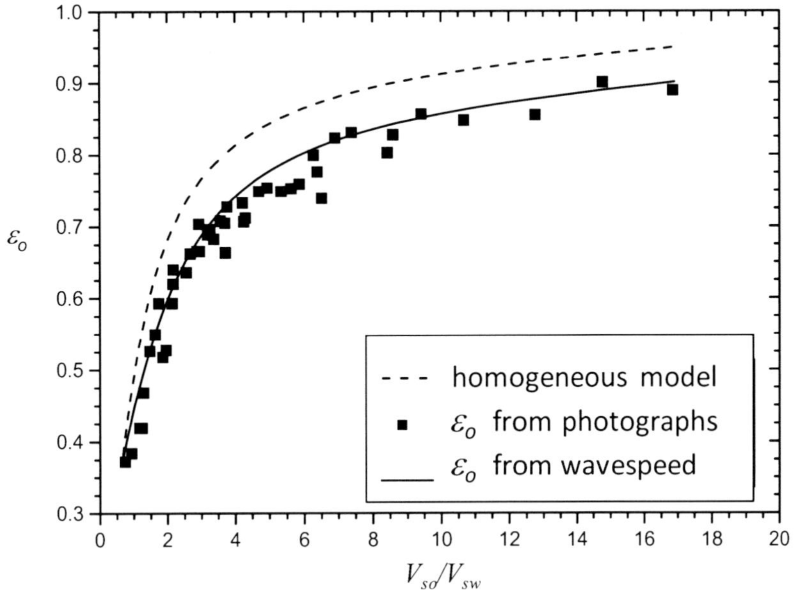

Figure 17: Comparison between photographic and wave speed based measurements of the oil holdup, ε_o, [55].

5. CONCLUDING REMARKS

From the discussion in this chapter, it can be concluded that the ultrasonic technique does seem promising. Even though research is still needed before it can be regularly used in the oil industry, several aspects of its use have been successfully demonstrated in the laboratory and a substantial amount of knowledge has been acquired. In broad terms, the main task ahead would be to get ultrasonic data for the actual flows and analyze it in the light of the knowledge already gained. The way to go from there would present itself more clearly.

In more specific terms, if the ultrasonic technique could be used to detect the prevailing flow pattern in a given situation, this information in itself would be extremely useful in MFM. Furthermore, if the ultrasonic technique can indeed provide direct information about the phase fractions in three- or four-phase flows the need for separators in the oil fields would be eliminated. In this case, the technique would fit the "Leave-as-it-is" MFM category, which aims at being totally non-intrusive to the flow, requiring no separation, no homogenization, and no alteration of the original flow regime. Even though this goal might be far off, the ultrasonic technique would be a much simpler method of measuring the phase fractions in two-phase flows, when the multiphase flow has to be separated, than many instruments discussed in [1]. The downside is, of course, that the ultrasonic technique requires extensive calibration. The ultrasonic technique could also be an alternative to more complex and expensive density measuring instruments.

In the case of homogeneous flows, the ultrasonic technique as a phase fraction measuring instrument could be used with a number of other instruments to provide the required phase flow rates. For horizontal and vertical slug flows, which are not homogeneous, the ultrasonic technique alone could provide rough estimates of the phases volume flow rate.

REFERENCES

[1] G. Falcone, G.F. Hewitt, and C. Alimonti, *Multiphase Flow Metering*. Amesterdam: Elsevier, 2010.

[2] R. Thorn, G.A. Johansen, and E.A. Hammer, "Three-phase flow measurement in the offshore oil industry: is there a place for process tomography?" in *1st World Congr. Ind. Process Tomography*, 1999, pp. 228-235.

[3] W.F.J. Slijkerman, A.W. Jamieson, W.J. Priddy, O. Oakland, and H. Moestue, "Oil companies′needs in multiphase flow metering" in *13th North Sea Flow Measuement Workshop*, 1995, pp. 228-235.

[4] R.L Powell, "Experimental techniques for multiphase flows," *Phys. Fluids*, vol. 20, pp. 1-20, 2008.

[5] A. Jaworek, A. Krupa, and M. Trela, "Capacitance sensor for void fraction measurement in water/steam flows," *Flow Measurement Instrumentation*, vol. 15, pp. 317-324, 2004.

[6] L.J. Bond, M. Morra, M.S. Greenwood, J.A. Bamberger, R.A. Pappas, "Ultrasonic technologies for advanced process monitoring, measurement, and control," in *20th IEEE Instrumentation & Measurement Technology*, 2003, pp. 1288–1293.

[7] J.A. Bamberger, and M.S. Greenwood, "Measuring fluid and slurry density and solids concentration non-invasively," *Ultrasonics*, vol. 42, pp. 563-567, 2004.

[8] J. G. Bouchard, P. A. Payne, and S. Szyszko, "Non-invasive measurement of process states using acoustic emission techniques coupled with advanced signal processing," *Trans. IChemE.*, vol. 72, Part A, p. 20025, 1994.

[9] M. K. Horne, and N. Roland, "Ultrasonic rate measurement in two-phase bubble flow," in *SPE Annual Technical Conference & Exhibition*, 2003, pp. 711-713.

[10] H. K. Kytömaa, "Theory of sound propagation in suspensions: a guide to particle size and concentration characterization," *Powder Tech.*, vol. 82, pp. 115-121, 1995.

[11] H. Murakawa, H. Kikura, and M. Aritomi, "Application of ultrasonic Doppler method for bubbly flow measurement using two ultrasonic frequencies," *Exp. Therm. Fluid Sci.*, vol. 29, pp. 843-850, 2005.

[12] M. Vatanakul, Y. Zheng, and M. Couturier, "Application of ultrasonic technique in multiphase flow," *Ind. Eng. Chem. Res.*, vol. 43, pp. 5681-5691, 2004.

[13] S. Wada, H. Kikura, and M. Aritomi, "Pattern recognition and signal processing of ultrasonic echo signal on two-phase flow," *Flow Measurement Instrumentation*, vol. 17, pp. 207-224, 2006.

[14] Warsito, M. Ohkawa, N. Kawata, and S. Uchida, "Cross-sectional distributions of gas and solid holdups in slurry bubble column investigated by ultrasonic computed tomography," *Chem. Eng. Sci.*, vol. 54, pp. 4711-4728, 1999.

[15] L. A. Xu, R. G. Green, A. Plaskowski, and M. S. Beck, "The pulsed ultrasonic cross-correlation flowmeter for two-phase flow measurements," *J. Phys. E.: Sci. Instrum.*, vol. 21, pp. 406-414, 1988.

[16] Y. Zheng, and Q. Zhang, "Simultaneous measurement of gas and solid holdups in multiphase systems using ultrasonic technique," *Chem. Eng. Sci.*, vol. 59, pp. 3505-3514, 2004.

[17] R.D.M. Carvalho, O.J. Venturini, E.I. Tanahashi, F. Neves Jr, and F.A. França, "Application of the ultrasonic technique and high-speed filming for the study of the structure of air-water bubbly flows," *Exp. Therm. Fluid Sci.*, vol. 33, pp. 1065-1086, 2009.

[18] L. Xu, Y. Han, L. Xu, and J. Yang, "Application of ultrasonic tomography to monitoring gas/liquid flow," *Chem. Eng. Sci.*, vol. 52, 13, pp. 2171-2183, 1997.

[19] L. J. Xu, and L. A. Xu, "Gas/liquid two-phase flow regime identification by ultrasonic tomography," *Flow Measurement Instrumentation*, vol. 8, 3/4, pp. 145-155, 1997.

[20] J. S. Chang, Y. Ichikawa, and G. A. Irons, "Flow regime characterization and liquid film thickness measurement in horizontal gas-liquid two-phase flow by an ultrasonic method," in *AIAA/ASME Joint Plasma Thermophysics*, 1982, HTC, vol. 2, pp. 12–17.

[21] S. Fiedler, S. Yildiz, and H. Auracher, "Determination of film thickness and flooding during reflux condensation in a small inclined tube with an ultrasonic transducer," *Int. J. Energy Res.*, vol. 4, pp. 315-325, 2003.

[22] C. Tavlarides, and L.L. Tsouris, "Volume fraction measurements of water in oil by an ultrasonic technique," *Ind. Eng. Chem. Res.*, vol. 32, pp. 998-1002, 1993.

[23] C. Tsouris, M.A. Norato, and L.L. Tavlarides, "A pulse-echo ultrasonic probe for in-situ volume fraction measurements in liquid-liquid dispersions," *Ind. Eng. Chem. Res.*, vol. 34, pp. 3154-3158, 1995.

[24] M.X. Su, X.S. Cai, M.H. Xue, L.L. Dong, and F. Xu, "Particle sizing in dense two-phase droplet systems by ultrasonic attenuation and velocity spectra," *Sci. China Ser. E: Technological Sci.*, vol. 52, pp. 1502-1510, 2009.

[25] V. Melnikov, V. Drobkov, and A. Shustov, "Simultaneous determination of multiphase flowrates and concentrations," *US 2003/0051558 A1*, March 20, 2003.

[26] Clampon Ultrasonic Intelligent Sensors [Online]. http://www.clampon.com/.

[27] S. W. Jones, A. Amblard, and C. Favreau, "Interaction of an ultrasonic wave with a bubbly mixture," *Exp. Fluids,* vol. 4, pp. 341-349, 1986.

[28] J.L. Gonçalves, P.S. Lopes, T.A. Paiva, A.B. Jorge, R.D.M. Carvalho, and O.J. Venturini, "Development of a multiphase flow metering procedure based on the ultrasonic technique and artificial neural networks," submitted for presentation and publication in *15th International Conference on Multiphase Technology*, 2011.

[29] J.L. Gonçalves, T.A. Paiva, G.C. Costa, R.I. Deperon, R.D.M. Carvalho, and O.J. Venturini, "Application of the ultrasonic technique for monitoring oil-gas mixtures and oil-solids supensions," submitted for presentation and publication in Brasil Offshore, 2011.

[30] T.J. Hanratty, "Gas-liquid flow in pipelines," U.S. Department of Energy, University of Illinois, Urbana, IL, Final Report Grant DEFG02-86ER13556, 2004.

[31] B.D. Hanratty, and T.J Woods, "Influence of Froude number on physical processes determining frequency of slugging in horizontal gas-liquid flow," *Int. J. Multiphase Flow,* vol. 25, pp. 1195-1223, 1999.

[32] Y. Barnea, and D. Taitel, "Two-phase slug flow," in *Advs. Heat Transfer,* Academic Press, Inc., pp. 83-132, 1990.

[33] A.E. Hubbard, and M.G. Dukler, "A model for gas-liquid slug flow in horizontal and near horizontal tubes," *Ind. Eng. Chem. Fund.,* vol. 14, pp. 337-347, 1975.

[34] L. Barnea, and D. Shemer, "Visualization of the instantaneous velocity profiles in gas-liquid slug flows," *Physicochem. Hydrodyn.*, vol. 8, pp. 243-253, 1987.

[35] R. van Hout, D. Barnea, and L. Shemer, "Evolution of statistical parameters of gas-liquid slug flow along vertical pipes," *Int. J. Multiphase Flow*, vol. 27, pp. 1579-1602, 2001.

[36] T. Masala, G. Harvel, and J.S. Chang, "Separated two-phase flow regime parameter measurement by a high speed ultrasonic pulse-echo system," *Rev. Scientific Instruments*, vol. 78, p. 114901, 2007.

[37] F. A. Grangeiro Filho, "Characterization of the horizontal intermittent air-water flow using the ultrasound technique aided by high-speed filming" (in Portuguese), M.Sc. thesis, Ciências e Engenharia de Petróleo, Universidade Estadual de Campinas (UNICAMP), Campinas, Brazil, 2010.

[38] E.I. Tanahashi, T.A. Paiva, F.A. Grangeiro, R.D.M. Carvalho, O.J. Venturini, A.C. Bannwart, and V.C.Bizotto, "Application of the ultrasonic technique for monitoring intermittent liquid-gas and liquid-solid flows," in *7th North American Conf. Multiphase Tech.*, 2010, Bannf, Canada, pp. 125-139.

[39] T.B. Benjamin, "Gravity currents and related phenomena," *J. Fluid Mech.*, vol. 31, pp. 209-248, 1968.

[40] J. Carlson, and P.-E. Martinssson, "A simple scattering model for measuring particle mass fractions in multiphase flows," *Ultrasonics*, vol. 39, pp. 585-590, 2002.

[41] R.J. Urich, "A sound velocity method for determining the compressibility of finely divided substances," *J. Acoustical Soc. Am.,* vol. 18, pp. 983-987, 1947.

[42] Y. Soong, I.K. Gamko, A.G. Blackwell, R.R. Schehl, and M.F. Zarochak, "Measurements of solids concentration in a three-phase reactor by an ultrasonic technique," *Chem. Eng. J.*, vol. 60, pp. 161-167, 1995.

[43] V. Stolojanu, and A. Prakash, "Hydrodynamic measurements in a slurry bubble column using ultrasonic techniques," *Chem. Eng. Sci.*, vol. 52, pp. 4225-4230, 1997.

[44] Y. Zheng, "Ultrasonic measurement for two/three phase flow detection," *Can. J. Chem. Eng.*, vol. 81, pp. 268-270, 2003.

[45] K.G. Foote, "Ultrasonic particulate sensing", US *4527420*, July 9, 1985.

[46] A. Shukla, and A. Prakash, "Ultrasonic technique to determine particle size and concentration in slurry systems," *Chem. Eng. Sci.,* vol. 61, pp. 2468-2475, 2006.

[47] T.A. Paiva, J.L. Gonçalves, O.J. Venturini, and R.D.M. Carvalho, "Study of ultrasonic signals in liquid-gas-solid mixtures and influence of particle size on acoustic dissipation," submitted for presentation and publication in *15th Int. Conf. Multiphase Tech.*, 2011.

[48] B.E. Wyslouzi, M.A. Kessick, and J.H. Masliyah, "Pipeline flow behavior of heavy crude oil emulsions," *Can. J. Chem. Eng.*, vol. 65, p. 353, 1987.

[49] M.M.M. Ribeiro, C. Gonçalves, P.F. Regueiras, M.M.L. Guimarães, J.J.C. Cruz Pinto, "Measurements of toluene-water dispersions hold-up using a noninvasive ultrasonic technique," *Chem. Eng. J.*, vol. 118, pp. 47-54, 2006.

[50] A. Havilicek, and H. Sovová, "Measurement of liquid holdup of the dispersed phase by an ultrasonic method," *Collection Czechoslovak Chem. Comm.*, vol. 49, p. 378, 1984.

[51] D. Barnea, and N. Brauner, "Holdup of the liquid slug in two-phase intermittent flow," *Int. J. Multiphase Flow*, vol. 11, pp. 43-49, 1985.

[52] A.C. Bannwart, "Wavespeed and volumetric fraction in core annular flow," *Int. J. Multiphase Flow*, vol. 24, pp. 961-974, 1998.

[53] O.M.H. Rodriguez, and A.C. Bannwart, "Experimental study of interfacial waves in vertical core flow," *J. Petroleum Sci. Tech.*, vol. 54, pp. 140-148, 2006.

[54] G.B. Wallis, *One-Dimensional Two-Phase Flow*, McGraw-Hill, 1969.

[55] A. C. Bannwart, "Modeling aspects of oil-water core-annular flows," *J. Petroleum Sci. Tech.*, vol. 32, pp. 127-143, 2001.

CHAPTER 5

Direct Contact Condensation of Steam Injected into Water – New Developments in Condensation Regime and Steam Plume Length Prediction

Anka de With*

Nuclear Research and Consultancy Group, Utrechtseweg 310, P.O. Box 9034, 6800ES Arnhem, The Netherlands

Abstract: Direct Contact Condensation (DCC) is a special mode of condensation where condensation occurs on the interface between steam and water. The condensation of steam injected into water is one of the least studied forms of condensation. Nevertheless, there is a range of devices which rely heavily on effective use of DCC, such as steam driven jet pumps, steam ejectors and safety valves in nuclear reactors. Therefore, correct prediction and modelling of DCC behaviour are crucial to obtain an optimised design of such devices. In this chapter the process of the DCC of steam injected into water is described in details, where different regions of the process will be presented. The process is compared with other modes of condensation and is described as one of the multi-phase flows. In addition, an analytical model for DCC is presented. Furthermore, the chapter focuses on the behaviour of injected steam, commonly termed as a regime. Different regimes of DCC are presented in details and parameters that determine the DCC regime are discussed. A three dimensional condensation regime diagram is presented, where the DCC regime is shown as a function of steam inflow rate, temperature of the water subcooling and steam injector diameter size. After that, the penetration distance of steam into water or a steam plume length is discussed and a two dimensional steam plume length diagram is presented. The diagram shows the length of the steam plume in relation to the steam Reynolds number and the condensation potential.

Keywords: Direct contact condensation, condensation regime, regime map, steam plume length.

1. INTRODUCTION

A condensation of steam injected into water is an example of multiphase flows in which fluids have different physical or thermodynamic properties and are separated by interfaces. Multiphase flows are the most common flows of fluids in nature and they appear in many industrial devices. Some examples of multiphase flows are boiling and condensing liquids, the drifting of clouds in the atmosphere and liquid sprays.

The frequent appearance and industrial significance of multiphase flows has resulted in intensive investigation of the subject, both experimentally and theoretically. However, despite the substantial developments in the field of multi-phase flows, accurate modelling remains out-of-reach when the phenomenon of phase transition is the dominant flow feature, as seen in condensation and cavitation.

The theme of this chapter is the process that takes place when steam is injected into water. The condensation of steam injected into water is a form of a Direct Contact Condensation (DCC) and is one of the least studied forms of condensation. Various experimental studies have been performed, investigating the behaviour of DCC [1-5]. While this has resulted in a substantial increase in experimental data, a generic understanding of condensation of steam in water is still very limited. Direct contact condensation is presented in this chapter as one of the modes of the condensation. After that, multi-phase flows are shortly described. The second section of the chapter describes the process of DCC in details. Different regions of the process are described. Furthermore, the behaviour of the injected steam, commonly named the condensation regime, are presented. Also shown is an analytical model for DCC in three-dimensions.

***Address correspondence to Anka de With:** Nuclear Research and Consultancy Group, Utrechtseweg 310, P.O. Box 9034, 6800ES Arnhem, The Netherlands; Tel: +31 26356 8579; Fax: +31 26356 8536; E-mail: dewith@nrg.eu

Lixin Cheng and Dieter Mewes (Eds)

Crucial in modelling the process of DCC is the condensation regime and the penetration distance of steam into water, the steam plume length. Therefore, following the analytical model for DCC, the three-dimensional condensation regime diagram and two-dimensional steam plume length diagram are presented. For a wide range of flow conditions, these are capable of predicting the behaviour of condensing steam, the regime, and the length of a steam plume. At the end of the chapter, the effect of the water flow on the behaviour of DCC is shortly described.

2. DIRECT CONTACT CONDENSATION

Condensation is the removal of heat from a system so that part or all of the vapour phase is converted into liquid [6, 7]. Different modes of condensation exist and they are presented in Fig. **1**.The most common mode is condensation of vapour on the cold surface of a solid. Film condensation (Fig. **1a**) occurs when steam comes in contact with a clean, uncontaminated surface and causes the formation of a liquid film covering the entire surface. If the surface is coated with a substance that inhibits wetting, dropwise condensation occurs (Fig. **1b**). Another common mode of condensation is homogeneous condensation (Fig. **1c**), where vapour condenses to form droplets suspended in a gas phase to form a fog. Direct contact condensation occurs when vapour comes in a direct contact with a cold liquid. (Fig. **1d** to **f**) shows three different cases of DCC. In (Fig. **1d**), vapour comes in contact with a liquid jet which is injected into a space filled with vapour. In the case shown in (Fig. **1e**), vapour is injected onto the surface of liquid and in (Fig. **1f**), vapour is injected into a pool of liquid. The latter case is the subject of this chapter.

2.1. Heat and Mass Transfer

A special case of DCC is the condensation of vapour on the interface of its own liquid. Here, the mass transfer can be described by kinetic theory as a difference between a rate of arrival of molecules from the vapour space to the interface and a rate of departure of molecules from the surface of the liquid into the vapour space [7], where molecular flux is:

$$|j| = \left(\frac{M}{2\pi R}\right)^{\frac{1}{2}} \frac{\Delta p}{T^{\frac{1}{2}}} \tag{1}$$

In order for condensation to occur, the arrival rate must exceed the departure rate.

The driving force for condensation of vapour on the interface of its own liquid is the saturation temperature, which is dependent on the vapour pressure. The heat transfer is fixed by the heat transfer coefficient, which incorporates information about the effect of the flow on the transport of heat. The interfacial heat transfer coefficient can be defined as $h_i = \Phi/\Delta T$, where the condensation energy flux Φ is a product of mass flux and latent heat ($\Phi = h_{fg} j$). This leads to an equation for a local heat transfer coefficient:

$$h_i = h_{fg} \left(\frac{M}{2\pi RT}\right)^{\frac{1}{2}} \frac{\Delta p}{\Delta T} \tag{2}$$

Eq. (2) is only an approximation and does not contain factors describing effects of turbulence, non-condensable gases and some others. Finding a more accurate equation for the heat transfer coefficient is one of the main difficulties in developing an accurate model for DCC. The other difficulty is a definition of the shape and position of the interface. The latter are connected with properties of steam and eddies around the interface and have strong influence on the heat transfer.

Heat transfer with DCC occurs in the liquid and the heat transfer coefficient for this process is higher than for other types of condensation process. With DCC the heat transfer coefficient is in the order of 10^6 W/(m^2 K), with filmwise condensation $6 \times 10^2 - 6 \times 10^3$ W/(m^2 K) and with dropwise condensation $6 \times 10^4 - 6 \times 10^5$ W/(m^2 K) [8].

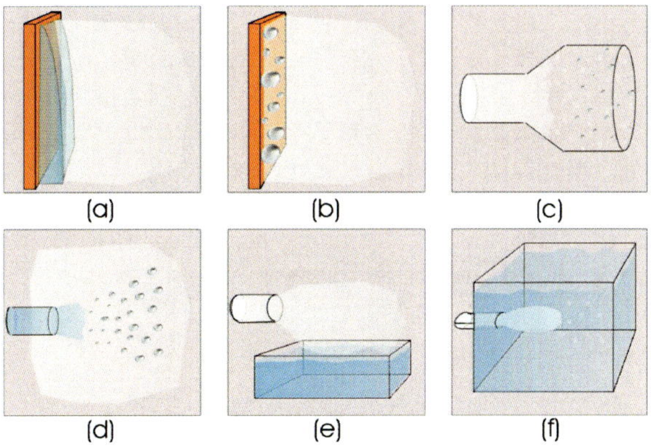

Figure 1: Modes of condensation: (a) Film condensation on a cold surface. (b) Dropwise condensation on a cold surface. (c) Homogeneous condensation resulting from the change in pressure. (d) DCC of vapour on the liquid jet. (e) DCC of vapour on the liquid surface. (f) DCC of vapour injected into a pool of liquid.

2.2. Multiphase Flows

Multiphase flows are combined flows of different phases such as bubbly flow, flow of water with solid particles, liquid droplets with dust particles in gas and many others. Special case of multiphase flows are flows with phase transition an example of which is also DCC.

Multiphase flows can be classified according to phase materials and interfacial structures in different systems and classes. The first classification gives three main systems of multiphase flows; liquid-gas, gas-solid and liquid-liquid system [8]. Classification according to interfacial structures gives three main classes of multiphase flows; separated flows, example of which is a jet flow, transitional flows, example of which is a slug flow, and dispersed flows, example of which is a bubbly flow [9, 10].

Different modelling techniques for multiphase flows can be used. The simplest is the homogeneous flow model [8, 11, 12], which assumes well mixed phases moving at the same velocity, and can only give satisfactory results for the void fraction and overall pressure gradient. The use of the model is therefore limited only to specific cases of multiphase flows.

Another model is a drift - flux model. The model, developed by Wallis [10], looks at the relative motion of phases and is therefore applicable to flows with well-defined velocities of different phases. As a result it gives a simplified description of some separated and dispersed flows such as bubbly and slug flow, but does not give information about the motion of the individual phases and can not be extended to more complicated flow regimes and processes such as boiling and condensation [13].

More complicated multiphase processes can be better described by separate continuity, momentum and energy equations for the different phases with additional jump conditions, which describe exchanges of mass, momentum and energy between different components on their interfaces. Therefore, jump conditions act as interfacial boundary conditions. Simplified governing equations describing multiphase flows are presented by Hetsroni [8], more complicated equations are described by Drew and Passman [14]. Drew and Passman's model is constructed from three conservation equations: mass, momentum and energy conservation, which can be used for different components involved in the process separately, and from three equations for jump conditions. Jump conditions describe the mass, momentum and energy transfer between phases and connect together separate conservation equations for different components of the process. Their model can be used to model all different types and processes of multiphase flows.

3. DIRECT CONTACT CONDENSATION OF STEAM INJECTED INTO WATER

The DCC of steam injected into water is a complicated process, because it appears in different regimes, depending on the environmental conditions, and it involves different regions. Independent of the condensation regime, the process of DCC of steam injected into water consists of four different regions: a steam plume, the interface between steam and water, the hot water layer and the bulk water. (Fig. **2**) shows a schematic picture of these regions. The first region consists of pure steam and is called the steam plume. The plume's outer surface is the interface where condensation takes place. Surrounding the interface is a hot water layer which contains of steam bubbles and is characterized by an increase in water temperature. The hot water layer is surrounded by bulk water. A two-phase jet is formed in the bulk water down-stream of the plume. The exact behaviour and the precise appearance of each region will differ depending on the regime. Nevertheless, each region is always present when there is a condensation of steam into water.

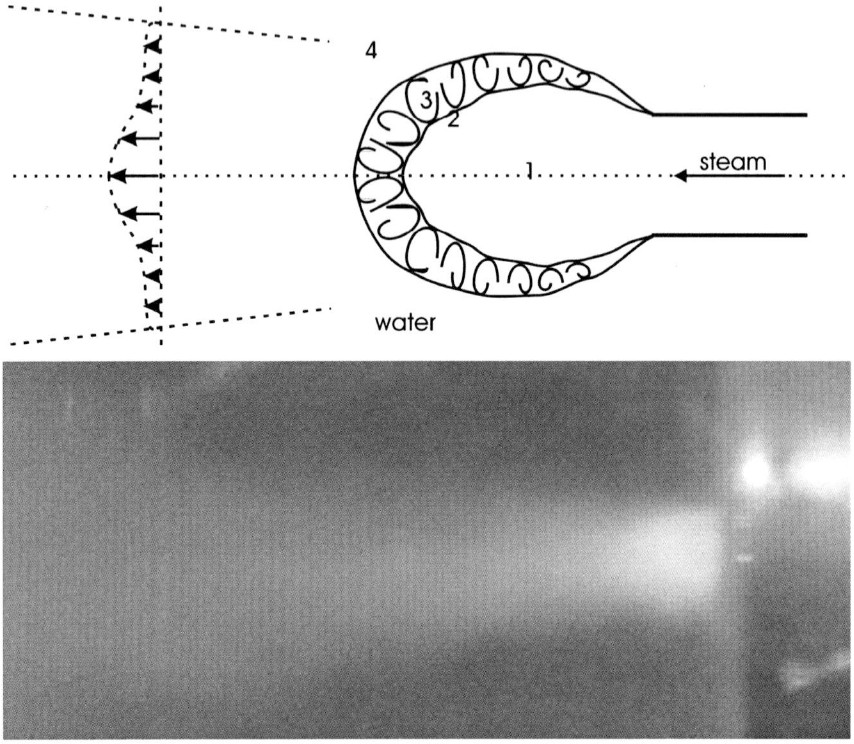

Figure 2: Schematic picture, showing the different regions of DCC of steam injected into water: steam plume (1), interface (2), hot water layer (3) and bulk water (4), and a photograph of DCC of steam injected into water.

3.1. Regions of DCC

3.1.1. Steam Plume

The steam plume is the first region in the condensation process and occurs at the steam pipe exit. In most computational models the steam plume is approximated by a pure steam region with temperature and velocity gradients in both downstream and radial directions [1-2]. The shape of the steam plume is predominantly dependent on the inflow conditions. These include steam inflow rate and water subcooling temperature. The steam inflow rate (G_0) is defined as a ratio between mass flow of steam and the steam injector exit area. The water subcooling is the temperature difference between steam and bulk water ($\Delta T=T_s-T_w$). A range of steam plume shapes can be observed during experimental observations and include: hemispherical, conical, ellipsoidal and divergent shapes. Alternatively, a steam plume in the shape of a bubble can be observed and with very low rates of steam inflow the plume is reduced to a small film covering the injector exit [15-17].

3.1.2. Interface

The surface of the steam plume or interface, divides the steam from the hot water layer. It is the region where DCC occurs with convective heat and mass transfer through the interface. The size of the interface is related to the amount of condensed steam. The exact shape and position of the interface are difficult to define. As suggested by Liang and Griffith [3], this shape depends on the interfacial eddies and the local temperature in the hot water layer that surrounds the interface.

The size of the interface depends on the rate of condensation and it also affects this rate. If the interface is small, heat and mass transfer can only take place across a small area and condensation will take place comparatively slowly. If the steam plume is large, the interface is larger and the condensation will be accelerated. Fast condensation across the interface of a plume increases the instability leading to a collapse of the bubble when the steam supply is insufficient to keep up with the rate of condensation.

3.1.3. Hot Water Layer

The hot water layer is a two-phase layer in the vicinity of the interface with a temperature close to the value of saturation. The water layer at the interface is mixed with steam bubbles and there is a large amount of turbulent motion. Turbulence is formed by the momentum subtracted from the condensing steam and has a large effect on the shape of the interface and the condensation rate across the interface [3, 18, 19].

3.1.4. Bulk Water

Stagnant or moving water surrounds the steam plume and hot water layer. As shown by Celata [18] the water velocity directly affects turbulence in the hot water layer and consequently the steam plume shape and the heat transfer coefficient. Steam, condensing in water, also generates a temperature distribution in the bulk water. However, the temperature of the water is significantly increased only in close proximity of the interface. Further downstream of the plume the increase of the water temperature is marginal [20].

Dependent on the inflow conditions, the bulk water can be simply water or a two-phase mixture of water and uncondensed steam bubbles. Downstream of the condensation region a self-similar jet is developed in the water [21].

3.2. Regimes of DCC

The dynamic behaviour and the geometrical appearance of the steam plume can be subdivided in three main regimes: chugging regime, jetting regime and bubbling regime.

3.2.1. Chugging Regime

The chugging regime is characterised by a flat or curved shape of the steam plume located in the injector or its vicinity, and the steam plume size is close to the injector's cross-sectional area. (Fig. **3**) shows a possible cycle of a chugging regime.

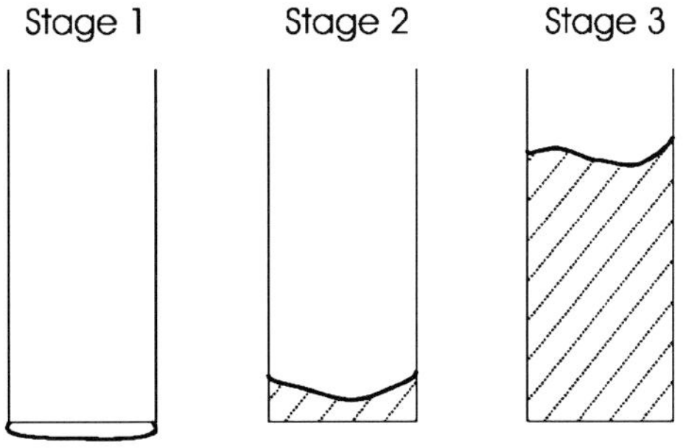

Figure 3: Chugging regime cycle as observed by Chan and Lee [4]. The regime was observed with water subcooling of 66 K. The shaded area represents water. Steam is injected vertically downwards.

The chugging regime occurs at steam inflow rates up to 80 kg/(m^{2}s), depending on the surrounding water temperature and steam injector diameter. The location of the steam plume in the injector may vary continuously, moving from the outer edge of the injector into the injector and backwards [4]. Chugging occurs because the steam inflow rate is smaller than the condensation rate which causes suction of surrounding water into the steam injector. If chugging occurs inside the injector in a pulsating manner, the process is called interfacial condensation oscillation. For this type of condensation the steam inflow rate is normally smaller than 5 kg/(m^{2}s).

3.2.2. Jetting Regime

For steam inflow rates higher than about 100 kg/(m^{2}s), steam forms a plume which holds approximately constant size and shape. This type of condensation is called the jetting regime. Jetting consists of three main plume shapes which include: conical, ellipsoidal and divergent shape (Fig. **4**). For some inflow conditions close to the bubbling regime, the plume can also take a hemispherical shape. The length of the steam plume (L) in the jetting regime can vary from one millimetre to more than fifteen centimetres. If the condensation occurs in a conical or ellipsoidal jetting regime, the plume will take a regular shape with a smooth surface (Fig. **4** (a) and (b)). For water temperatures higher than 343 K to 353 K, the plume will take an irregular divergent shape (Fig. **4** (c)). This regime is named divergent jetting. The length of a divergent plume is usually bigger than two centimetres.

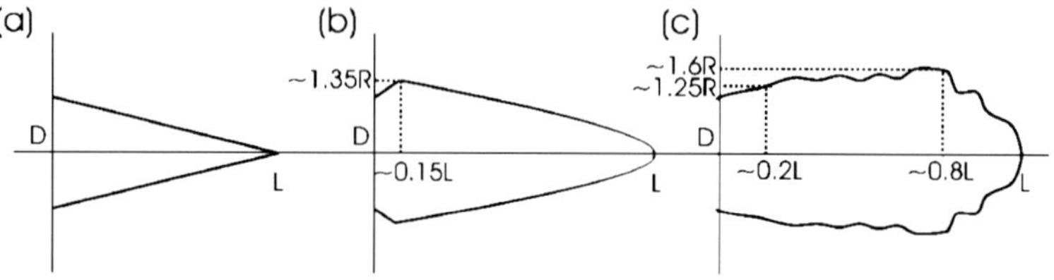

Figure 4: Jetting regime plume shapes observed by Chun *et al.* [16]. (a) Conical plume, ΔT=76.7 K, G_0=286 kg/(m^{2}s), D=10.85 mm, L=20 mm, (b) Ellipsoidal plume, ΔT=58.5 K, G_0=312 kg/(m^{2}s), D=7.65 mm, L=26 mm, (c) Divergent plume, ΔT=27.1 K, G_0=530 kg/(m^{2}s), D=7.65 mm, L=34 mm. Steam injection is to the right.

3.2.3. Bubbling Regime

For steam inflow rates between chugging and jetting, bubbling occurs. Here, injected steam generates a regular or an irregular bubble at the edge of the injector. The generated bubble grows and when a maximum size of the bubble is reached, the whole or a part of the bubble is detached from the injector. The remaining part of the original bubble starts to grow until a maximum bubble size is reached [17]. In certain cases there is a collapse of the original bubble after detachment. Detached parts of the bubble are dragged away from the steam injector and continue to condense, generating a trace of smaller steam bubbles in the downstream flow [22]. (Fig. **5**) shows a possible cycle of a bubbling regime of DCC.

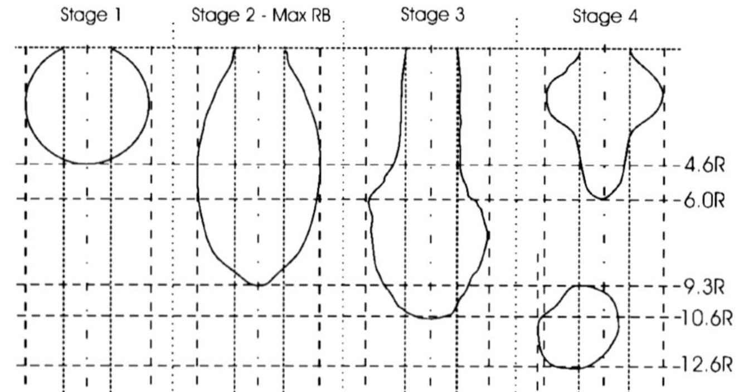

Figure 5: Bubbling regime cycle observed by Simpson and Chan [17]. ΔT=37 K, G_0=305.5 kg/(m^{2}s), D=0.635 cm. Steam is injected vertically downwards.

3.3. Analytical Model for DCC

An appropriate analytical model describing DCC of steam injected into water must incorporate fundamental aspects of two-phase flow, flow dynamics and condensation. The equations are based on the general equations for three-dimensional unsteady two-phase model derived by Drew and Passman [14]. It consists of balance equations and jump conditions for both phases and has incorporated important aspects of condensation.

The first equation of the model is the mass conservation or continuity equation:

$$\frac{\partial}{\partial t}(\alpha_k \rho_k) + \nabla \cdot (\alpha_k \rho_k \vec{v}_k) = \Gamma_k \quad \textbf{(3)}$$

where the index k is replaced by the index s for the steam phase or w for the water phase for each separate equation. In addition to Equation (3) a mass jump condition exists:

$$\Gamma_s = \Gamma_w \quad \textbf{(4)}$$

where interfacial mass exchange is preserved and the system does not have any additional sources or sinks on the interface.

The second equation of the system is a momentum equation for condensation (Eq. (5)) followed by a jump condition (Eq. (6)).

$$\frac{\partial}{\partial t}(\alpha_k \rho_k \vec{v}_k) + \nabla \cdot (\alpha_k \rho_k \vec{v}_k \vec{v}_k) = -\nabla(\alpha_k p_k) - \nabla \cdot (\alpha_k \underline{T_k}) + \Gamma_k (\vec{v}_{in} - \vec{v}_k) + M_k \quad \textbf{(5)}$$

$$M_s + M_w + \vec{v}_{is}\Gamma_s + \vec{v}_{iw}\Gamma_w = m \quad \textbf{(6)}$$

The energy equation for DCC is:

$$\frac{\partial}{\partial t}\left[\alpha_k \rho_k \left(e_k + \frac{\vec{v}_k \vec{v}_k}{2}\right)\right] + \nabla \cdot \left[\alpha_k \rho_k \vec{v}_k \left(e_k + \frac{\vec{v}_k \vec{v}_k}{2}\right)\right] = -\nabla(\alpha_k p_k \vec{v}_k) - \nabla \cdot (\alpha_k \underline{T} \cdot \vec{v}_k) + $$
$$+ Q_k + W_k + \Gamma_k \left(e_{ki} + \frac{\vec{v}_{ki} \vec{v}_{ki}}{2}\right) \quad \textbf{(7)}$$

and the energy jump condition is:

$$Q_s + Q_w + W_s + W_w + \left(e_{si} + \frac{\vec{v}_{si} \vec{v}_{si}}{2}\right)\Gamma_s + \left(e_{wi} + \frac{\vec{v}_{wi} \vec{v}_{wi}}{2}\right)\Gamma_w = \varepsilon \quad \textbf{(8)}$$

For more details about the model see [23, 24].

To compute the unknowns of each of the systems: -pressure, steam and water density, steam and water velocity, and steam and water internal energy,- the system of equations have to be solved by approximation methods. Crucial to the modelling of DCC of steam condensing in water is the approximation of the variables such as Q, W and Γ as well the geometry of the steam plume and the heat transfer coefficient. These quantities are computed simultaneously with the general equations describing processes involved. Geometry of the plume, the shape and the length of the plume, is important in order to specify the position, where equations of each of the phases have to be applied during the calculation. Hence, the condensation regime and the length of the plume have to be known prior to calculation. The diagrams for the prediction of the condensation regime and the length of the steam plume are presented in the next section.

4. PREDICTION OF THE DCC FEATURES

4.1. Prediction of the DCC Condensation Regime

Reliable and accurate prediction of the regime is essential for applications where DCC is the dominant flow feature. The DCC condensation regime can be found in regime maps. Regime maps were proposed by Liang and Griffith [3], for steam injected through an injector with a diameter D=0.02 m. Aya and Nariai proposed a regime map for steam injected through injectors with diameters in the range of D=0.009 m to D=0.038 m [25, 26]. Furthermore, Chan and Lee [4] constructed a map for a steam injector with diameter D=0.051 m. Block [22] presented a regime map, which was originally taken from Chan and Lee [4], but used a different notation for the regimes. The notation was based on the pressure variation in the water into which steam was injected. Also the regime map, presented by Youn *et al.* [27] is based on pressure variation in the water. All these regime maps accurately predict the regimes, however, as they were designed for a single or a limited number of injectors, show only a limited accuracy.

A comparison of the two-dimensional regime maps shows that all maps contain similar features. However, steam and water conditions, at which the regime pattern is observed, can vary significantly [23, 24]. Furthermore, distinct variations are seen when comparing the regime maps based on different injector diameters. Therefore, the regime pattern is not scalable from one injector size to another if there is a substantial change in injector size. For example, while the regime map provides a distinct chugging regime with large injector diameters, chugging is marginal with a comparatively small injector diameter. This observation is very significant and implies that the regime map cannot be normalised with respect to the injector size; hence, a single two dimensional regime map is unsuitable for a wide range of injector sizes.

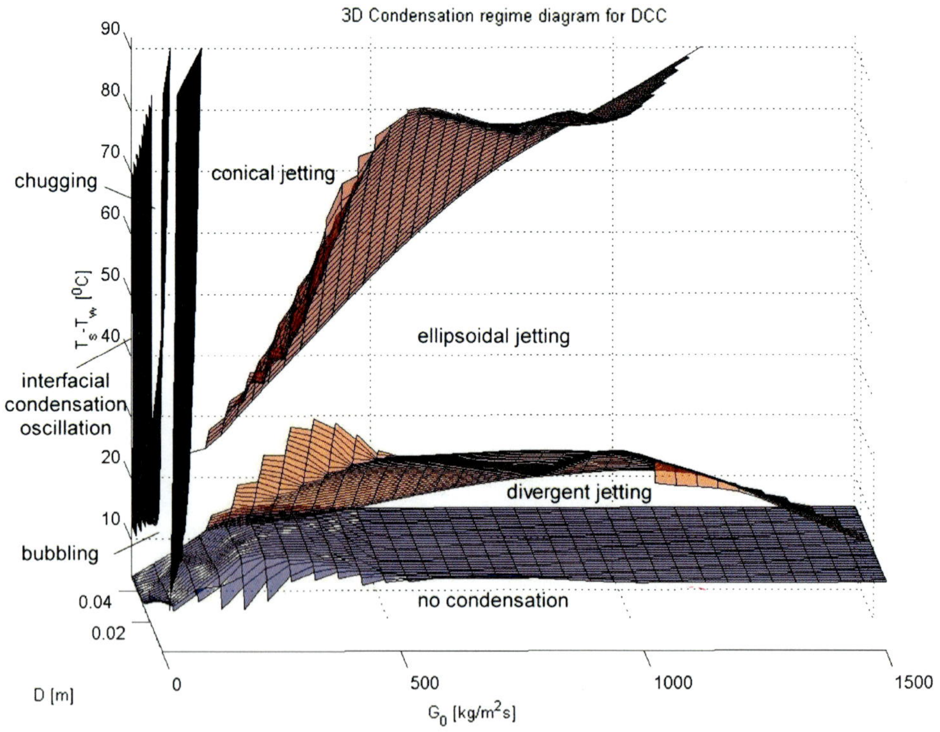

Figure 6: The 3DCR diagram viewed from a side with small steam injector diameters. Three different jetting regimes are clearly seen on the diagram. At small steam inflow rates, bubbling, chugging and interfacial condensation oscillation can be identified.

More general is a Three-Dimensional Condensation Regime (3DCR) diagram [24] and [28]. The diagram has been developed from existing two-dimensional regime maps, placed on one three-dimensional diagram, and from a series of additional data, showing the condensation regime at different steam and water conditions. The traditional dependencies of steam mass inflow rate (G_0) and water subcooling ($\Delta T=T_s-T_w$) were used. Based on the above observation regarding the injector diameter, the injector diameter (D) was selected as a third dependency. Other selections of dependency have been studied but provided very poor correlation. The three-dimensional condensation regime diagram is shown in Figs. **6** and **7**. It presents areas of different regimes for different steam inflow rates in the range from 0 kg/(m^{2}s) to 1500 kg/(m^{2}s), for steam injector diameter sizes between 0.00135 m to 0.05 m and for water subcooling ranging from 0 K to 90 K. (Figs. **6** and **7**) show two different views of the 3DCR diagram. (Fig. **6**) shows a side view from a side with small steam injector diameters and (Fig. **7**) shows side view from a side with large steam injector diameters.

The diagram shows areas of different jetting regimes: conical jetting, ellipsoidal jetting and divergent jetting. Jetting regimes cover most of the 3DCR diagram and they appear at steam inflow rates around 200 kg/(m^{2}s) and extend to high steam inflow rates. The divergent jetting appears above the no-condensation area and covers a section of the jetting area that ranges from approximately 10 K to 25 K water subcooling, depending on a steam injector diameter. Above the divergent jetting area, an ellipsoidal jetting area can be observed. It expands to high water subcooling at steam inflow rates higher than 1000 kg/(m^{2}s).

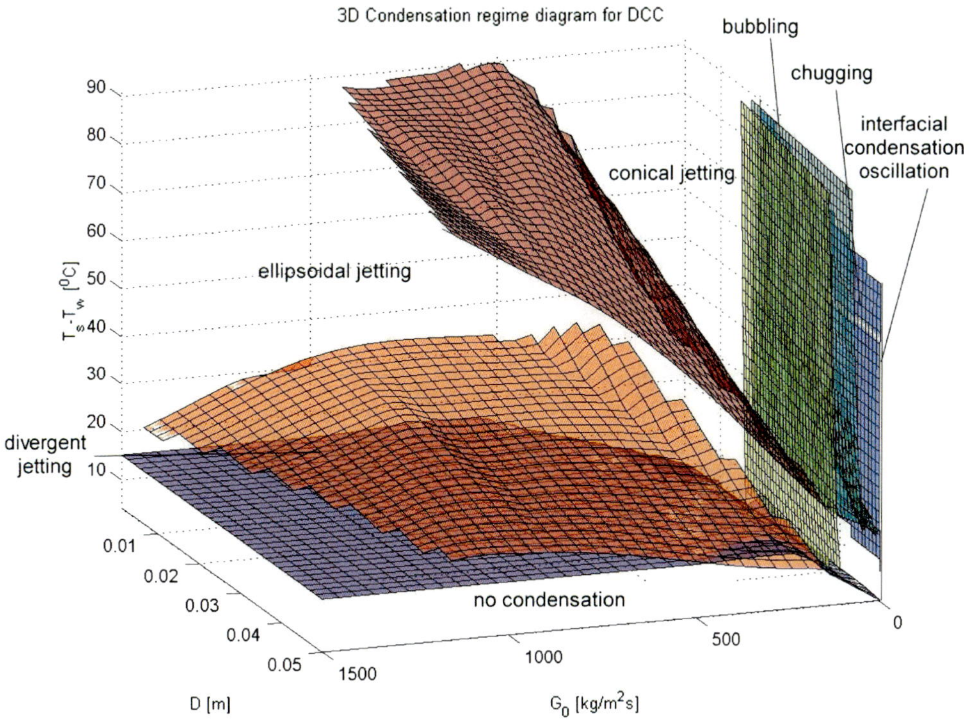

Figure 7: The 3DCR diagram viewed from a side with large steam injector diameters.

Atlower steam inflow rates, this area is diagonally cut with an area of conical jetting, which occurs at a water subcooling around 25 K at 200 kg/(m^{2}s) and expands to 90 K at 200 kg/(m^{2}s). For steam inflow rates smaller than 200 kg/(m^{2}s) bubbling, chugging and interfacial condensation oscillation occurs. The diagram also shows that for very small steam inflow rates up to 2 kg/(m^{2}s) interfacial condensation oscillation ranges from 15 K to 60 K of water subcooling. The chugging regime starts when the water subcooling is 20 K and extends to 90 K and covers an area of the 3DCR diagram up to a steam inflow rate of 50 kg/(m^{2}s). With lower water subcooling a bubbling regime is observed. For steam inflow rates between 50 kg/(m^{2}s) and 120 kg/(m^{2}s) a bubbling regime occurs with water subcooling up to 90 K.

The 3DCR diagram has been validated and validation showed that there is a general agreement between the new 3DCR diagram and the data. A detailed validation of the diagram can be found in [24, 28].

4.2. Prediction of the DCC Steam Plume Length

In addition to the precise nature of the condensation regime, another important feature of DCC of steam injected into water is the length of a steam plume. However, the length of a steam plume depends on the flow conditions of the process and is therefore difficult to predict. Researchers have performed experiments in which the steam plume length was measured. The data was used to derive correlations, predicting the dimensionless steam plume length as a function of condensation potential $B = (c_p/h_{fg})\Delta T$ and dimensionless steam mass inflow rate G_0/G_m ($G_m = 275$ kg/(m^{2}s)). Different correlations for dimensional steam plume length are shown in Table **1**.

Table 1: Different correlations for dimensional steam plume length.

Author	Correlation	Validity
Kerney *et al.* [1]	$L/D=0.2588B^{-1}(G_0/G_m)^{0.5}$ $L/D=0.3583B^{-0.8311}(G_0/G_m)^{0.6446}$	D=0.00495 m G_0 from 338 to 1240 kg/(m^{2}s) B from 0.0028 to 0.135
Weimer *et al.* [29]	$L/D=17.75B^{-1}(G_0/G_m)^{0.5}(\rho_s/\rho_\infty)^{0.5}$ $L/D=10.285B^{-0.801}(G_0/G_m)^{0.713}(\rho_s/\rho_\infty)^{0.384}$	D=0.00317 m G_0 from 321 to 1136 kg/(m^{2}s) B from 0.0025 to 0.063 ρ_∞/ρ_s from 3980 to 27700
Chun *et al.* [16]	$L/D=0.5923B^{-0.66}(G_0/G_m)^{0.3444}$	D=0.00135 m G_0=1488 kg/(m^{2}s) B from 0.035 to 0.15
Kim *et al.* [5]	$L/D=0.503B^{-0.70127}(G_0/G_m)^{0.47688}$	D=0.005 m G_0=1188 kg/(m^{2}s) B from 0.037 to 0.12

The correlations are only valid for a narrow range of flow conditions and, as they are based on a limited amount of data, are also not very accurate.

More general prediction of the steam plume length can be found on the Two-Dimensional Steam Plume Length (2DSPL) diagram [24, 30]. This diagram has been created from a large number of data. All collected data was plotted on single graphs in attempt to understand the underlying trends. First, the dimensionless steam plume length data was plotted in a graph using condensation potential B and dimensionless steam mass inflow rate G_0/G_m as axes. The results showed some structure but generally no clearly defined trends.

Further graphs were then generated where the dimensionless steam plume length (L/D) and the steam plume length (L) were presented as functions of steam mass inflow rate (G_0), condensation potential (B), injector diameter (D) and steam Reynolds number ($Re= DG_0/\eta$). In addition to physical parameters, known to affect the process of DCC, the Reynolds number was chosen as one of potential parameters. The Reynolds number is directly related to the inertia of injected steam and hence it should be related to the penetration length of the steam.

A coherent plot of the experimental data was found when using the steam plume length (L) as a function of Reynolds number (Re) and condensation potential (B). Therefore, the steam plume length data as a function of Reynolds number and condensation potential was used to develop a two-dimensional steam plume length diagram. The diagram was produced from the available data through interpolation and filtering. The two-diagram steam plume length diagram is shown on Fig. **8**. It shows the length of the steam plume in metres (L[m]) in relation to the steam Reynolds number (Re) in the range between 0.24×10^5 and 13.74×10^5, and the condensation potential (B) in the range between 0.029 and 0.155. Different colours on the diagram represent the length of the steam plume in metres.

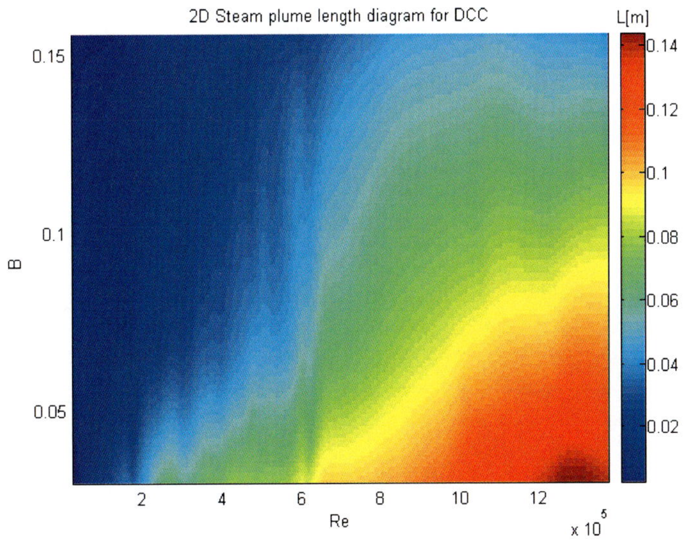

Figure 8: 2DSPL diagram for DCC. Colours indicate predicted length of the steam plume in metres. The longest plumes are to be expected at high Reynolds numbers and low condensation potentials; short plumes at low Reynolds numbers.

The 2DSPL diagram shows that steam develops long plumes when injected at high Reynolds numbers and short plumes at low Reynolds numbers. Furthermore, the longest plumes are to be expected at high Reynolds numbers and low condensation potentials. This is in accordance with the physics of the process which suggests that the plume is longer if the inertia of the injected steam is higher, but is shortened with increasing condensation rate. The 2DSPL diagram has been validated and validation showed that error of the diagram is 13.7%. A detailed validation of the diagram can be found in [24, 30].

4.3. Effect of Water Flow on DCC

When steam is injected into water flow, the condensation of steam is affected by the movement of the surrounding water. The shear layer between a developing steam plume and a water flow may cause a steam plume to extend. On the other hand a constant supply of cold water around a steam plume may cause steam to condense faster compared with DCC into stagnant water at the same conditions. Hence, the regime of a DCC of steam into a water flow as well as shape and length of a steam plume is highly likely to be dependent on the velocity of the water flow.

Experiments were performed, where steam was injected into a water flow, flowing in the direction of the injected steam [23]. Experiments showed, that the three dimensional condensation regime diagram for DCC of steam injected into a water flow has similar features to the condensation regime diagram for stagnant water. However, the location of the various regimes in the diagram differs. For example, more steam is needed to achieve jetting at DCC in a water flow than in stagnant water. It has also been shown, that the length of the steam plume decreased by 60 to 65% for all tested Reynolds numbers when steam was injected into a water flow.

From these results follows, that the effect of increased condensation due to the constant supply of fresh water is much larger than the momentum effect of the water flow. Hence, heat and mass transfer due to condensation seem to affect the steam plume more than the momentum of the injected steam.

5. CONCLUSIONS

A special mode of condensation, a direct contact condensation of steam injected into water, has been presented in this chapter. Main features of the process, regions and regimes of DCC, were described in details. Furthermore, the analytical model of DCC has been presented.

Important for modelling of the DCC process are the condensation regime and the length of the developed steam plume. Therefore, the three-dimensional condensation regime diagram and the two-dimensional steam plume length diagram were shown in this chapter. The diagrams can be used for the prediction of the DCC regime and the length of the steam plume for a wide range of conditions. The diagrams are based on the observations that the physical parameters with which the DCC process can be described, are: the steam inflow rate, dimensions of the steam injector, the water subcooling and the velocity of the water flow. As a result, the two diagrams give a better prediction for the regime and the length of a steam plume compared to other models, which use only steam inflow rate and the water subcooling as defining parameters for DCC.

At the end of the chapter, a condensation behaviour found in DCC of steam injected into a water flow was shortly discussed. The DCC behaviour in water flow resembles most of the trends seen with DCC into stagnant water. However, specific characteristics occur at different flow conditions as compared with DCC of steam injected into stagnant water.

NOMENCLATURE

B	condensation potential
c_p	specific heat at constant pressure
D	diameter of steam injector
e	internal energy
G	steam flow rate
h	heat transfer coefficient
h_{fg}	latent heat
j	molecular flux
L	plume length
M	molecular weight
M	momentum transfer
m	surface tension force
p	pressure
Q	heat exchange
R	gas constant
R	injector radius
Re	steam Reynolds number
T	temperature
$\underline{T}$	stress tensor
ΔT	temperature of water subcooling
v	velocity
W	work exchange
α	volume fraction
Γ	volumetric mass transfer rate
ε	energy source
η	viscosity
ρ	density
Φ	condensation energy flux

Subscript

i, in	interface
k	phase k
ki	contact between phase k and interface
m	mixture
m	mean value
s	steam
w	water

0 at the injector
∞ free stream conditions

REFERENCES

[1] P.J. Kerney, G.M. Faeth, D.R. Olson, "Penetration characteristics of a submerged steam jet", *AIChE J.*, vol. 18 (3), pp. 548-553, 1972.
[2] Y.S. Kim and J.W. Park, "Determination of the steam-water direct contact condensation heat transfer coefficient using interfacial transport models", *Proc. Am. Nucl. Soc. - Thermal Hydraulics Div.*, vol. 10, pp. 110-117, 1997.
[3] K.S. Liang and P. Griffith, "Experimental and analytical study of direct contact condensation of steam in water", *Nucl. Eng. Des.*, vol. 147, pp. 425-435, 1994.
[4] C.K. Chan and C.K.B. Lee, "A regime map for direct contact condensation", *Int. J. Multiphase Flow*, vol. 8(1), pp. 11-20, 1982.
[5] H.Y. Kim, Y.Y. Bae, C.H. Song, J.K. Park, S.M Choi, "Experimental study on stable steam condensation in quenching tank", *Int. J. Energy Res.*, vol. 25, pp. 239-252, 2001.
[6] P.B. Whalley, *Boiling, Condensation and Gas-Liquid Flow*, New York: Oxford University Press Inc., 1987.
[7] J.G. Collier and J.R. Thome, *Convective Boiling and Condensation 3ed*, Oxford, Oxford Science Publication, Clarendon Press, 1996.
[8] G. Hetsroni, *Handbook of Multiphase Systems*, Washington, Hemisphere Publishing Corporation, 1982.
[9] N. Andrianov, "Analytical and numerical investigation of two-phase flows", PhD thesis, Fakultät für Mathematik der Otto-von-Guericke Universität Magdeburg, DE, 2003.
[10] G. B. Wallis, G.B, *One-dimensional Two-Phase Flow*, New York, McGraw-Hill, Inc., 1969.
[11] G.F. Hewitt and N.S. Hall-Taylor, *Annular Two-Phase Flow*, New York, Pergamon Press, 1970.
[12] P.B. Whalley, *Two-Phase Flow and Heat Transfer*, New York, Oxford University Press Inc., 1996.
[13] C. Kleinstreuer, *Two-Phase Flow: Theory and Applications*, London, Teylor and Francis, 2003.
[14] D.A. Drew and S.L. Passman, *Theory of Multicomponent Fluids*, York, PA, Springer-Verlag New York, Inc., Printed by Maple-Vail Book Manufacturing Group, 1999.
[15] L.D. Chen and G.M. Faeth, "Condensation of submerged vapor jets in subcooled liquids", *J. Heat Transfer*, vol. 104, pp. 774-780, 1982.
[16] M.H. Chun, Y.S. Kim, J.W. Park, "An investigation of direct condensation of steam jet in subcooled water", *Int. Comm. Heat Mass Transfer*, vol. 23(7), pp. 947-958, 1996.
[17] M.E. Simpson and C.K. Chan, "Hydrodynamics of a subsonic vapor jet in subcooled liquid", *J. Heat Transfer*, vol. 104, pp. 271-278, 1982.
[18] G.P. Celata, M. Cumo, G.E. Farello, G. Focardi, "Direct contact condensation of steam on slowly moving water", *Nucl. Eng. Des.*, vol. 96, pp. 21-31, 1986.
[19] K.S. Liang, "Experimental and analytical study of direct contact condensation of steam in water", PhD thesis, Department of Nuclear Engineering, Massachusetts Institute of Technology, USA, 1991.
[20] R.J.E. van Wissen, K.R.A.M. Schreel, C.W.M. van der Geld, J. Wieringa, "Turbulence production by a steam-driven jet in a water vessel", *Int. J. Heat Fluid Flow*, vol. 25, pp. 173-179, 2004.
[21] R.J.E. van Wissen, K.R.A.M. Schreel, C.W.M. van der Geld, "Particle image velocimetry measurements of a steam-driven confined turbulent water jet", *J. Fluid Mech.*, vol. 530, pp. 353-368, 2005.
[22] J.A. Block, "Condensation-driven fluid motions", *Int. J. Multiphase Flow*, vol. 6, pp. 113-129, 1980.
[23] A. Petrovic, "Analytical study of flow regimes for direct contact condensation based on parametrical investigation", *ASME J. Pressure Vessel Tech.*, vol. 127 (1), pp. 20-25, 2005.
[24] A. Petrovic – de With, "Characterisation and modelling of flow mechanisms for direct contact condensation of steam injected into water", PhD thesis, University of Hertfordshire, Hatfield, UK, 2006.
[25] I. Aya and H. Nariai, "Boundaries between regimes of pressure oscillation induced by steam condensation in pressure suppression containment", *Nucl. Eng. Des.*, vol. 99, pp. 31-40, 1987.
[26] I. Aya and H. Nariai, "Evaluation of heat-transfer coefficient at direct-contact condensation of cold water and steam", *Nucl. Eng. Des.*, vol. 131, pp. 17-24, 1991.
[27] D.H. Youn, K.B. Ko, Y.Y. Lee, M.H. Kim, Y.Y. Bae, J.K. Park, J.K, "The direct contact condensation of steam in a pool at low mass flux", *J. Nucl. Sci. Tech.*, vol. 40(10), pp. 881-885, 2003.
[28] A. Petrovic - de With, R.K. Calay, G. de With, "Three dimensional regime map for direct contact condensation of steam injected into water", *Int. J. Heat Mass Transfer*, vol. 50, pp. 1762-1770, 2007.

[29] J. C. Weimer, G.M. Faeth, D.R. Olson, "Penetration of Vapor Jets Submerged in Subcooled Liquids", *AIChE J.*, vol. 19(3), pp. 552-558, 1973.

[30] A. de With, "Steam Plume Length Diagram for Direct Contact Condensation of Steam Injected into Water, *Int. J. Heat Fluid Flow*, vol. 30, pp. 971-982, 2009.

CHAPTER 6

Drag Reduction with Surfactants and Polymeric Additives in Multiphase Flow

Lixin Cheng[1,*], Lei Liu[2] and Dieter Mewes[3]

[1]School of Engineering, University of Portsmouth, Anglesea Building, Anglesea Road, Portsmouth, PO1 3DJ, United Kingdom; [2]State Key Laboratory of Multiphase Flow in Power Engineering, Xi'an Jiaotong University, Xi'an 710049, China and [3]Institute of Process Engineering, University of Hanover, Callinstr. 36, D-30167, Hanover, Germany

Abstract: Addition of small amount of surfactants or polymeric additives may reduce the drag in multiphase flow systems. This chapter presents an overview on the available studies of drag reduction with surfactants and polymeric additives in multiphase flow including gas-liquid two-phase flow, liquid-liquid two-phase flow and gas-liquid-liquid three-phase flow. First, the research background of drag reduction technologies with additives is presented. Then, basic knowledge of drag reduction surfactants and polymeric additives is briefly described. Measured physical properties of surfactant and polymer solutions such as surface tension and viscosity are shown. Next, the drag reduction mechanisms of polymer and surfactant solutions are discussed. The, comprehensive review on the available studies on drag reduction with polymers and surfactants in multiphase flow is presented. The mechanisms are also reviewed. Based on the present review, future research needs are discussed and recommended.

Keywords: Multi-phase flow, drag reduction, surfactant, polymer, additive, flow pattern, gas-liquid two phase flow, liquid-liquid two phase flow, gas-liquid-liquid three-phase flow, mechanisms.

1. INTRODUCTION

The use of additives such as surfactants and polymeric additives for liquids appears to be quite viable and has attracted a lot of research over the past decades. Addition of very small amount surfactants or polymeric additives in a fluid can either enhance the boiling heat transfer [1] or reduce the drag of fluid flow including single phase flow [2-18, 41-43] and multiphase flow [53-81]. Cheng *et al.* [1] presented a comprehensive review on boiling phenomena with surfactants and polymeric additives. They also pointed out that study of drag reduction in two phase flow and flow boiling by the addition of surfactants and polymeric additives should be developed and be connected to the heat transfer mechanisms of flow boiling with surfactants and polymeric additives. The present chapter focuses on an overview of drag reduction technologies with additives in multiphase flow.

Drag reduction is a phenomenon by which small amount of additives (ppm) can greatly reduce the frictional pressure drop or friction factor of a fluid flow. The aim for the drag reduction is to improve the fluid-mechanical efficiency using active agents, known as Drag Reduction Agents or Additives (DRA). Tom [2] first reported such a phenomenon that the addition of small concentrations of high molecular weight polymer to water or other solvent can produce large reductions in frictional pressure drop for turbulent flows past a surface, leading to the possibility of increased pipeline capacities and faster ships. Since then, many research studies and publications have been devoted to the subject over the past 60 years.

Percent drag reduction or effectiveness of a drag reduction agent or additive is defined as the ratio of reduction in the frictional pressure drop when the flow rates are held constant to the frictional pressure drop without DRA, as shown in Eq. (1). effectiveness of DRA ε can be defined as follows:

$$\varepsilon = \frac{\Delta p_{without\,DRA} - \Delta p_{with\,DRA}}{\Delta p_{without\,DRA}} \times 100 \times \% \quad (1)$$

*Address correspondence to Lixing Cheng: School of Engineering, University of Portsmouth, Anglesea Building, Anglesea Road, Portsmouth, PO1 3DJ, United Kingdom; E-mail: lixincheng@hotmail.com.

where $\Delta p_{without\,DRA}$ is the frictional pressure drop in the absence of DRA and $\Delta p_{with\,DRA}$ is the frictional pressure drop in the presence of DRA.

The effectiveness of DRA ε may also be defined according to friction factor as:

$$\varepsilon = \frac{f_{without\,DRA} - f_{with\,DRA}}{f_{without\,DRA}} \times 100 \times \% \tag{2}$$

where $f_{without\,DRA}$ is the friction factor in the absence of DRA and $f_{with\,DRA}$ is the friction factor in the presence of DRA.

The drag reduction technologies in fluid flow have found a wide range of applications in industrial and military applications. Many studies on these technologies, including theoretical, experimental, and numerical approaches, have been conducted over the past years. Various drag reduction technologies such as diluted polymer solutions, surfactants, micro-bubbles and compliant coating as drag reducers have been investigated in the literature. For the polymer solutions, it is possible to achieve up to 80% drag reduction with only a few parts per million of polymer. Ionic and non-ionic surfactants can also offer similar drag reduction as polymer solutions but at a high concentration of few percent. Micro-bubbles are perhaps the cheapest and non-polluted drag reducer. However, the control of the bubble size and the angle of ejection can impose technical challenges. This method can be combined with other drag reduction technologies such as polymers to enhance its effectiveness. Compliant coatings can be designed to behave in a similar fashion to the skin of a dolphin. The development of the compliant coating is theoretically complicated, although the optimized coating can offer a drag reduction up to 50%. Despite the extensive research in the area of drag reduction over the past decades, for each drag reduction technology there is no universally accepted model that explains its mechanisms. Of the available drag reduction technologies, polymeric additives are the most effective drag reducers although other DR technologies have been proved to be effective. Surfactants are also effective in some cases but less investigated than polymeric additives. For both surfactants and polymeric additives in fluid flow, unified drag reduction mechanisms have not yet developed. Instead, there are various mechanisms available to explain the phenomena.

Drag reduction technologies can be applied for both Newtonian fluids and non-Newtonian fluids including single turbulent flow and multiphase flows such as gas-liquid two-phase flow, liquid- liquid two-phase flow and gas–liquid-liquid three-phase flow systems. It has been widely observed that addition of a small amount of macromolecular polymer to a Newtonian fluid may lead to a dramatic reduction in fluid resistance. Similar studies have been reported in regard to non-Newtonian fluids but less investigated than those for Newtonian fluids.

Over the past years, a number of studies have focused on single phase flow and some mechanisms have been reached. Studies on drag reduction with additives in multiphase flows have also been conducted but less understood than those in single phase flow. Multiphase flows such as gas–liquid two-phase flow, liquid-liquid two-phase flow, gas-liquid-liquid flows, gas-liquid-solid multiphase flows are frequently encountered in many industrial units such as distillation columns, pipelines, boiler tubes, condensers, evaporators, chemical reactors, oil and gas production processes *etc.* For example, drag reduction is important to the increase of oil and gas production rates and to the reduction of transportation costs. Offshore production has necessitated transportation of both gas and liquid phases over long distances before separation. It involves very complex multiphase flow phenomena. Reducing the pressure drop in the transportation of oil-gas products will results in great energy and cost savings.

It can be seen from the literature survey that although some studies have been done on drag reduction by polymer solution in single phase flow, but a few attempts have been made to study the effect of polymer solutions in reducing the high axial pressure drop in two-phase flow. Especially for the mechanisms of drag reduction it is far from understanding. Experimental evidences show that the polymer increases the thickness of the viscous sub layer and the transition zone. The mechanism of this boundary layer effect is not yet fully understood, but supporting experimental evidences have been given by various researchers [2-18]. Single phase drag reduction due to polymer based additives has been thoroughly studied and the mechanisms are beginning to be understood to be related to dampening of turbulent bursts and reduction of Reynolds stresses [16, 17]. In addition to mechanisms available to single-phase flow, there are also a very large number of experimental studies on single phase flow drag reduction with additives [2-18]. Although understanding of single phase flow drag reduction mechanisms may be of help in understanding the drag reduction mechanisms in complex multiphase flow, the complexity of multiphase flow phenomena which involve phase interface changes still need new mechanisms and theory to explain such phenomena. They are also the basis in developing and investigating new drag reduction technologies. Both polymeric additives and surfactants should be focused on as their mechanisms may be quite different in reducing drag in multiphase flow.

This chapter presents an overview of studies on drag reduction with surfactants and polymeric additives in multiphase flow including gas-liquid two-phase flow, liquid-liquid two-phase flow and gas-liquid-liquid three- phase flow. Future research needs are also identified according to this review.

2. DRAG REDUCTION AGENTS OR ADDITIVES (DRAS)

For multiphase flow, both surfactants and polymers may be used as Drag Reduction Agents (DRAs). The behaviors of DRAs are affected by several factors, such as pipe diameter, flow direction, fluid temperature, fluid viscosity, surface tension and pipe inclination and the flow regimes effects. In what follows, fundamental of surfactants and polymeric additives are first presented. Then, the effects of surfactants and polymeric additives on the physical properties are also presented. Some performances are summarized and their behaviors are analyzed.

2.1. Fundamental of Surfactants

Surface active substances or surfactants are amphiphilic compounds having a lyophilic, in particular hydrophilic part (polar group) and a lyophobic, in particular hydrophobic part (often hydrocarbon chain). The amphiphilic structure of surfactants is responsible for their tendency to concentrate at interfaces and to aggregate in solutions into various supramolecular structures, such as micelles and bilayers. According to the nature of the polar group, surfactants can be classified into nonionic and ionic surfactants, which may be of anionic, cationic, and amphoteric or zwitterionic nature. Nonionic surfactants have no charge, anionic surfactants have a negative molecular charge, cationic surfactants have a positive molecule charge, and amphoteric or zwitterrionic surfactants have both positive and negative charges [19, 20]. Anionic and nonionic surfactants provide most of industrial surfactant requirements and are the most common.

Surfactants greatly reduce the surface tension of solvents, water and water-based solutions, inks, fountain solutions, adhesives and other coating formulations. To reduce the surface tension, however, surfactant molecules have to migrate to the interface, and this takes some finite amount of time. The formulation will eventually reach equilibrium (static) surface tension after certain time. This takes several seconds or even several hours depending on the type of surfactant and the concentration of solutions. During this dynamic process before reaching equilibrium, the surface tension changes with the time and thus is defined as dynamic surface tension. In general, surfactants with smaller (lighter) molecule mass (short hydrophobic tail) diffuse more rapidly to the interface than that with larger (higher) molecule mass. Higher molecular weight surfactants affect a higher equilibrium surface-tension depression compared to lower molecular weight surfactants [19, 20]. In addition, most surfactants at higher concentrations cause change of the physical properties of the surfactant solutions and cause strong surface films between adjacent molecules, the strength of which determines surface properties of the surfactant solutions. In general, surface tension decreases with increasing surfactant solution concentration, and dynamic surface tension is usually higher than equilibrium surface tension at a fixed concentration. Higher solution temperature results in lower surface tension in both equilibrium and dynamics conditions. Additionally for all surfactants, surface tension decreases asymptotically with increasing concentration. The asymptotic limit is commonly referred to as the critical micelle concentration (cmc) of the surfactants. Critical micelle concentration (cmc) is characterized by micelle formation, or micellization, which is the property of surface-active solutes that lends to the formation of colloid-sized clusters, *i.e.* at a particular concentration, additives form aggregates in the bulk phase or a surfactant cluster in solution that are termed micelles. Different shapes and sizes of micelles such as globular or spherical, rod-shaped or cylindrical, and lamellar or plate-like exist depending upon the surfactant type and its concentration, solution temperature, presence of other ions and water-soluble organic compounds in the solutions. The micelle is a dynamic entity and its structure and shape can change with time. Critical micelle concentration (cmc) may be determined by many different techniques. A survey of methods for cmc determination is summarized by Lange [20]. Popular techniques include surface tension, turbidity, self diffusion, conductivity, osmotic pressure, solubilization, surfactant selective electrodes and fluorescence methods. Nearly all of these methods involve plotting a measurement as a function of surfactant concentration or as a function of the logarithm of surfactant concentration. The cmc is then deduced as a breakpoint. Critical micelle concentration is an important parameter in boiling phenomena with surfactants.

2.2. The Effect of Surfactants on the Physical Properties (Surface Tension and Viscosity) of Aqueous Surfactant Solutions

The surface tension and viscosity of aqueous surfactant solutions are possibly the primary thermo-physical properties that are influenced by the presence of surfactants, whereas all other physical properties are remained unchanged. The extent of surface tension reduction and/or viscosity variation of aqueous surfactant solutions (possibly at higher concentrations) depend upon the chemistry of the surfactant type as well as the concentration of the aqueous solutions. Addition of surfactants to water reduces the surface tension of the aqueous surfactant solutions considerably. The reduction in the surface tension is dependent upon several factors such as surfactant bulk concentration, surfactant type and molecular weight, solution temperature, interfacial conditions and so on.

In order to determine the effect of surfactants on the surface tension of aqueous surfactant solutions, we measured the equilibrium (static) surface tensions of aqueous solutions with two surfactants in the Institute of Thermodynamics at the Leibniz University of Hanover. The surfactants are SDS ($C_{12}H_{25}NaO_4S$) (≥ 95%, anionic) and Triton X-100 ($C_{33}H_{60}O_{10}$) (reinst, nonionic) supplied by Carl Roth GmbH in Germany. An optical contact angle measurement instrument manufactured by Dataphysics in Germany was used to measure the equilibrium surface tensions with the pendant-drop-method. (Fig. **1**) shows the variation of the measured equilibrium surface tension versus the concentration of SDS and Triton X-100 solutions at 23°C. The surface tensions were measured by static pendant liquid drops contacted by air. The measured equilibrium surface tensions decrease with increasing of concentrations for both surfactants. The concentration of the solutions has a great effect on the surface tension of the solutions. When the concentration reaches at a critical value, the surface tension will tend to a constant value which corresponds to the critical micelle concentration (cmc) for each surfactant. However, the critical value of Triton X-100 solution is much smaller than that of SDS solution. The critical value of SDS solution is more than 1000 ppm while that of Triton X-100 solution is only about 200 ppm. This manifests that the surfactant type has a great effect on the surface tension and the cmc value.

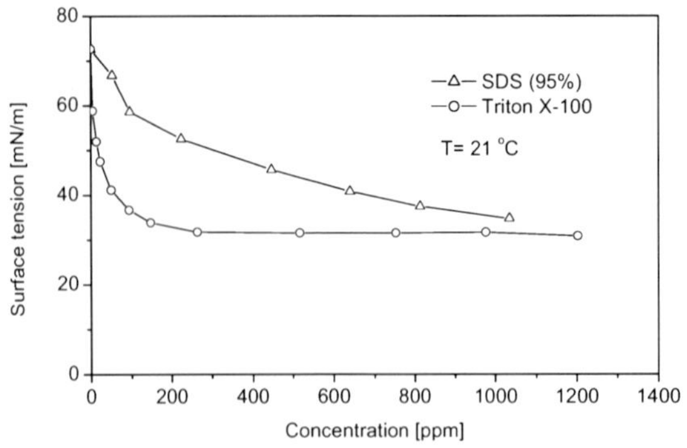

Figure 1: Variation of the measured equilibrium surface tension versus concentration [1].

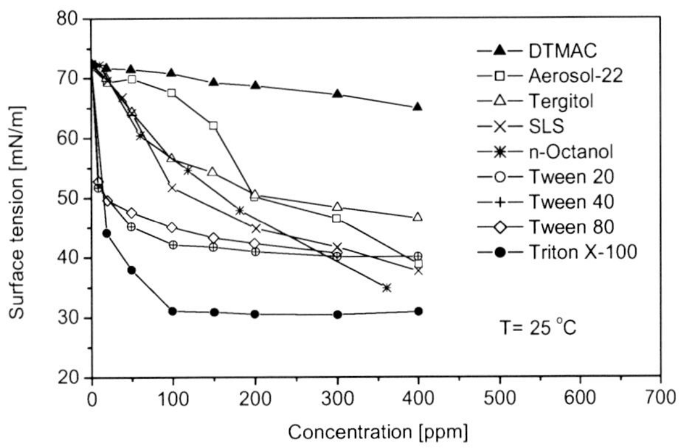

Figure 2: Variation of the measured equilibrium surface tension versus concentration presented by Wu *et al.* [21].

Several studies have focused on the effect of surfactant types and concentrations on the equilibrium surface tension of aqueous surfactant solutions in the literature [21-28] and similar conclusions have been reached. One typical example is the measurement of the equilibrium surface tension of solutions reported by Wu *et al.* [21]. They measured the equilibrium surface tension of aqueous solutions for nine additives including three anionic surfactants: SLS (sodium lauryl sulfate), Tergitol and Aerosol-22; one cationic surfactants: DTMAC (dodecyl trimethyl ammonium chloride) and five nonionic surfactants: Triton X-100, n-Octanol, Tween 20, Tween 40 and Tween 80 (polyoxyethylene sorbitan mono-oleate) at 25°C. (Fig. **2**) shows the variation of the measured equilibrium surface tension versus the solution concentration which was measured by Wu *et al.* [21]. It can be concluded that surface tensions decrease with increasing additive concentrations for all the nine surfactants. In addition, it can also be concluded that the depression of equilibrium surface tensions increases in the order: nonionic > anionic > cationic. The molecular weights of these surfactants considered in their study also vary in the same order, which suggests that higher molecular weight surfactants affect a higher surface-tension depression in the aqueous surfactant solutions. For nonionic surfactants: Triton X-100 > Tween 20, 40, 80 > n-Octanol; and for anionic surfactants: SLS > Aerosol-22 > Tergitol. For nonionic surfactants, the depression of equilibrium surface tensions increases with increasing the molecular weight of surfactants while it is contrary for anionic surfactants. Obviously, the surfactant type and the molecular weight (chemistry structure) have a great effect on the surface tension change of the aqueous surfactant solutions. Similar results have been obtained by Zhang and Manglik [27], and Wasekar and Manglik [28] as well. In addition, the surface tension decreases asymptotically with increasing the concentration of the aqueous surfactant solutions. This asymptotic limit is the critical micelle concentration (cmc) of surfactants.

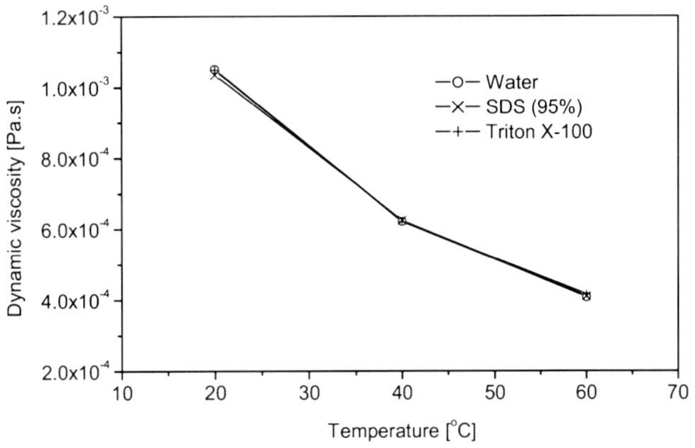

Figure 3: Variation of the measured dynamic viscosity versus temperature [1].

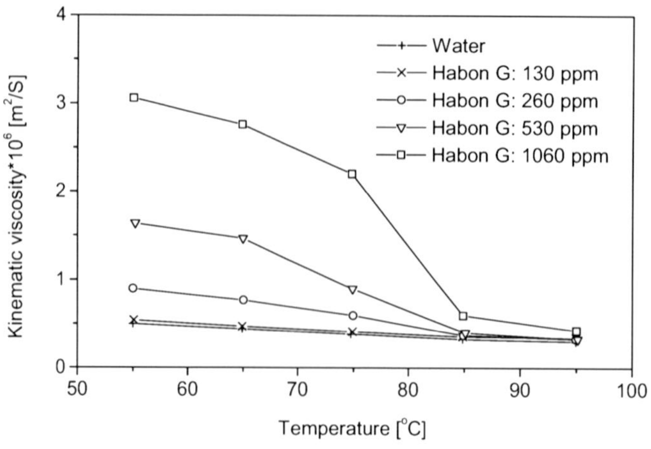

Figure 4: Variation of the measured kinematice viscosity versus temperature by Hetsroni *et al.* [26].

The viscosity of aqueous surfactant solutions may also be affected by the addition of surfactants in some cases. In order to determine the effect of surfactants on the viscosity of aqueous surfactant solutions, we measured the dynamic viscosities of aqueous surfactant solutions for SDS and TritonX-100 at the maximum solution concentrations for respective surfactants in the Institute of Process Engineering at the Leibniz University of Hanover. We also measured the dynamic viscosity of deionized water which is used as a reference for comparison to the measured dynamic viscosities of the surfactant solutions. A Rheometrics Fluids Spectrometer II (RFS) manufactured by Rheometric Scientific GmbH in Germany was used to measure the dynamic viscosities of aqueous SDS solution at the concentration of 1035 ppm, aqueous Triton X-100 solution at the concentration of 1203 ppm and deionized water at three different temperatures at atmospheric pressure. (Fig. **3**) shows the measured dynamic viscosities of SDS solution, Triton X-100 solution and deionized water at three different temperatures. There is no obvious difference among the dynamic viscosities for the three fluids at the same temperature. This manifests that both SDS and Triton X-100 solutions reveal the Newtonian fluidic behavior as that of water. Especially for aqueous Triton X-100 solution, even the test concentration of 1203 ppm is much higher than the cmc (about 300 ppm), the solution still reveals the Newtonian fluidic behavior. Thus, it can be concluded that the addition of Triton X-100 does not change the viscosity of the aqueous solutions at higher concentrations. This is different from the some measured results by other researchers in the literature, for example, the measured results of the aqueous Habon G solutions by Hetsroni *et al.* [26] as shown in (Fig. **4**) where the solution viscosity increases with increasing the solution concentration at the same temperature. The aqueous Habon G solutions reveal Non-Newtonian fluidic behavior except for the case at very low solution concentrations. The reason is possibly due to the effect of the surfactant type. Several studies have also concluded that surfactants with pre-micellar or dilute concentrations cause no significant change in the dynamic viscosity of solution [28, 29]. The measured viscosities for both SLS and Tween-80 aqueous solutions at room temperature were constant and independent of shear rates. Therefore, these two surfactant solutions show the Newtonian fluidic behavior in a wide concentration range of 125 to 500 ppm for polymeric surfactant solutions [30]. Post-micellar solutions, however, may show appreciable increases in dynamic viscosity [31] and tend to be viscoelastic [32, 33], which behave like polymers as discussed in the next section. Hetsroni *et al.* [34] measured the shear viscosities of aqueous Alkyl (8-16) Glucoside (Plantacare 818 UP) solutions at 25 and 60 °C. The increased trend of the measured shear viscosity with increasing shear rate indicates that aqueous Plantcare 818 solutions reveal the non-Newtonian fluidic behavior at higher concentrations. The measured concentrations ranging from 600 ppm to 7000 ppm, mostly exceed the cmc of the surfactant. It is reasonable that the solutions behave like the non-Newtonian fluids. In addition, Hetsroni *et al.* [26] studied the effect of temperature on the dynamic shear viscosity of aqueous Habon G solution over a temperature range of 25 to 60°C. The concentration of the aqueous Habon G solution was around its cmc. For a given shear rate, the viscosity decreases with the increase of solution temperature. Moreover, the change in shear viscosity at a temperature of 333 K over the shear rate range of 1 to 1000 per second was negligible. They also measured the kinematic viscosities for various concentrations of Habon G over a temperature of 55 to 95°C, as shown in (Fig. **4**). At the same temperature, the kinematic viscosity increases with the increases of the concentration of the surfactant solution. This is contradiction to our measured results presented in (Fig. **4**) and also does not follow the general law if the solutions are the Newtonian fluids. The reason is unclear but could be the effect of the surfactant type. Furthermore, at the near boiling temperature, the kinematic viscosities approach the value of pure water for both pre-micellar and micellar solutions. Such a behavior is more pronounced with decreasing the solution concentrations.

2.3. The Effect of Polymeric Additives on the Physical Properties of Aqueous Polymeric Additive Solutions

Polymers are typically large molecules, macromolecules, or agglomerates of small chemical units called monomers, and are broadly classified as biological and non-biological macromolecules. Addition of polymers in water primarily increases the solution viscosity, which tends to increase with increasing the concentration of aqueous polymeric additive solutions as well as the molecular weight of the polymers. The polymeric additive solutions often display a shear-rate depended shear-thinning rheology [32] with the exception of some surface-active polymers (or polymeric surfactants) such as hydroxyethyl cellulose (HEC) and polyethylene oxide (PEO), most aqueous polymeric additive solutions do not show any significant change in surface tension [22, 31, 36]. The viscosity of aqueous polymeric additive solutions, however, can influence the surface tension measurement considerably, especially at higher viscosities and bubble frequencies. The reduced surface tensions in aqueous polymeric additive solutions are largely brought about by the molecular adsorption of the polymeric additives to the vapor-liquid interface. The time scales of this process vary from the order of seconds to minutes depending upon the polymer chemistry and the solution concentration, which is possibly due to the slow process of diffusion transport of polymer molecules to the interface and their subsequent reorientation. This dynamic adsorption process, along with time scales of 10-100 ms for the boiling bubble dynamics in water, thus results in a rather complex interfacial behavior, which significantly alters the nucleate boiling in polymeric additive solutions. Zhang and Manglik [37] measured the surface tensions at different bubble frequencies for the various polymeric concentrations of HEC QP-300 solutions at 23°C and 80°C. (Fig. **5**) shows the variation of his measured dynamic surface tensions of HEC QP-300 solutions at various bubble frequencies at a temperature of 23°C versus the concentration of the aqueous solutions. Similar to that of surfactants, the dynamic surface tension of the polymeric additive solutions is found to increase with increasing the bubble frequency and decreasing the solution concentration. A critical polymer concentration (cpc) akin to cmc of surfactants is observed. It must be pointed out that HEC is a kind of polymeric surfactant which can depress the surface tension of aqueous polymeric additive solutions. However, most polymers generally have little effect on the surface tensions of solutions. Therefore, it is important to select those polymers which can depress the surface tension of aqueous polymeric additive solutions when study the boiling phenomena with polymeric additives.

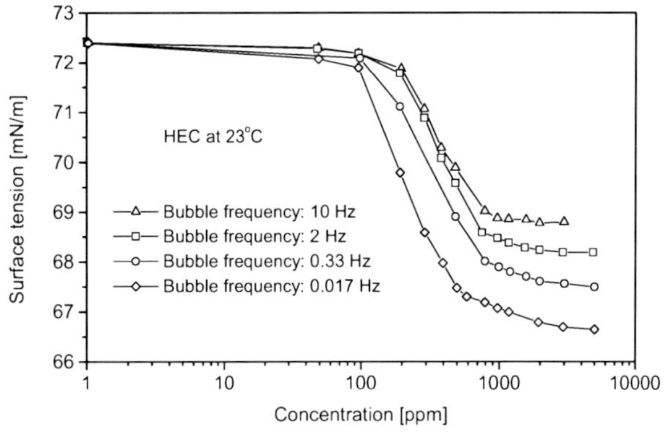

Figure 5: Variation of the measured dynamic surface tension versus concentration by Zhang and Manglik [37].

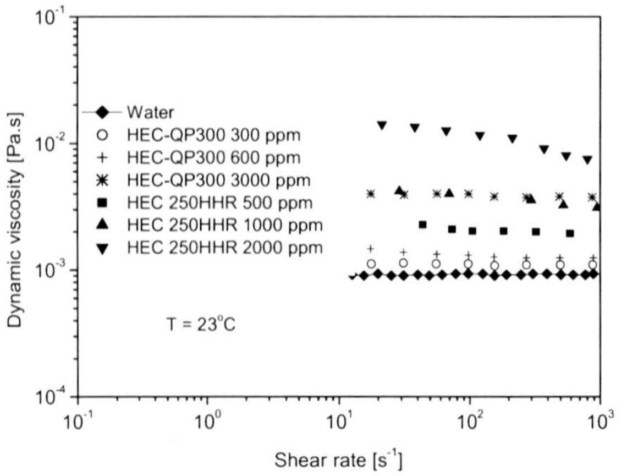

Figure 6: Variation of the measured dynamic surface tension versus concentration for HEC-QP300 by presented by Zhang [38] and for HEC250HHR presented by Hu *et al.* [31].

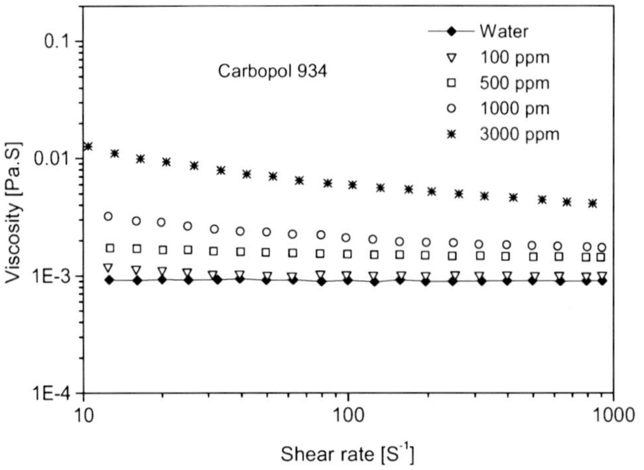

Figure 7: Variation of the measured viscosity versus shear rate by Zhang and Manglik [37].

Many researchers measured the viscosity of aqueous polymeric additive solutions in their studies of the boiling phenomena of aqueous polymeric additive solutions. (Fig. **6**) shows the measured viscosities for HEC 250HHR at various solution concentrations by Hu *et al.* [31] and that for HEC-QP300 at various solution concentrations and water by Zhang [38]. It can be seen that aqueous HEC solutions are more viscous than pure water, and the viscosity of the aqueous polymeric additive solutions increases with increasing concentration for each polymer. Actually the polymeric additive solutions behave like the Newtonian fluids at lower solution concentrations as the viscosities at solution concentrations remain constant with increasing shear stresses. By comparison, the HEC-QP300 solutions at the concentration of 3000 ppm show the Newtonian fluidic shear-constant behavior while the HEC 250HHR solutions at the concentration of 1000 ppm show the non-Newtonian fluidic shear-thinning behavior. This attributes to the effect of molecular weights and degrees of polymerization of the HEC family polymers. HEC 250HHR has a higher molecular weight ($M = 1.3 \times 10^6$) and a higher degree of polymerization as compared with HEC-QP300 ($M = 4\text{-}6\times10^6$). Similar results were obtained by Maestro *et al.* [39]. Their data for the lower molecular weight HEC9 ($M = 9\times10^4$) reveal the Newtonian fluidic characteristics even at a very high concentration (10% by weight), and the non-Newtonian fluidic shear-thinning behavior for the higher molecular weight HEC130 ($M = 1.3\times10^6$) solution at a concentration of 0.75% by weight. The results clearly indicate that molecular weight and the degree of polymerization play a role in the rheological behavior of polymeric additive solutions. As mentioned above, HECs are polymeric surfactants, they behavior like Newtonian fluids at low concentrations. Zhang and Manglik [37] also measured the viscosities of Carbopol 934 solutions at various concentrations and compared the results with those of water. As shown in (Fig. **7**), the polymeric additive solutions show the non-Newtonian fluidic shear-thinning behavior apparently, even at lower solution concentrations. The reason is that Carbopol 934 has a higher molecular weight ($M = 3\times10^6$) and a higher degree of polymerization as compared with HEC QP-300. Therefore, it has confirmed that a higher viscosity of a polymeric solution reflects a high degree of polymerization of the polymer.

3. DRAG REDUCTION WITH POLYMER AND SURFACTANT SOLUTIONS

3.1. Drag Reduction with Polymer Solutions

Polymer solutions are the most widely studied and most often employed of the drag reducing systems in single phase and multiphase flows although surfactants are sometimes used. Several typical polymer drag reducing solutions are shown in Table **1**.

It has been shown by experimental results that the higher the molecular weight, the more effective a given polymer as a drag reducer [40]. Polymers with a molecular weight below 100000 seem to be ineffective. As the average molecular weight of poly(ethylene oxide) (PEO) is increased from 2×10^5 to above 5×10^6, the solution concentration to achieve about 70% drag reduction on a rotating disk is reduced from 600 to 100 ppm [40]. The longer polymer chain provides more chance for entanglement and interaction with the flow. The extension of the polymer chain is critical for drag reduction. The most effective drag reducing polymers are essentially in linear structure, with maximum extensivity for a given molecular weight. Polyethylene oxide, polyisobutylene and polyacrylamide are typical examples of linear polymers. Polymers lacking linear structure, such as gum Arabic and the dextrans are ineffective for drag reduction.

Table 1: Drag reducing polymer solutions [40].

Water-soluble polymers	Solvent-soluble polymers
Poly(ethylene oxide)	Polyisobutylene
Polyacrylamide	Polystyrene
Guar gum	Poly(methyl methacrylate)
Xanthan gum	Polydimethylsideoxane
Carboxymethyl cellulose	Poly(cis-isoprene)
Hydroxyethyl cellulose	

3.2. Drag Reduction Mechanisms and Applications with Polymer Solutions

Over the past decades, a number of theories have been developed to explain how polymer molecules interfere with production, growth or transport of turbulent disturbances. The polymer hydrodynamic coil interacts with and disrupts eddies and micro-vortices in turbulent flows. These theories may explain many of the observed changes in flow structure during drag reduction [10-19]. However, the complexity of the drag reduction phenomenon has led some researchers to believe that more than one drag reduction mechanism must exist. Experimental evidence may be construed to support several theoretical models. It is possible that more than one mechanism may be required to adequately explain the phenomenon. A number studies on single phase flow drag reduction have been conducted. Some studies on multiphase flow have also been performed but much less than those for single phase flow.

In the application aspect, research on polymers originated with the development of the chemical industry in the 20th century. Polymer solutions are now widely used for many applications, *i.e.* the oil, food, cosmetics and chemical industry. General guidelines for the selection of a Drag Reduction Agent (DRA) for a given multiphase flow application do not exist. The most important requirement is that the DRA is soluble in the liquid. Table **2** lists several commercial drag reduction agents.

Table 2: Some commercial drag reduction agents [41]

Drag reduction agents	Solvent	Commercial source
Poly(isobutylene) (PIB)	Crude oil, most crude oil fractions	Shell Oil Co.
CDR®	Crude oil, gasoline, fuel oil	Conoco Inc.
FLO®	Crude oil, gasoline, fuel oil	ChemLink Petroleum
Poly(1,2-butyleneoxide)	Crude oil	Dow Chemical Co.
Poly(ethylene oxide), PolyOx®	Water, sewerage systems, fire hoses, mist suppressant	Union Carbide Corp.
SEPARAN® AP-30	Water	Dow Chemical Co.

3.3. Drag Reduction with Surfactant Solutions

One of the earliest studies on the effect of surfactant solutions on drag reduction was reported by Mysels [42]. However, the relevant research on surfactant solutions has received less attention than polymer solutions. Surfactant solutions have become a favourite drag reduction agent due to their chemical and mechanical stability which is an important factor in practical applications. Furthermore, surfactant solution systems exhibiting drag reduction at concentrations similar to dilute polymer solutions (< 100 ppm) have also been reported [43].

3.4. Drag Reduction Mechanisms with Surfactant Solutions

Formation of micelles and their shapes are main factors influencing the drag reduction ability of the surfactant solutions. The main characteristics feature of the friction behaviour of surfactant solutions is the disappearance of drag reduction when a critical wall stress is reached. The spherical micelle is generally conceived as a small ball-like particle of colloidal dimensions and fairly constant in size for a given surfactant. These spherical micelles exist only in relative dilute solutions. In concentrated solutions, however, the lamellar micelle is favoured. Under the influence of an electrolyte, spherical micelles can rearrange into cylindrical or rod-like micelles [43].

The drag reduction behaviours of surfactant solutions are different from those of polymer solutions. While the soap solution exhibits drag reduction low wall shear stress values, the polymer solutions show relatively small drag reduction at low Reynolds numbers and increasingly large reduction at high Reynolds numbers. These two types of behaviour are obviously a consequence of the morphological difference between micellar and polymeric structures [44]. It can be assumed that: (i) the flexible polymer molecule needs to be elongated by a large velocity gradient before its full drag reducing ability is developed, and (ii) the surfactant particles are oriented much more easily at lower velocity gradients, but the micelles collapse at high shear stresses associated with large velocity gradients [44]. In terms of equivalent molecular weight, micelles are known to have larger values than polymers and thus they would shift the onset of drag reduction to a lower shear stress value [40, 44].

4. MULTIPHASE FLOW DRAG REDUCTION STUDIES

The drag reduction mechanisms are much more complex in multiphase flow than those in single phase flow due to the different flow regimes and the interfacial behaviors. Flow patterns are very important in multiphase flow dynamics as pointed out by Cheng *et al.* [45, 46]. Some available studies of multiphase flow drag reduction mechanisms have shown that the flow regime transition has a significant effect on drag reduction characteristics. Therefore, it is of significant help very important to understand the two phase flow regimes for various situations. In this section, gas-liquid two phase flow patterns in vertical, horizontal and inclined pipes are described at first. Flow patterns for highly viscous and non-Newtonian fluids are also presented. Then, summary of the available studies on multiphase flow drag reduction are reviewed. Based on the review, future research needs are discussed.

4.1. Two Phase Flow Patterns

4.1.1. Vertical Adiabatic Two-Phase Flows

(Fig. **8a**) shows the most commonly observed two-phase flow patterns in a vertical tube. Bubbly flow occurs when a relatively small quantity of gas or vapor is mixed with a moderate flow rate of liquid. Increasing the gas flow rate may lead to plug flow, which some observers call Taylor bubble flow. With a further increase in gas flow rate, one may observe slug flow which consists of a regular train of large bubbles separated by liquid slugs. Each of these bubbles occupies nearly the entire channel cross section except for a thin liquid layer on the wall and their length is typically 1 to 2 times the channel diameter. An increase in both gas and liquid flow rates will lead to an unstable flow pattern which is called churn flow. A relatively higher gas flow rate generates a wispy-annular flow pattern, which is not observed or recognized as such in many studies. Very high gas flow rates may cause some of the liquid flow to be entrained as droplets carried along with the continuous gas phase in annular flows. At even higher gas flow rates, all the liquid is sheared from the wall to form the mist flow regime.

4.1.2 Horizontal Adiabatic Two-Phase Flows

(Fig. **8b**) shows the most commonly observed flow patterns for co-current flow of gas and liquid in a horizontal tube. Two-phase flow patterns in a horizontal tube are similar to those in a vertical tube but distribution of the liquid is influenced by gravity. In general, most flow patterns in horizontal tubes show a non-symmetrical structure which is due to the effect of gravity on the different densities of the phases. This generates a tendency of stratification in the vertical direction, with the liquid having a tendency to occupy the lower part of the channel and the gas the upper part. Stratified flow is usually observed at relatively low flow rates of gas and liquid. As the gas and liquid flow rates are increased, the smooth interface of the liquid becomes rippled and wavy. This pattern is called a stratified wavy flow. If the liquid flow rate is further increased while the vapor flow is maintained low, an intermittent flow pattern will develop in which gas pockets or plugs are entrapped in the main liquid flow and then a plug flow will develop. If flow rates of gas and liquid are increase together, a so called slug flow regime will develop. The main distinction between slug and plug flow is in the more pronounced nature of intermittent liquid mass separated by a larger gas bubble. With further increase in the gas flow alone, annular flow will develop. The gas flow in the core of an annular flow may entrain a portion of the liquid phase in the form of droplets and in some cases the liquid film may also entrain some small bubbles. At relatively large liquid flow rates, with little gas flow, one would observe the so called dispersed bubble flow in which the liquid phase is in the dispersed form of the scattered bubbles. At very high gas flow rates, the mist flow is reached, which can begin at the top perimeter where the annular film is the thinnest and then progress downstream to the bottom perimeter.

4.1.3. Flow Patterns in Inclined Channels

There are relatively few experimental observations on flow patterns in inclined tubes, notwithstanding the technical importance of such flows. Hewitt [47] has summarized this topic briefly. In short, flow patterns in inclined channels seem to have the same basic structures as in vertical and horizontal flows except for the limitation or total suppression of the churn regime.

4.1.4. Highly Viscous and Non-Newtonian Fluids

For highly viscous Newtonian liquids and non-Newtonian fluids flow patterns like those in (Fig. **9**) are observed by Diziubinski *et al.* [48]. The difference between these flow patterns and those for normal Newtonian fluids is in the flow structures. For example, a bubble may become larger in a highly viscous Newtonian bubbly flow due to the effect of viscosity on the liquid and gas velocities. There is very little research on highly viscous Newtonian gas-liquid flows. Just to mention two studies, one study is for horizontal tubes by Chhabra and Richardson [49] and the other one is for a vertical tube by Diziubinski *et al.* [48]. These studies have provided a good starting point for further work on these topics. Further extensive experiments should be done over a wide range of conditions to develop a generalized flow map for highly viscous Newtonian fluids.

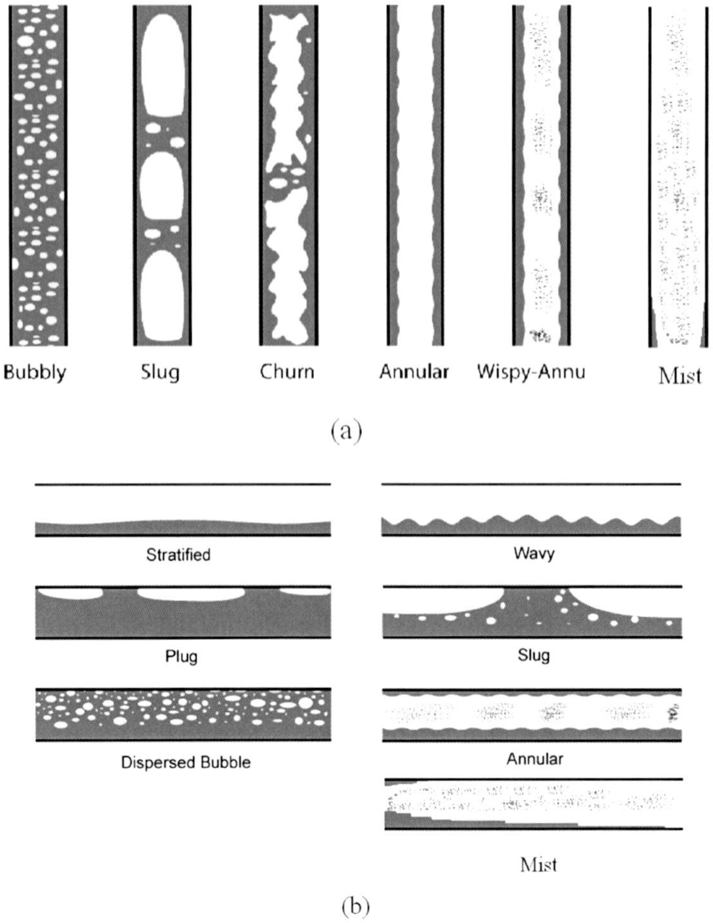

Figure 8: (a) Schematic of flow patterns in vertical upward gas liquid cocurrent flow; (b) Schematic of flow in horizontal gas liquid concurrent flow [45].

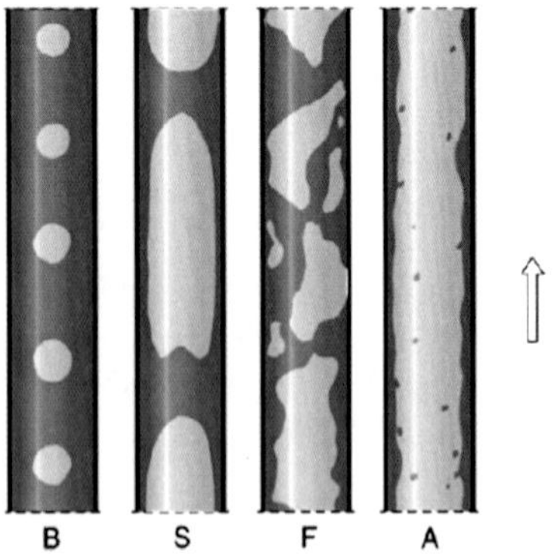

Figure 9: Schematic flow patterns of now-Newtonian two-phase flow in vertical channels: *B*—bubble flow, *S*—slug flow, *F*—froth flow, *A*—annular flow, observed by Dziubinski *et al.* [48].

4.1.5. Flow Patterns in Countercurrent Flows

There is very little published information regarding flow patterns in countercurrent flow situations. In general, flow regimes in countercurrent flows have a very limited range of existence due to the fact that a continuous increase in the flow rate of either phase would lead to so called flooding, which means that the passage of the other phase would be blocked and co-current flow would be established. In horizontal channels countercurrent flow may exist only as a stratified smooth or stratified wavy flow. In vertical channels, it exists only for downward liquid flow against rising vapor. The observed flow patterns are limited to annular, churn and plug flow. So far, there is a scarcity of information on this topic.

4.2. Two Phase Flow Pattern Maps

Flow pattern maps are generally used to identify the two phase flow regimes. Flow pattern maps may be classified primarily into two types: empirical flow pattern maps which are generally fit to the observed flow pattern database and theoretical or semi-theoretical flow pattern maps whose transitions are predicted from physical models of the flow phenomena. Cheng *et al.* have presented a comprehensive review on two phase flow patterns and flow pattern maps. One of the best known empirical flow pattern maps for horizontal flow is that of Baker [50] shown in (Fig. **10**). It was based on observations of co-current flow of gaseous and condensate petroleum products in horizontal pipes and was constructed with two parameter groups G_G/λ and $G_L\psi$ (where G_G and G_L are gas and liquid mass velocities, respectively), taking into account their physical properties by introducing the following parameters:

$$\lambda = \left(\frac{\rho_G}{\rho_A}\frac{\rho_L}{\rho_W}\right)^{\frac{1}{2}} \tag{3}$$

$$\psi = \left(\frac{\sigma_W}{\sigma}\right)\left[\left(\frac{\mu_L}{\mu_W}\right)\left(\frac{\rho_W}{\rho_L}\right)^2\right]^{\frac{1}{3}} \tag{4}$$

where ρ_A, ρ_W, σ_W and μ_W are the density of air, the density of water, the surface tension of water and the dynamic viscosity of water, respectively, at one atmosphere pressure and room temperature while ρ_L, ρ_G, σ and μ_L are the liquid density, gas density, surface tension and liquid viscosity, respectively. The correction factors λ and ψ had previously been used for correlating data on flooding points in distillation columns. Although Baker's flow map coordinates includes these apparently relevant variables for scaling a variety of different conditions, later investigations have showed that this map does not adequately predict horizontal flow regimes in numerous situations as:

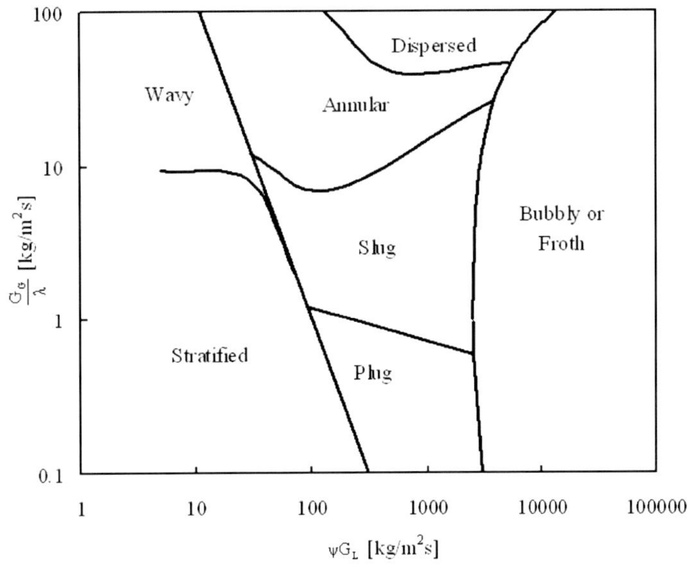

Figure 10: The Baker flow pattern map for horizontal gas liquid cocurrent flow [50].

Perhaps the most comprehensive treatment of flow pattern transitions in horizontal flow on a semi-theoretical basis is that of Taitel and Dukler [51]. It has been proved successful in predicting a fairly wide range of system conditions. (Fig. **11**) shows the Taitel and Dukler flow pattern map. The parameter groups, which are based on semi-theoretical derivations for different flow pattern transitions in horizontal or slightly inclined channels (θ is the angle of inclination) are as follows:

$$X = \left[\frac{(dp/dz)_L}{(dp/dz)_G} \right]^{\frac{1}{2}} \quad \textbf{(5)}$$

$$Fr = \frac{G_G}{\left[\rho_G (\rho_L - \rho_G) Dg \cos\theta \right]^{\frac{1}{2}}} \quad \textbf{(6)}$$

$$T = \left[\frac{\left| (dp/dz)_L \right|}{g(\rho_L - \rho_G)\cos\theta} \right]^{\frac{1}{2}} \quad \textbf{(7)}$$

$$K = Fr \left[\frac{G_L D}{\mu_L} \right]^{\frac{1}{2}} \quad \textbf{(8)}$$

where X is the Martinelli parameter, $(dp/dz)_L$ is the frictional pressure gradient as if the liquid in the two phase flow were flowing alone in the tube, $(dp/dz)_G$ is the frictional pressure gradient as if the gas in the two-phase flow were flowing alone in the tube, Fr is the Froude number, D is the tube diameter, g is the acceleration due to gravity, ρ_L is the liquid density, ρ_G is the gas density and μ_L is the liquid viscosity. They suggested the K versus X coordinate with a theoretically derived boundary curve, C, for transition from stratified smooth to stratified wavy flow. The Fr versus X relationship was proposed for the transitions between stratified wavy, annular-dispersed (droplets), dispersed bubble and intermittent (plug or slug) flows. The theoretically determined transition curves A and B (at X = 1.6) between the said regimes were also given in those coordinates. Finally, T versus X was proposed for defining the transition between dispersed bubble and intermittent (plug or slug) flow regimes with the transition line D. The transition curves shown in (Fig. **8**) are for the case of zero inclined-angle (horizontal). All the transition criteria used by Taitel and Dukler have some theoretical basis, although sometimes rather tenuous. There are also a number of other flow maps such as Hewitt and Roberts vertical flow map [52] and Mandhane *et al.* horizontal flow map [53] *etc.* A comprehensive review on flow regimes and flow pattern maps including macro-scale and micro-scale channels is presented by Cheng *et al.* [45].

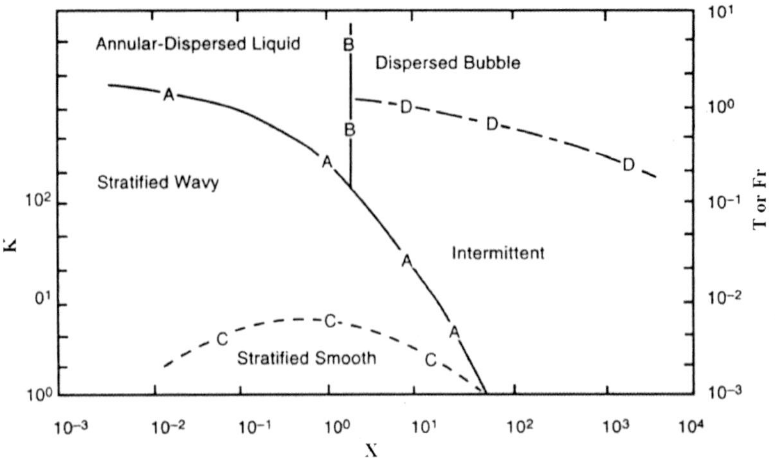

Figure 11: The Taitel and Dukler [51] flow pattern map for horizontal gas-liquid co-current flow: Coordinates of curves A and B are Fr vs. X, coordinates of curve C are K vs. X and coordinates of curve D are T vs. X.

4.3. Multiphase Flow Drag Reduction Research

Addition of small amounts of surfactants or polymeric additives in single phase liquid flow largely affects a wide range of phenomena (*e.g.* stability of laminar flow, transition to turbulence) and thus may reduce the drag in the single phase liquid flow. The drag reduction phenomenon has also been observed in for multiphase flow such as gas-liquid two-phase flow, liquid-liquid two-phase flow and gas-liquid-liquid three-phase flow. However, the effect of drag reduction additives on multiphase phase flow is much more complex than that on single phase liquid flow. As different gas-liquid or liquid-liquid interfacial structures (flow regimes) may occur in multiphase flow, addition of a drag reduction agent may result in flow regime changes which give quite different pressure drop behavior. The drag reduction performance in multiphase flow may also be affected by other factors such as pipe diameter, pipe inclination and fluid properties *etc.* studies of multiphase flow drag reduction are not as many as those of single phase liquid flow in the literature. Furthermore, the corresponding physical mechanisms of multiphase flow drag reduction are much less understood than those of single liquid flow. There are some explanations on the drag reduction mechanism in multiphase flow. However there are no generally unified mechanisms so far. In this section, available studies on multiphase phase flow drag reduction are summarized.

More than 10 years ago, Manfield *et al.* [54] performed a literature review of drag reduction with additives in multiphase flow. As pointed out by them that despite the large amount of work published on drag reduction in single-phase flow, limited work has been published on the effect of additives in multiphase flow. Especially, there is very limited work in the areas of surfactant drag reduction agents in multiphase flow.

For using a surfactant as a drag reduction agent in multiphase flow, Spedding and Hand [57] in a review paper summarize the studies that examine the effect of surfactant additives on two-phase pipe-flow. However, all the aforementioned studies principally examine the surface and interfacial effects of surfactants and not the reduction of frictional pressure drop caused by the turbulence modification within the liquid layer.

Furthermore, the effects of surfactants or polymeric additives on two phase flow regimes are not yet well investigated. It is obvious that flow regimes are very important in the study of drag reduction with additives. In fact, the available flow pattern maps cannot apply to the two phase flow with additives if flow regime changes occur. In this case, a new flow pattern map considering the additive effects is needed. Apparently there is no such a flow map available in the literature.

Jubran *et al.* [55] presented a review on drag reduction agents in multiphase flow. Their review includes recent work on drag reduction agents in single phase flow and multiphase flow. They focused on the theories of drag reduction, the influences of drag reduction agent types and hydrodynamic and heat transfer characteristics of flows in the presence of drag reduction additives with particular reference to the oil industry. They pointed out that drag reduction phenomena in multiphase flow are still far from being well understood in spite of the numerous investigations available in the literature. This is due to the dependence of such phenomena on a large number of parameters such as liquid viscosity, pipe diameter, liquid and gas velocities, composition of oil such as the wax content, pipe surface roughness, water cut, pipeline inclination, drag reduction additive concentration, types of drag reduction agents, shear degradation of drag reduction agents, temperature and pH. Therefore, much work is needed to understand these aspects.

Most recently, Al-Sarkhi [56] presented a literature review on drag reduction with polymers in gas-liquid and liquid-liquid flows in pipes, focusing on the characteristics of two-phase flow with drag reduction additives and the research approaches and methodology concerning drag reduction with additives in multiphase flow. He also summarized the suggested mechanisms for drag reduction phenomena of two phase flows from the available studies in the literature. However, there are quite different mechanisms explaining the drag reduction results. Therefore, there is a need to develop a unifying interpretation for the effect of drag reduction agents in multiphase flow. He also discussed some of the industrial application of using drag reduction additives in two phase flow. According to his review, he suggested some potential research areas in the relevant aspects.

Table **3** presents some of the available studies including surfactants and polymeric additives. Most of the studies are related to polymeric additives as already mentioned in the reviews. Some studies are on the surfactants but less than polymeric additives. The available studies include gas-liquid, gas-liquid-liquid three phase flows with different flow directions such as horizontal flow, vertical flow, downward flow and inclined flow. The available studies are reviewed according to classifications of polymeric additives, surfactant additives, gas-liquid two phase flow with additives, liquid-liquid two phase flow with additives, and gas-liquid-liquid three phase flow with additives in the following.

Table 3: Summary of several studies on drag reduction with polymeric additives and surfactants in multiphase flow.

Source	Pipe diameter (mm)	Fluids	Flow direction	Additive	Remarks
Greskovich and Shrier [58, 59]	1.5 in and 0.0381 m	Air-water, nitrogen-kerosene	Horizontal flow	Polymer solutions (Polyethylene oxide coagulant (Polyox); polyisobutylene (Vistanex L-200)	For air-water flows, pressure drop reduction as high as 50% for slug flow. For hydrocarbon-nitrogen 25 to 30%
Sylvester *et al.* [62]	1, 2 and 3 inch	Natural gas-hexane	Horizontal flow	Ploymerica	Drag reduction in annular-mist flow,
Rosehart *et al.* [60]	0.0254 m		Horizonatal flow	Polyacrylamide	
Sylvester and Brill [61]	0.0127 m	Air-water		polyethylene oxide	
Mowla *et al.* [63]		Air-water	Horizontal flow	Polymer additive	
Kang *et al.* [64]	0.01 m	Oil-water-CO_2	Horizontal and inclined upward flow	DRA	The effect of DRA on pressure gradient and flow regime was studied.
Al-Sarkhi *et al.* [56]	0.0254 m	Air-water	Horizontal flow	Co-polymer of polyacrylamide and sodium acrylate	The influence of the polymers on the transition to slugging, the transition to roll waves, the liquid holdup and the interfacial drag.
Al-Sarkhi and Hanrattyl. [68, 69]	0.0953 m and 0.0254	Air-water	Horizontal flow	Co-polymer of polyacrylamide and sodium acrylate	Drag reduction of 48% and up to 63% were realized.
Soleimani *et al.* [71]	0.0254 m	Air-water	Horizontal flow		
Fernandes *et al.* [72]		Air-water			
Mowla and Naderi [73, 74]	0.0254 m and 0.0127 m	Air-crude oil	Horizontal flow	Polyalpha-olefin (Polyisobutylene)	40% of pressure drop reduction was obtained
Kang *et al.* [65]	0.01 m	Cardon dioxide and two types of oil	Horizontal flow	DRA, CDR WS 500M flow improver	The effect of viscosity on pressure drop reduction were experimentally studied.
Wilens and Thomas [77] and Wikens *et al.* [76]	0.052 m	Air-water	Horizontal flow	Sodium dodecyl sulfate (SDS)	The pressure drop reduction is due to solely to the suppression of intermittent flow patterns
Lioumbass *et al.* [79]	0.024 m	Gas-liquid	Inclined downward flow up to 15°	Non-ionic surfactant Tween 80®	The surfactant has a dramatic effect characteristics of stratified-wavy flow.
Al-Wahaibi *et al.* [80]	0.0127 m	Air-water	Horizontal and inclined upward flow	MAGNAFLOC 110L (polymer)	Drag reduction of 71% was obtained.
Duangprasert *et al.* [78]		Air-water	Vertical flow	SDS surfactant solutions	The effect of SDS on flow regimes.
Al-Yaari *et al.* [81]	0.0254 m	Oil-water	Horizontal flow	Water soluble polymer solution Magnafloc 1011 and polyethylene oxide.	Effect of polymer concentrations and molecular weights on flow patterns and pressure drops.

4.3.1. Polymeric Additives

As shown in Table **3**, a number of studies deal with polymeric additives in two phase flow. Some studies focus on the effects of polymeric additives on flow regime changes which result in drag reduction in the two phase flow. Some studies are described here.

Greskovich and Shrier [58, 59] studied polymer drag reduction agent in a 0.038-m diameter horizontal pipe. They reported drag reduction of 40–50% in air–water slug flow. They attributed this to a reduction in frictional losses as opposed to accelerational changes within the slug.

Rosehart *et al.* [60] studied polymer drag reduction agent (polyacrylamide) in slug flow in a 0.0254-m diameter horizontal pipe. They found no changes to the slug characteristics. There was no change in the slug translational velocity. At the highest additive concentrations there was a decrease in slug frequency, primarily due to increased viscosity. The greatest drag reduction occurred at the highest slug velocities. This is due not only to a reduction in skin frictional losses, but also due to lost energy (accelerational). Increasing the additive concentration also increased the drag reduction up to a saturation concentration beyond which there was no additional drag reduction.

Sylvester and Brill [61] studied polymer drag reduction additive in annular flow of air-water two phase flow in a 0.0127-m diameter horizontal pipe. They found drag reduction of up to 37% at high flow rates if the additive was not recycled.

Sylvester *et al.* [62] reported the effect of liquid flow rate and gas flow rate on drag reduction in horizontal natural gas–hexane pipe flow in three different diameters (1, 2 and 3 in.). Drag reduction of 34% were obtained. It was found that drag reduction increased with decreasing gas rate.

Mowla *et al.* [63] and considered the effects of polymer additives on two-phase flow drag reduction for air–water system. Mowla and Naderi [74] reported experimental results of drag reduction in annular flow of oil and air two-phase flow.

Kang *et al.* [64, 65] and Kang and Jepson [66] studied drag reduction in horizontal or slightly inclined slug flow and annular entrained flow of oil and carbon dioxide.

Otten and Fayed [67] reported drag reduction for no-Newtonian slug flow. Their results are similar to those of Rosehart *et al.* [60]. Furthermore, they observed pressure drop was always lower that predicted by the Lockhart-Martinelli correlation. It should be mentioned here that studies on drag reduction of non-Newtonian fluid flow with additives. It is difficult to understand the phenomena with limited studies.

Al-Sarkhi and Hanratty [68, 69] studied polymer DRA in a horizontal, 0.0953-m diameter multiphase pipe. They compared air/water flow with and without the DRA at equal flow rates for high gas flow conditions. They found that annular flow became stratified with the injection of the DRA. This was due to damped disturbance waves which led to reduced atomization. They found a reduced pressure gradient that reached a plateau with increasing DRA concentration. They also found sensitivity to injection technique. The reduced pressure gradient observed was a combined effect of flow pattern change and a liquid phase mechanisms as their polymer solution presented single phase drag reduction. They reported that the injection of a polymer solution into horizontal annular air–water pipe flow reduces drag by 48%.

Soleimani *et al.* [71] studied drag-reducing polymers on pseudo-slugs-interfacial drag and transition to slug flow in a horizontal, 0.0254-m diameter pipe. They compared air/water flow with and without drag reduction agent at equal flow rates in the range of pseudo-slug flow. (Figs. **12** and **13**) show their comparisons. Furthermore, they found the transition to slug flow to occur at higher superficial liquid velocities in the presence of the drag reduction additives; The DRA led to a decrease in the liquid shear stress at the wall. Increasing the concentration decreased wave frequency, decreased wave velocity, decreased the small wavelength waves, dampened the high frequency waves, and increased the holdup. The pressure drop reduction was not monotonic with increasing DRA concentration due to competing effects (*e.g.*, holdup, interfacial drag) at each concentration. They suggested that the drag reduction observed was due to the elimination of small wavelength waves on roll waves rather than damping of the roll waves.

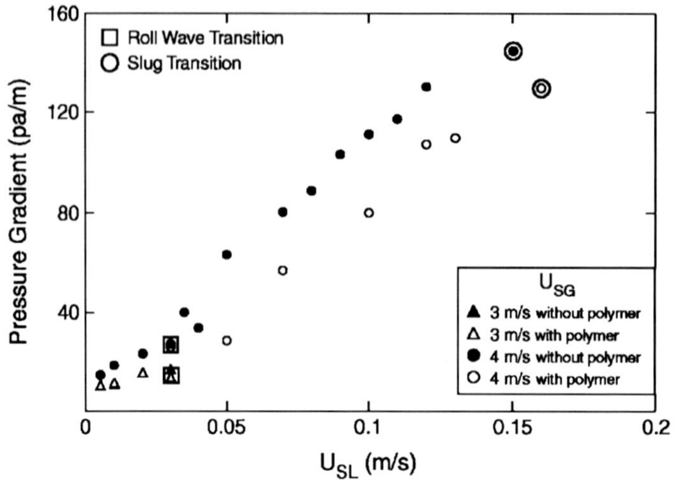

Figure 12: Measurements of the effect of U_{SL} on the pressure gradient for U_{SG} = 3 and 4 m/s with (C_H = 100 ppm) and without polymer additives.

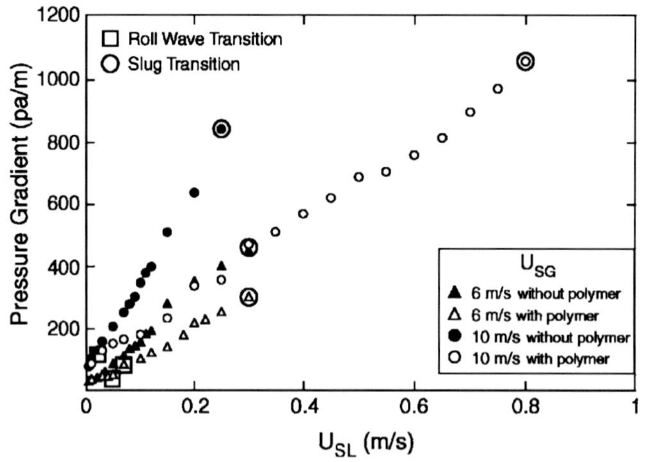

Figrue 13: Measurement of the effect of U_{SL} on the pressure gradient for U_{SG} = 6 and 10 m/s with (C_H = 100 ppm) and without polymer additive.

Fernandes *et al.* [72] studied polymer DRA in a high pressure (1.9 MPa) horizontal, 0.019-m diameter multiphase pipe. They compared methane/condensate flow with and without DRA for high gas flow rate conditions. They found drag reduction in annular flow due to both reduction of entrainment and reduced interfacial friction due to a reduced liquid film roughness, with a negligible contribution of liquid turbulence. They proposed a model for estimating the drag reduction.

More recently Fernandes *et al.* [82] conducted experimental study of drag reduction n vertical two phase annular flow. They concluded that the drag reducers have two main effects on the flow: they reduce the frictional pressure drop and increase the liquid holdup. These phenomena have opposing effects on the overall pressure gradient.

It should be mentioned that most of the available studies deal with two phase flow in horizontal pipes. Some deal with two phase flow in inclined pipes. However, only few deal with two phase flow in vertical flows.

4.3.2. Surfactant Additives

Spedding and Hand [57] reported observations of the addition of 100 ppm of Chemtreat 271 surfactant in air–water two-phase flow in a 0.0935-m diameter horizontal pipe. They reported that the surfactants delayed the transition from stratified smooth to stratified wavy flow, and thus resulted in two phase flow drag reduction. Furthermore, the drag reduction led to increased liquid holdup. Wave damping was also found in the liquid film in annular flow. They further found that the surfactant caused a modification to the slug front region, leading to drag reduction. Several studies on drag reduction with surfactants are also presented in this section.

Hart *et al.* [75] examined the influence of the interfacial tension on the wavy-stratified and annular flow regime characteristics using an aqueous surfactant solution (Tween 80®) and they noticed that the surfactant additives cause higher pressure drop values when they are present in the flow. However, the concentration of the solutions used was lower than the critical micelle concentration (CMC) of the Tween® solution (CMC=0.13% w/w), *i.e.* the concentration above which the surfactant molecules tend to assemble to aggregates (micelles) and start to behave as DRA.

Wilkens *et al.* [76] and Wikens and Thomas [77] performed experiment studies on the flow pattern suppression in horizontal air water pipe flow with surfactant additive. Their results suggest that addition of the surfactant to the gas-liquid flow significantly reduces the occurrence of slug flow. In addition, previously unreported flow patterns were observed to exit between slug and dispersed bubble flows. They concluded that new mechanisms for slug flow transition are needed.

Duangpasert *et al.* [78] studied two phase gas liquid flows in vertical pipes with SDS surfactant solutions. They focused on the influence of surfactant addition on the flow regimes, the corresponding pressure gradients and the bubble sizes and velocities.

Lioumbas *et al.* [79] investigated the influences of a surfactant additive on the interfacial structure and on the transition from the smooth to the wavy stratified flow regimes in inclined pipes. Their measurement of liquid layer thickness and axial velocity has confirmed that the addition of small amounts of non-ionic surfactant additives during co-current downward flow strongly influences both the interfacial characteristics and the flow field within the liquid layer and thus reduce the pressure drops of the fluid.

As indicated in section 2, Cheng *et al.* [1] measured surface tension and viscosity of two surfactant solutions and found surface tension was decreased with surfactant additives. The physical property changes should be considered in the flow regimes changes in the presence of surfactants. However, so far no such consideration has been reported. Therefore, in future, systematic studies of drag reduction mechanisms should consider the physical property changes with addition of surfactants.

4.3.3. Gas-Liquid Two Phase Flow with Polymeric Additives and Surfactants

Most of the available studies are relevant to gas-liquid two phase flow. Air-water is the most used fluids. Some studies used oil-air. The relevant studies are presented in Table **3**.

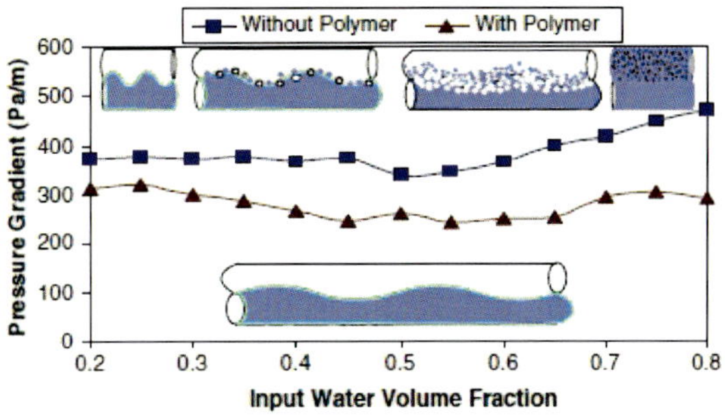

Figure 14: Measurement of the effect of 50 ppm PDRA and water fraction on oil-water pressure drop at mixture velocity of 1 m/s [81].

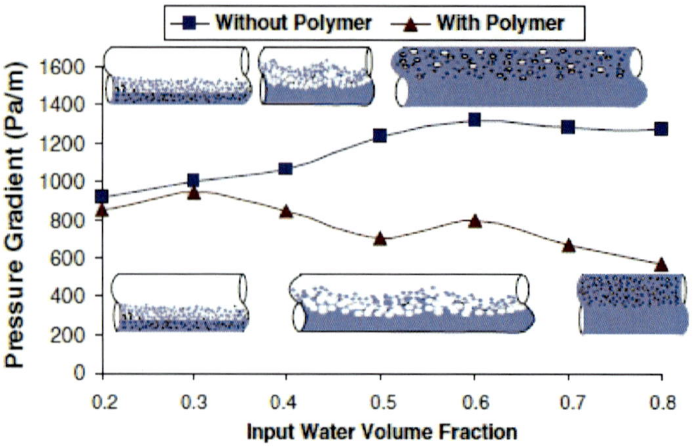

Figure 15: Measurement of the effect of 50 ppm PDRA and water fraction on oil-water pressure drop at mixture velocity of 2 m/s [81].

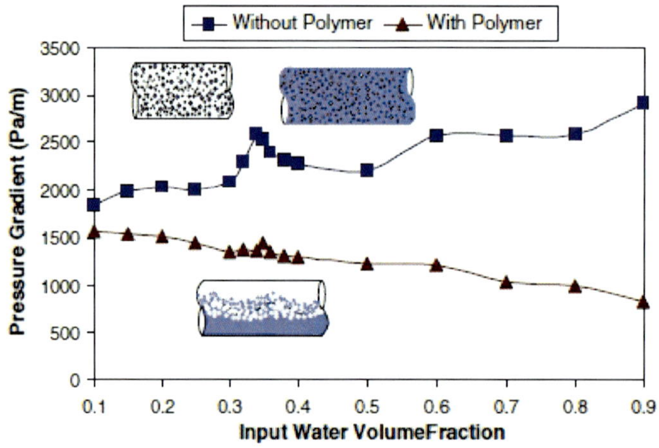

Figure 16: Measurement of the effect of 50 ppm PDRA and water fraction on oil-water pressure drop at mixture velocity of 3 m/s [81].

4.3.4. Liquid-Liquid Two Phase Flow with Additives

There are several studies on drag reduction of liquid-liquid two phase flow with additives in the literature but such studies are very few. Al-Wahaibi *et al.* [80] studied the effect of a drag-reducing polymer on oil-water flow in a 14 mm ID acrylic pipe. Oil and a co-polymer (Magnafloc 1011) of polyacrylamide and sodium acrylate were used. Their results showed a strong effect of the Drag Reduction Polymer (DRP) on flow patterns. The presence of DRP extended the region of stratified flow and delayed transition to slug flow. The addition of the polymer clearly damped interfacial waves. The DRP cuased a decrease in pressure gradient and a maximum drag reduction of about 50% was found when the polymer was introduced into annular flow. The height of the interface and the water holdup increased with DRP.

Al-Yaari *et al.* [81] measured drag reduction for oil-water flowing in a horizontal 0.0254 mm pipe. The injection of water soluble polymer solution in some cases produced drag reduction of about 65% with concentration of only 10-15 ppm. They also observed different oil-water configurations and phase inversion in dispersed flow regime. Injection of polymer drag reduction additive into water continuous layer or the water continuous dispersed flow regimes in oil-water immiscible flow changes the flow pattern and causes a higher degree of stratification. Pressure drops are greatly affected by the polymer additives as shown in (Figs. **14** to **16**).

4.3.5. Gas-Liquid-Liquid Three Phase Flow with Additives

Kang *et al.* [65], who conducted experiments in inclined pipes, observed that the presence of a Drag Reducing Agent (DRA) (whose type is not explicitly stated) in a three-phase CO_2–oil–water flow has a significant effect on the flow pattern transitions and results to an important drag reduction (92%) during stratified pipe downward flow. However, studies on gas-liquid-liquid three phases are very limited. More experimental research work is needed. Furthermore, reasonable mechanisms are also needed to better explain the drag reduction phenomena.

4.4. Mechanisms and Theoretical Analysis

Multiphase pipe flow has numerous additional unique mechanisms available to reduce pressure loss, including but not limited to: interfacial stress reduction, holdup change, flow pattern change, and reduction of effective density (vertical flow). Two critical issues exist with polymer DRAs in terms of hydrocarbon transport; dry out and shear degradation. Polymer DRA requires a co-solvent. During high gas–liquid ratio multiphase flow these co-solvents have a tendency to vaporize (*i.e.*, dry out). This leads to precipitation of the DRA and a loss of drag reduction benefit. Shear degradation is a phenomenon that has been well documented. Under high shear conditions, long polymer DRA chains are permanently broken, thus permanently reducing or eliminating the drag reduction capabilities of the DRA.

These critical issues occurred in polymeric additives could be eliminated with the use of surfactant drag reducing agents (SDRA). Surfactants are soluble in most liquid phases and therefore do not require a co-solvent and therefore they are not subject to dry out. SDRA also have much smaller molecular chains than those of the polymers used in DRA. With surfactants the mechanism of single phase drag reduction is related to the formation of micelles. When surfactants are added to a liquid they go to the liquid surface and act to reduce the surface tension. Above the Critical Micelle Concentration (CMC) additional surfactant molecules no longer migrate to the surface and they begin to form micelle structures within the liquid. It has been proposed that in single-phase flow, surfactant drag reduction occurs if the micelle structures formed are rod-like. These rod-like micelles then mimic the performance of long chain polymers in dampening turbulent bursts near the pipe wall. The benefit of these self-assembling structures is that they break apart when subjected to conditions of high shear but then reassemble downstream when the high-shear condition is gone. Thus it does not need to be re-injected following a high-shear condition, such as a pumping station or a choke. Elimination of intermittent flow, in particular slug flow, is seen as a key to increasing throughput in gas/liquid flow due to the high pressure losses associated with these flow patterns. Ability to control formation of intermittent flow may also provide key insights into the mechanism of slug formation and pave the way towards a comprehensive model for gas/liquid flow in pipes. The focus of this work will be to quantify the flow pattern effect on pressure drop. More specifically, it will focus on the effect of eliminating slug flow on pressure drop through the use of SDRA.

So far, understanding the mechanisms of multiphase flow drag reduction with additives is still unsatisfactory. As already mentioned, flow regimes are very important in understanding the mechanisms. However, most available studies used air-water two phase flows in horizontal pipe. Several mechanisms in the available studies are summarized as follows:

- Annular flow: the destruction of disturbance waves (because of wave damping) and, as a consequence, the discontinuance of atomization and of the spreading of the liquid layer up the walls of a horizontal pipe. A change to stratified flow occurs.

- Stratified flow: polymers damp waves. This can have two effects: a decrease of interfacial drag and an increase of gas velocity because of the increase of the height of the stratified liquid layer. These have opposite influences on the pressure gradient.

- Slug flow: a damping of turbulence in the liquid decreases the wall drag and changes the behavior of the gas bubble just behind the back of the slug. PDRA also reduces the slug frequency.

- Bubbly flow: polymers have the possibility of affecting turbulence and bubble size. The maximum drag reduction usually occurs when a slug, pseudoslug or annular flow changes to stratified flow by adding drag reducing polymers. The effectiveness of a drag-reducing polymer is sensitive to the technology used to introduce polymer into the flow and the concentration of the injected master polymer solution. The aggregates of polymers have a more important effect on the turbulence and drag reduction than individual molecules. Therefore, degradation occurs by the destruction of these groups or aggregates.

It should be pointed out that there are few studies on inclined and vertical two phase flows. In vertical flows this could affect the effect the transition from bubbly flow to slug or churn flow. PDRA also resulted in a significant increase in the liquid holdup [82]. Liquid–liquid flows may be investigated in similar way. Differences related to the use of oil soluble or water soluble DRP should be identified. Obviously, more experimental work is needed to multiphase flow drag reduction. Further understanding the mechanisms is needed. Due to the flow regime change, the available flow pattern maps will possibly not apply. Thus, new flow regime maps are needed for multiphase flow with additives. So far no such a map is available. Systematic studies on multiphase flow pressure drop according to flow regimes are also necessary. Apparently no such studies are available. In the end, unified mechanisms and prediction methods for flow regimes and pressure drop should be targeted.

Furthermore, nearly all the available studies are relevant to large diameter pipes. In recent years, small diameter tubes are also very important but no such studies are available [46]. It is therefore very essential to perform such studies. Also studies on no-Newtonian fluid flow are very limited and thus it is also very important to conduct such studies.

5. CONCLUSIONS

This chapter presents an overview on drag reduction with surfactants and polymeric additives in multiphase flow. The following conclusions have been obtained through this review:

1. Most of the available studies are relevant to drag reduction with polymeric additives while some studies are relevant to drag reduction with surfactants. Studies of drag reduction with surfactants are less popular than those with polymeric additives.

2. Most of the available studies deal with gas-liquid two phase flow. There are limited studies on liquid-liquid two phase flow and gas-liquid-liquid three phase flow with drag reduction additives. More studies on those aspects are needed.

3. Most of the available studies deal with two phase flow with drag reduction additives in horizontal pipes. There are few studies on inclined and vertical two phase flows. Flow regimes are greatly affected by the drag reduction additives and drag reduction can be achieved due to the flow regime changes. For example, in vertical flows this could affect the effect the transition from bubbly flow to slug or churn flow. This may also result in a significant increase in the liquid holdup Liquid–liquid flows may be investigated in similar way.

4. Obviously, more experimental studies are needed to investigate multiphase flow drag reduction for various conditions and pipe arrangement.

5. Further understanding the drag reduction mechanisms is needed. Due to the flow regime change, new flow maps are needed for multiphase flow with additives. So far no such a flow pattern map is available.

6. Systematic studies on multiphase flow pressure drop according to flow regimes are also necessary. Apparently no such studies are available. In the end, unified mechanisms and prediction methods for flow regimes and pressure drop should be targeted.

7. Nearly all the available studies are relevant to large diameter pipes. In recent years, small diameter tubes are also very important but no such a study on drag reduction with additives in small diameter channels is available. It is therefore very essential to conduct studies on this aspect, considering pipe diameter effect together with other main affecting factors.

8. Studies on no-Newtonian fluid flow are very limited and thus it is also very important to conduct studies on non-Newtonian two phase flow drag reduction with surfactants and polymeric additives. Furthermore, the physical mechanisms related to various flow regimes and pressure drop reductions should be systematically investigated. Furthermore, pressure drop prediction methods should also be developed.

ACKNOWLEDGEMENTS

Dr. Lixin Cheng was supported by a visitor grant (as a renowned expert) awarded by Xi'an Jiaotong University in China for his collaboration with Prof. Lei Liu at the State Key Laboratory of Multiphase Flow in Power Engineering at Xi'an Jiaotong University in December 2010.

NOMENCLATURE

C_H	concentration, ppm
D	inner tube diameter, m
$(dp/dz)_L$	frictional pressure gradient as if the liquid in the two phase flow were flowing alone in the tube, Pa/m
$(dp/dz)_G$	frictional pressure gradient as if the gas in the two-phase flow were flowing alone in the tube, Pa/m
Fr	Froude number, Eq. (6)
$f_{without\,DRA}$	friction factor in the absence of DRA
$f_{with\,DRA}$	friction factor in the presence of DRA
G	mass velocity, kg/m^{2}s
g	acceleration due to gravity, 9.81 m/s^2
K	parameter defined by Eq. (8)
T	parameter defined by Eq. (7)
X	Martinelli number, Eq. (5)

Greek Symbols

$\Delta p_{without\,DRA}$	frictional pressure drop in the absence of DRA, Pa
$\Delta p_{with\,DRA}$	frictional pressure drop in the presence of DRA, Pa
λ	correction factor defined by Eq. (3)
ε	effectiveness of DRA
θ	inclination angle, rad.
μ	dynamic viscosity, kg/ms
ρ	density, kg /m^3
σ	surface tension, N /m
ψ	correction factor defined by Eq. (4)

Subscripts

A	air
G	gas phase
L	liquid phase
W	water
with DRA	in the presence of DRA
without DRA	in the absence of DRA

REFERENCES

[1] L. Cheng, D. Mewes, and A. Luke, "Boiling phenomena with surfactants and polymeric additives: a state-of-the-art review," *Int. J. Heat Mass Transfer*, vol. 50, pp. 2744-2771, 2007.

[2] B. A. Toms, "Some observation on the flow of linear polymer solution through straight tube at large Reynolds number," In: *Proc. 1st Int. Congr. Rheology*, North Holland, Amsterdam, vol. 2, 1948 pp. 135-141.

[3] J. G. Savins, "Drag reduction characteristics of solutions of macromolecules in turbulent pipe flow," *Society of Petroleum Engineers J.*, vol. 2, pp. 203-214, 1964.

[4] F. A. Seyer and A. B. Metzner, "Turbulent phenomena in drag reduction systems," *AIChE J.*, vol. 15, pp. 426, 1969.

[5] G. Astria, "A phenomenological interpretation and correlation of drag reduction," *AIChE J.* , vol. 15, pp. 564-567, 1969.

[6] M. J. Rudd, "Laser Dopplermeter and polymer drag reduction," *AIChE Symp. Sers.* Vol. 67 (111), pp. 21-26, 1971.

[7] S. M. Kumor and N. D. Sylvester, "Effect of a drag reducing polymer on the turbulent boundary layer," *AIChE Symp. Sers.* Vol. 69, pp. 130, 1972.

[8] Z. Chara, J. L. Zakin, M. Severa, and J. Myska, "Turbulence measurements of drag reducing surfactant systems," *Exp. Fluids*, vol. 16, pp. 36-41, 1993.

[9] G. Hetsroni, J. L. Zakin, and A. Mosyak, "Low-speed streaks in drag-reduced turbulent flow," *Phys. Fluids*, vol. 9, pp. 2397-2404. 1997.

[10] B. Lu, X. Li, J. L. Zakin, and Y. Talmon, "A non-viscoelastic drag reducing cationic surfactant system," *J. Non-Newtonian Fluid Mech.*, vol. 71, pp. 59-72, 1997.

[11] H. K. Yoon and A. J. Ghajar, "Heat eddy diffusivity for viscoelastic turbulent pipe flows," *Int. Comm. Heat Mass Transfer*, vol. 14, pp. 237-249, 1987.

[12] H. K. Yoon and A. J. Ghajar, "A method for correlating the diameter & concentration effect on friction & heat transfer in drag reducing flow," *AIAA Thermophysics, Plasmadynamics & Lasers Conf.*, San Antonio, Texas, 1988.

[13] H. K. Yoon and A. J. Ghajar, "A heat transfer correlation for viscoelastic turbulent pipe flows," *Chem. Eng. Comm.*, vol. 78, pp. 167-177, 1989.

[14] G. C. Liaw, J. L. Zakin, and G. K. Patterson, "Effect of molecular characteristics of polymer on drag reduction," *AIChE J.*, vol. 17, pp. 391-397, 1971.

[15] C. B. Lester, "The basics of drag reduction," *Oil & Gas J.*, pp. 51-56, 1985.

[16] P. S. Virk, "Drag reduction fundamentals," *AIChE J.*, vol. 21, pp. 625-656, 1975.

[17] P. S. Virk, "Turbulent kinetic energy profile during drag reduction," *Phys. Fluids.*, vol. 18, pp. 415-419, 1975.

[18] C. M. White and M. Godfrey Mungal, "Mechanics and prediction of turbulent drag reduction with polymer additives," *Ann. Rev. Fluid Mech.*, vol. 40, pp. 235-256, 2008.

[19] V. B. Fainerman, D. Möbius, and R. Miller, *Surfactant: Chemistry, Interfacial Properties, Applications*, Elsevier Science BV, Amsterdam, 2001.

[20] L. Robert Lange, *Surfactants: A Practical Handbook*, Hanser Publishers, Munich, 1999.

[21] W. T. Wu, Y. M. Yang, and J. R. Maa, "Enhancement of nucleate boiling heat transfer and depression of surface tension by surfactant additives," *J. Heat Transfer*, vol. 117, pp. 526-529, 1995.

[22] R. M. Manglik, V. M. Wasekar, and J. Zhang, "Dynamic and equilibrium surface tension of aqueous surfactant and polymeric solutions," *Exp. Therm. Fluid Sci.*, vol. 25, pp. 55-64, 2001.

[23] W. T. Wu, Y. M. Yang, and J. R. Maa, "Nucleate pool boiling enhancement by means of surfactant additives," *Exp. Therm. Fluid Sci.*, vol. 18, pp. 195-209, 1998.

[24] P. D. Jontz and J. E. Myers, "The effect of dynamic surface tension on nucleate boiling coefficient," *AIChE J.*, vol. 6, pp. 34-38, 1960.

[25] R. M. Manglik, V. M. Wasekar, and J. Zhang, "Dynamic and equilibrium surface tension of aqueous surfactant and polymeric solutions," *Exp. Therm. Fluid Sci.*, vol. 25, pp. 55-64, 2001.

[26] G. Hetsroni, J. L. Zakin, Z. Lin, A. Mosyak, E. A. Pancallo, and R. Rozenblit, "The effect of surfactants on bubble growth, wall thermal patterns and heat transfer in pool boiling," *Int. J. Heat Mass Transfer*, vol. 44, pp. 485-497, 2001.

[27] J. Zhang and R. M. Manglik, "Effect of ethoxylation and molecular weight of cationic surfactants on nucleate boiling in aqueous solutions," *J. Heat Transfer*, vol. 126, pp. 34-42, 2004.

[28] V. M. Wasekar and R. M. Manglik, "The influence of additive molecular weight and ionic nature on the pool boiling performance of aqueous surfactant solutions," *Int. J. Heat Mass Transfer*, vol. 45, pp. 483-493, 2002.

[29] A. T. A. Wang and J. P. Hartnett, "Influence of surfactants on pool boiling of aqueous polyacrylamide solutions," *Wärme und Stoffubertragung*, vol. 27, pp. 245-248, 1992.

[30] R. P. Chhabra and J. F. Richardson, *Non-Newtonian Flow in the Process Industries: Fundamentals and Engineering Applications*, Butterworth-Heinemann, Oxford, 1999.

[31] R. Y. T. Hu, A. T. A Wang, and J. P. Hartnett, "Surface tension measurement of aqueous polymer solutions," *Exp. Therm. Fluid Sci.*, vol. 4, pp. 723-729, 1991.

[32] F. Tokiwa and K. Ohki, "Micellar properties of a series of sodium dodecylpolyoxyethylene sulfates from hydrodynamic data," *J. Physical Chemistry*, vol. 71, pp. 1343-1348, 1967.

[33] H. Hoffmann and H. Rehage, "Rheology of surfactant solutions," In: R. Zana (Ed.), *Surfactant Solutions*, Marcel Dekker, New York, 1987, pp. 209-239.

[34] S. Hernandez-Acosta, "Capillary rheometry of micellar aqueous solutions," *J. Non-Newtonian Fluid Mech.*, vol. 85, pp. 229-247, 1999.

[35] G. Hetsroni, M. Gurevich, A. Mosyak, R. Rozenblit, and Z. Segal, "Boiling enhancement with environmentally acceptable surfactant," *Int. J. Heat Fluid Flow*, vol. 25, pp. 841-848, 2004

[36] P. Kotchaphakdee and M. C. Williams, "Enhancement of nucleate pool boiling with polymeric additives," *Int. J. Heat Mass Transfer*, vol. 13, pp. 835-848, 1970.

[37] J. Zhang and R. M. Manglik, "Nucleate pool boiling of aqueous polymer solutions on a cylindrical heater," *J. Non-Newtonian Fluid Mech.*, vol. 125, pp. 185-196, 2005.

[38] J. Zhang, *Experimental and Computational Study of Nucleate Pool Boiling Heat Transfer in Aqueous Surfactant and Polymer Solutions*, PhD thesis, University of Cincinnati, Cincinnati 2004.

[39] A. Maestro, C. Gonalez, and J. M. Guitierrez, "Shear thinning and thixotropy of HMHEC and HEC water solutions," *J. Rheol.*, vol. 46, pp. 1445-1457, 2002.

[40] J. W. Hoyt, "Drag reduction," in *Encyclopedia of Polymer Science and Engineering*, vol. 5, 1986.

[41] M. J. O. D. Zohourin and A. Pourjavadi, "Oil soluble drag-reducing polymer," *J. Polymer Mats.*, vol. 11, pp. 239-247, 1994.

[42] K. J. Mysels, *US Patent* 2 492 173, 1949.

[43] A. Gyr and H. W. Bewersdorff, *Drag Reduction of Turbulent Flows by Additives*, Kluwer Academic Publishers, 1995.

[44] A. V. Shenoy, "A review of drag reduction with special reference to micellar system," *Colloid & Polym. Sci.*, vol. 262, pp. 319-337, 1984.

[45] L. Cheng, G. Ribatski, and J. R. Thome, "Gas-liquid two-phase flow patterns and flow pattern maps: fundamentals and applications," *ASME Appl. Mech. Rev.*, vol. 61, 050802, 1-28, 2008.

[46] L. Cheng and D. Mewes, "Review of two-phase flow and flow boiling of mixtures in small and mini channels," *Int. J. Multiphase Flow*, vol. 32, pp. 183-207, 2006.

[47] G. F. Hewitt, Liquid-Gas Systems, in: Hetsroni, G., *Handbook of Multiphase Systems*, Hemisphere Publishing Corporation, Washington DC. 1982

[48] M. Dziubinski, H. Fidos, and M. Sosno, "The flow pattern map of a liquid-gas flow in the vertical pipe," *Int. J. Multiphase Flow*, vol. 30, pp. 551-563, 2004.

[49] R. P. Chhabra and J. F. Richardson, "Prediction of flow patterns for cocurrent flow of gas and non-Newtonian liquid in horizontal pipes," *Can. J. Chem. Eng.*, vol. 62 pp. 449–454, 2004.

[50] O. Baker, "Simultaneous flow of oil and gas," *Oil Gas J.*, vol. 53, pp. 185-190, 1954.

[51] Y. Taitel and A. E. Dukler, "A model for predicting flow regime transitions in horizontal and near horizontal gas-liquid flow," *AIChE J.*, vol. 22, pp. 47-55, 1976.

[52] G. F. Hewitt and D. N. Roberts, "Studies of two-phase flow patterns by simultaneous X-ray and flash photography," *Report AERE-M 2159*, 1969, Atomic Energy Research Establishment, Harwell.

[53] J. M. Mandhane, G. A. Gregory, and K. Aziz, "A flow pattern map for gas-liquid flow in horizontal pipes: predictive models," *Int. J. Multiphase Flow*, vol. 1, pp. 537-553, 1974.

[54] P. D. Manfield, C. J. Lawrence, and G. F. Hewitt, "Drag reduction with additives in multiphase flow: a literature survey," *Multiphase Sci. Technol.*, vol. 11, pp. 197-221, 1999.

[55] B. A. Jubran, Y. H. Zurigat, and M. F. A. Goosen, "Drag reducing agents in multiphase flow pipelines: recent trends and future needs," *Petroleum Sci. Technol.*, vol. 23, pp. 1403-1424, 2005.

[56] A. Al-Sarkhi, "Drag reduction with polymer in gas-liquid/liquid-.iquid flows in pipes: a literature review," *J. Natural Gas Sci. Eng.*, vol. 2, pp. 41-48, 2010.

[57] P. L. Spedding and N. P. Hand, "A review analysis of the effect of surfactants on two-phase phenomena in horizontal air-water pipe flow," *Dev. Chem. Eng. Process*, vol. 5, pp. 267-279, 1997.

[58] E. J. Greskovich and A. L. Shrier, "Drag reduction in two-phase flows," *Ind. Eng. Chem. Fundam.*, vol. 10, pp. 646-648, 1971.

[59] E. J. Greskovich and A. L. Shrier, "Pressure drop and holdup in horizontal slug flow," *AIChE J.*, vol. 17, pp. 1214-1219, 1971.

[60] R. G. Rosehart, D.S. Scott, and E. Rhodes, "Gas-liquid slug flow with drag-reducing polymer solutions," *AIChE J.*, vol. 18, pp. 744-750, 1972.

[61] N. D. Sylvester and J. P. Brill, "Drag reduction in two-phase annular mist flow of air and water," *AIChE J.*, vol. 22, pp. 615-617, 1976.

[62] N. D. Sylvester, R. H. Dowling, and J. P. Brill, "Drag reduction in co-current horizontal gas-hesane pipe flow," *Polym. Eng. Sci.*, vol. 20, pp. 485-492, 1980.

[63] D. Mowla, M. S. Hatamipour, and M. Moshfeghian, "The effect of dilute polymer solution on drag reduction in horizontal two-phase flow," *Int. J. Engr. Analsis Design*, vol. 2, pp. 97-105, 1995.

[64] C. Kang, R. M. Vancko, Jr., A. S. Green, H. Kerr, and W. P. Jepson, "Effect of drag-reducing agents in multiphase flow pipelines," *J. Energy Resour. Technol.*, vol. 120, pp. 15-19, 1998.

[65] C. Kang, W. P. Jepson, and M. Gopal, "Effect of drag-reducing agent on slug characteristics in multiphase flow in inclined pipelines," *J. Energy Resour. Technol.*, vol. 121, pp. 86-90, 1999.

[66] C. Kang and W. P. Jepson, "Effect of drag-reducing agents in multiphase oil/gas horizontal flow," *Society of Petroleum Engineers*, SPE 58976, 2000.

[67] L. Otten and A.S. Fayed, "Pressure drop and drag reduction in two-phase non-Newtonian slug flow," *Can. J. Chem. Eng.*, vol. 54, pp. 111-114, 1976.

[68] A. Al-Sarkhi and T. J. Hanratty, "Effect of drag-reducing polymers on annular gas-liquid flow in a horizontal pipe," *Int. J. Multiphase Flow*, vol.27, pp. 1151-1162, 2001.

[69] A. Al-Sarkhi and T. J. Hanratty, "Effect of pipe diameter on the performance of drag-reducing polymers in annular gas-liquid flows," *IChemE, Part A*, vol. 27, pp. 1151-1162, 2001.

[70] A. Al-Sarkhi and A. Soleimani, "Effect of drag-reducing polymer two-phase gas-liquid flows in a horizontal pipe," *Trans. IChemE, Part A*, pp. 1583-1588, 2004.

[71] A. Soleimani, A. Al-Sarkhi, and T. J. Hanratty, "Effect of drag-reducing polymers on pseudo-slugs-interfacial drag and transition to slug flow," *Int. J. Multiphase Flow*, vol. 28, pp. 1911-1927, 2002.

[72] R. L. J. Fernandes, B. M. Jutte, and M. G. Rodriguez, "Drag reduction in horizontal annular two-phase flow," *Int. J. Multiphase Flow*, vol. 30, pp. 1051-1069, 2004.

[73] D. Mowla and A, Naderi, "Experimental study of drag reduction by a polymeric additive in slug two-phase flow of crude oil and air in horizontal pipes," *Chem. Eng. Sci.*, vol. 61, pp. 1549-1554, 2006.

[74] D. Mowla and A. Naderi, "Experimental investigation of drag reduction in annular two-phase flow of oil and air," *Iranian J. Sci. & Techni., Trans. B, Engineering*, vol. 32, pp. 601-609, 2008.

[75] J. Hart, P. J. Hamersma, and J. M. H. Fortuin, "Correlations predicting frictional pressure drop and liquid holdup during horizontal gas–liquid pipe flow with a small liquid holdup," *Int. J. Multiphase Flow*, vol. 15, pp. 947–964, 1989.

[76] R. J. Wilkens, D.K. Thomas, and S. R. Glassmeyer, "Surfactant use for slug flow pattern suppression and new flow pattern types in horizontal pipe," *J. Fluids Eng.*, vol. 128, pp. 164-169, 2006.

[77] R. J. Wilkens and D. K. Thomas, "Multiphase drag reduction: Effect of eliminating slugs," *Int. J. Multiphase Flow*, vol. 22, pp. 134-146, 2007.

[78] T. Duangprasert, A. Sirivat, K. Siemanond, and J. O. Wilkes, "Vertical two-phase flow regimes and pressure gradients under the influence of SDS surfactant," *Exp. Therm. Fluid. Sci.*, vol. 32, pp. 808-817, 2008.

[79] J. S. Lioumbas, A. A. Mouza, and S. V. Paras, "Effect of surfactant additives on co-current gas-liquid downflow," *Chem. Eng. Sci.*, vol. 61, pp. 4605-4616, 2006.

[80] T. Al-Wahaibi, S. Smith, and P. Angeli, "Effect of drag-reducing polymers on horizontal oil water flows," J. *Petri. Sci. Eng.*, vol. 57, pp. 334-346, 2007.

[81] A. Al-Yaari, A. Soleimani, B. Abu-Sharkh, U. Al-Mubaiyedh, and A. Al-sarkhi, "Effect of drag reducing polymers on oil-water flow in a horizontal pipe," *Int. J. Multiphase Flow*, vol. 35, pp. 516-524, 2009.

[82] R. L. J. Fernandes, B. A. Fleck, T.R. Heidrick, L. Torres, and M. G. Rodriguez, "Experimental study of DRA for vertical two-phase annular flow," *J. Energy Resour. Technl.*, vol. 131, 023002-1-023002-5, 2009.

CHAPTER 7

Void Fraction and Flow Patterns of Two-Phase Flow in Upward and Downward Vertical and Horizontal Pipes

Afshin J. Ghajar* **and Clement C. Tang**

School of Mechanical and Aerospace Engineering, Oklahoma State University, Stillwater, Oklahoma, 74078, USA

Abstract: A comparison of the performance of 54 void fraction correlations based on unbiased experimental data set of 3385 data points. A comprehensive literature search was undertaken for the available void fraction correlations and experimental void fraction data for upward and downward vertical and horizontal two-phase flows. The performance of the correlations in correctly predicting the diverse data set was evaluated. Comparisons between the correlations were made and appropriate recommendations were drawn. The analysis showed that most of the correlations developed are very restricted in terms of handling a wide variety of data sets. Based on this analysis void fraction correlations with the best predictive capability are highlighted.

Keywords: Two-phase flow, void fraction, flow pattern, flow visualization, vertical and horizontal pipes.

1. INTRODUCTION

Practical applications of gas-liquid flow, of two different components or a single substance, are commonly encountered in the petroleum, nuclear and process industries. The two gas and liquid phases may exist in flow of different components (*e.g.*, air and water) and/or in the event of phase change due to evaporation and condensation of a single fluid. Void fraction (α) is one of the most important parameters in characterizing two-phase flow. Void fraction is the key physical parameter for determining other two-phase parameters, namely two-phase density, gas and liquid velocities. In addition, void fraction plays an important role in the modeling of two-phase pressure drop, flow pattern transition, and heat transfer.

For industrial applications where two-phase flow is involved, the task of sizing the equipment for gathering, pumping, transporting and storing such a two-phase mixture requires the formidable task of predicting the phase distribution in the system from given operating conditions. The ability to quantify void fraction is of considerable importance in systems involving two-phase flow. For example in nuclear reactor technology, Boiling Water Reactor (BWR) uses light water as neutron moderator and coolant, and void fraction is significant in estimating the reactivity of the nuclear reactor.

The seemingly benign issue of determining the phase distribution from input conditions in a given pipe turns out to be a formidable task due to the slippage between the gas and the liquid phases. Currently, there is a plethora of void fraction correlations available in the literature. The fact that there are numerous correlations available would not be a concern had it not been for the fact that most of the correlations have some form of restrictions attached to them. For instance, one of the most common restrictions to the correlations, flow pattern dependency, is sometimes a purely subjective judgment of the investigator especially for those points on or near flow pattern boundaries. Another pitfall is that many void fraction correlations have only been validated with experimental data that is limited to specific conditions, such as pipe orientation, flow pattern, and gas-liquid combination. As a result, engineers are faced with the daunting task of choosing the appropriate correlation among the plethora of correlations available.

Further advancement in the development of better and more accurate void fraction correlations would not be proper if the tedious process of scrutinizing available correlations is disregarded or circumvented.

***Address correspondence to Afshin J. Ghajar:** School of Mechanical and Aerospace Engineering, Oklahoma State University, Stillwater, Oklahoma, 74078, USA; Tel: 405-744-5900; Fax: 405-744-7873; E-mail: afshin.ghajar@okstate.edu,

Lixin Cheng and Dieter Mewes (Eds)

Comparison of the correlations with the measured results provides a stepping stone for further developments of more accurate and improved correlations. The purpose of this work is to identify void fraction correlations that could acceptably predict most of the experimental data collected for upward and downward vertical and horizontal two-phase flows. A comprehensive search has been made in the literature to collect available void fraction correlations as well as the measured void fraction data from various sources. Experimental void fraction data was also measured rather than be solely dependent upon the data from other sources. Based on the analysis, the best performing correlations for all three pipe orientations were selected and recommendations were drawn based on their strengths and weaknesses. Since the main focus of this study is not on the development of void fraction correlations but rather in identifying the best performing correlations, detail discussions on the physics that affect void fraction are not within the scope of this study. The physics behind most of the correlations cited in this manuscript are available in the referenced sources.

2. EXPERIMENTAL SETUP FOR VOID FRACTION MEASUREMENT AND FLOW VISUALIZATION

The experimental setup is equipped for measuring heat transfer, pressure drop, void fraction, and also conducting flow visualization in air-water flow for all major flow patterns and inclination angles from 0° (horizontal) to ±90° (upward and downward vertical). The capabilities of the new experimental setup allow an undertaking that combines the study of heat transfer, flow patterns, pressure drop, void fraction, and inclination effects. The schematic of the overall experimental setup for flow pattern visualization and void fraction for the present study is given in Fig. **1**. The details of the test section are illustrated in Fig. **2**. The flow visualization section is the central portion of the void fraction section and it is constructed from a polycarbonate tube with an inner diameter of 12.7 mm. Detail discussions on the design, construction and functionality of this experimental setup are documented by Cook [1].

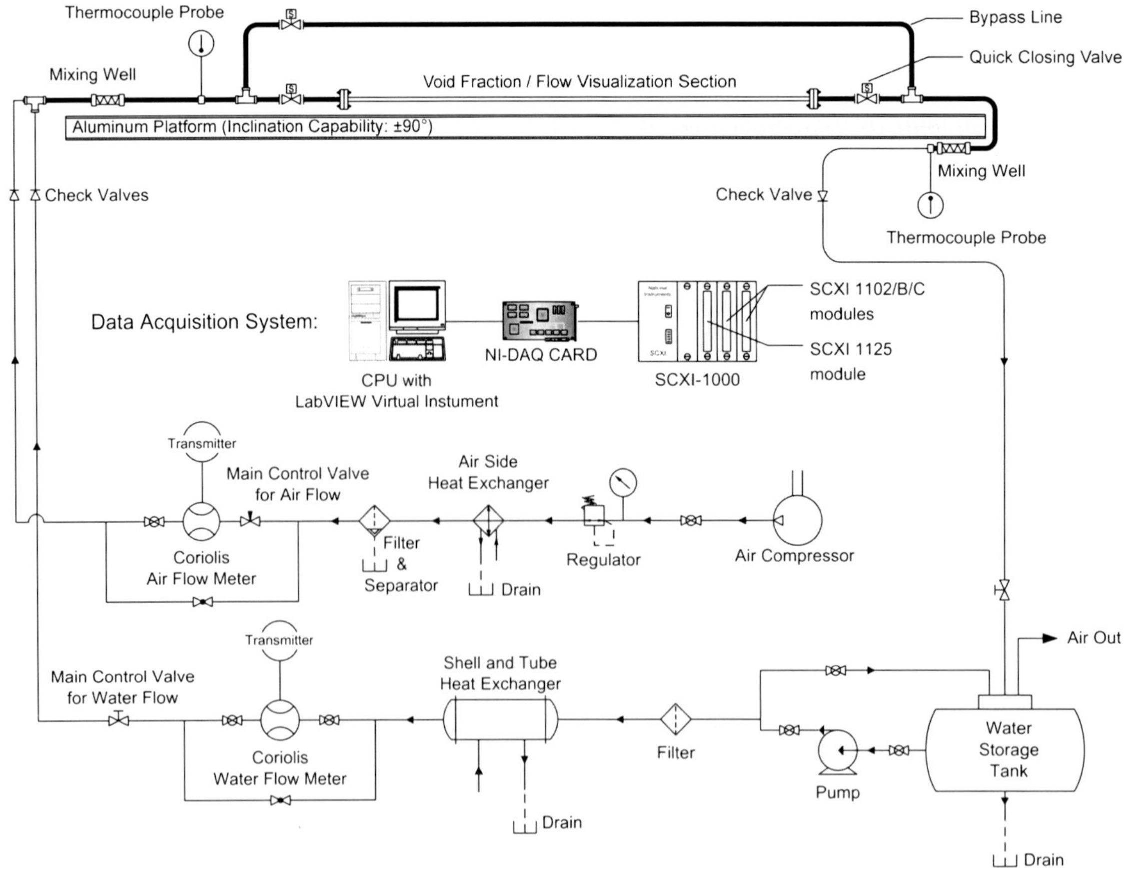

Figure 1: Schematic of experimental setup.

The fluids used in the test loop are air and water. The water is distilled and stored in a 55-gallon cylindrical polyethylene tank. A Bell & Gosset series 1535 coupled centrifugal pump was used to pump the water through an Aqua-Pure AP12T water filter. An ITT Standard model BCF 4063 one shell and two-tube pass heat exchanger removes the pump heat and the heat added during the test to maintain a constant inlet water temperature. From the heat exchanger, the water passes through an Emerson Micro Motion Coriolis flow meter (model CMF100) connected to a digital Field-Mount Transmitter (model RFT9739) that conditions the flow information for the data acquisition system. From the Coriolis flow meter it then flows into the test section.

Air is supplied *via* an Ingersoll-Rand T30 (model 2545) industrial air compressor. The air passes through a copper coil submerged in a vessel of water to lower the temperature of the air to room temperature. The air is then filtered and condensation removed in a coalescing filter. The air flow is measured by Emerson Micro Motion Coriolis flow meters (model CMF025 for high flow rates and model LMF3M for low flow rates). Both flow meters are connected to a Micro Motion 1700 transmitter. Air is regulated by a needle valve and is delivered to the test section by flexible tubing.

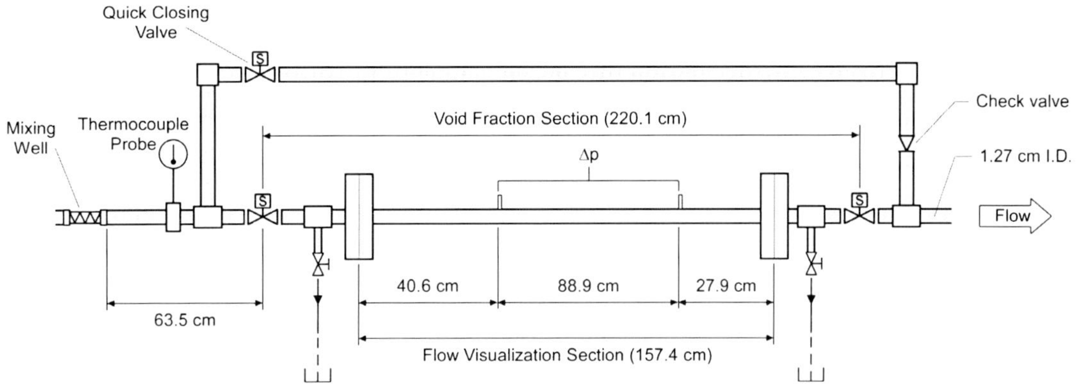

Figure 2: Test section for void fraction measurement and flow visualization.

The inlet liquid and gas temperatures and the exit bulk temperature were measured by Omega TMQSS-06U-6 thermocouple probes. Calibration of thermocouple probes showed that they were accurate within ±0.5°C. Two static mixers, one at the inlet and another at the outlet of the test section, are used to ensure that air and water are properly mixed such that accurate temperature of the mixture can be measured by the thermocouple probes. Upon exiting the test section, the water and air mixture is returned to the reservoir where it is separated and the water recycled.

The void fraction section is constructed to trap mixture of two-phase flow in order to measure the volume of the liquid portion. With the known volume of the void fraction section and the measured volume of the liquid portion, the value of the void fraction can be determined. To trap the two-phase mixture in the void fraction section, three quick closing valves are used. Two normally open valves are used for controlling fluid movement at the inlet and exit of the void fraction section, while a normally closed valve is for controlling the entry of fluid into a bypass line. The quick closing valves are W. E. Anderson Model ABV1DA101 pneumatic ball valves and they exhibit a positive seal when closed and have a closing time of 0.03 seconds.

When the valves are triggered, the two normally open valves close and the normally closed valve opens simultaneously. In this manner, a two-phase sample is trapped in the void fraction section while the air-water mixture is allowed to continue flowing through the bypass line. Air-water mixture trapped in the void fraction section was drained, and the volume of the liquid is measured. With the measured volume of the liquid (V_l), the void fraction (α) can be determined from the total volume of the test section (V_t) using $\alpha = 1 - V_l / V_t$. The uncertainty associated with the volume of the liquid is ±0.5 cm^3, and the uncertainty associated with the total volume of the test section is ±2 cm^3. Using the procedure prescribed by [2], the uncertainties associated with the measured void fraction results were estimated to be between ±1.25 and ±4.16%.

Flow patterns observations in upward and downward vertical and horizontal two-phase air-water flows were conducted as well. All observations for the flow pattern judgments were made at the flow visualization section (see Fig. **1**). By fixing the water flow rate, flow patterns were observed by varying air flow rates. Using visual observation and digital photography, distinctive flow patterns were observed.

3. EXPERIMENTAL FLOW PATTERN AND VOID FRACTION RESULTS

3.1. Flow Pattern Results

The main purpose of conducting experiment on flow patterns was to verify that the experimental setup has the capability of observing similar flow patterns as reported in the literature. By fixing the liquid flow rate, flow patterns were observed by varying gas flow rates. Using visual observation and digital photography, distinctive flow patterns were observed for upward and downward vertical and horizontal pipes. Digital images of each flow pattern were taken using a Nikon D50 digital camera with Nikkor 50 mm f/1.8D lens.

In horizontal pipe, the distinctive major flow patterns for gas-liquid two-phase flow are stratified, plug, slug, wavy and annular. Other flow patterns, such as slug/wavy, slug/bubbly, and wavy/annular, may be considered transitional flow patterns. The major flow patterns observed in this study for horizontal flow confirm the major flow patterns observed by Mandhane *et al.* [3] and Barnea *et al.* [4]. Flow patterns that are observed in horizontal gas-liquid two-phase flow are illustrated in Fig. **3**.

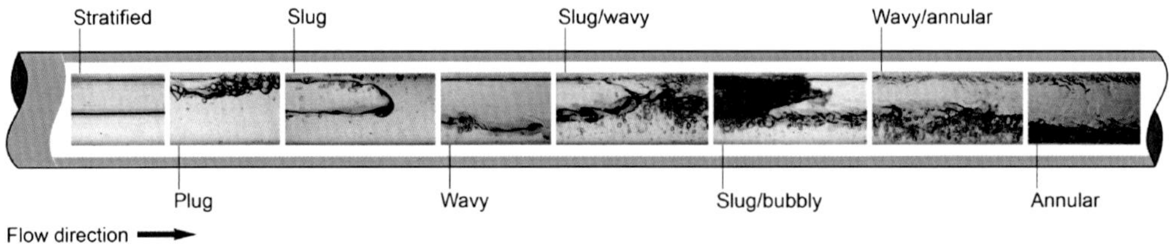

Figure 3: Flow patterns in horizontal two-phase air-water flow.

In upward vertical gas-liquid two-phase flow, the distinctive major flow patterns observed are bubble, slug, churn/froth, and annular. The major flow patterns observed in this study for upward vertical flow confirm the major flow patterns observed by Taitel *et al.* [5] and Weisman and Kang [6]. Flow patterns that are observed in upward vertical gas-liquid two-phase flow are illustrated in Fig. **4**.

In downward vertical gas-liquid two-phase flow, the distinctive major flow patterns observed are falling film, bubble, slug, froth, and annular. The major flow patterns observed in this study for downward vertical flow confirm the major flow patterns observed by Oshinowo and Charles [7]. Flow patterns that are observed in downward vertical gas-liquid two-phase flow are illustrated in Fig. **5**. For bubble flow at constant superficial gas velocity, increase in superficial liquid velocity causes the number of air bubble to increase while the bubbles decrease in size, thus allowing air bubbles to be more evenly distributed across the pipe cross-section. The observed bubble flow behavior at constant superficial gas velocity (U_{sg}) with varying superficial liquid velocity (U_{sl}) is illustrated in Fig. **6**.

The observed bubble flow behavior at constant superficial liquid velocity with varying superficial gas velocity is illustrated in Fig. **7**. For bubble flow at constant superficial liquid velocity, increase in superficial gas velocity causes the number of air bubble to increase as well as the sizes of the bubbles (see Fig. **7**). Further increase in superficial gas velocity would cause the coalescence of air bubbles that would eventually form slug flow.

3.2. Void Fraction Results

The void fraction experiments were conducted to meet several objectives. First, the measured void fraction results were to be used for validation of experimental data from other sources. Since the uncertainties associated with the measured void fractions from this study are determinable, they can be used to validate experimental data compiled from other sources. This validation is especially informative, since not all the sources reported detailed experimental uncertainties. Second, the measured void fraction results were to be used to fill in the gaps in regions that were not comprehensively measured by other sources. For example, in upward vertical flow, the experimental results from this study have provided 26% of data points to the data compiled from other sources for $0.25 \leq \alpha \leq 0.5$. Also, experimental results from this study for upward vertical flow have added 67 data points to the 58 data points compiled from other sources for churn/froth flow. Third, the measured void fraction results were to be used to evaluate variation of void fraction with superficial velocities of gas and liquid and groupings of various flow patterns for $0 < \alpha < 1$.

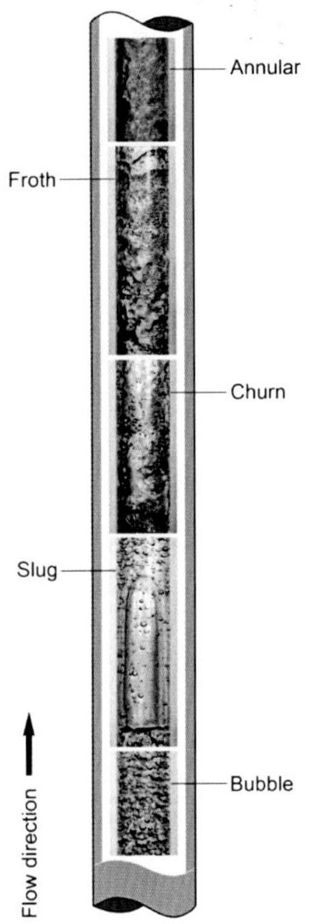

Figure 4: Flow patterns in upward vertical two-phase air-water flow.

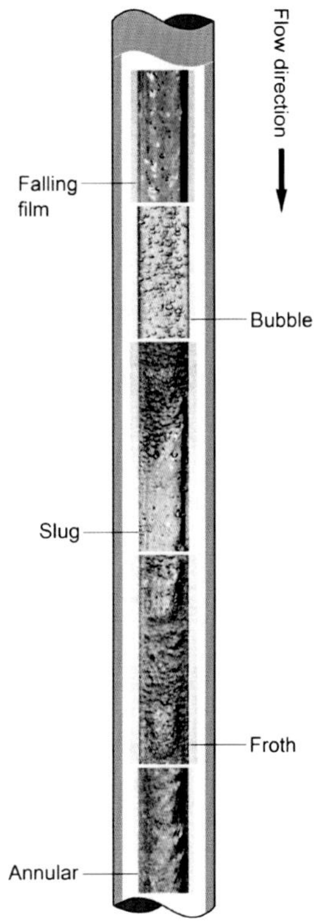

Figure 5: Flow patterns in downward vertical two-phase air-water flow.

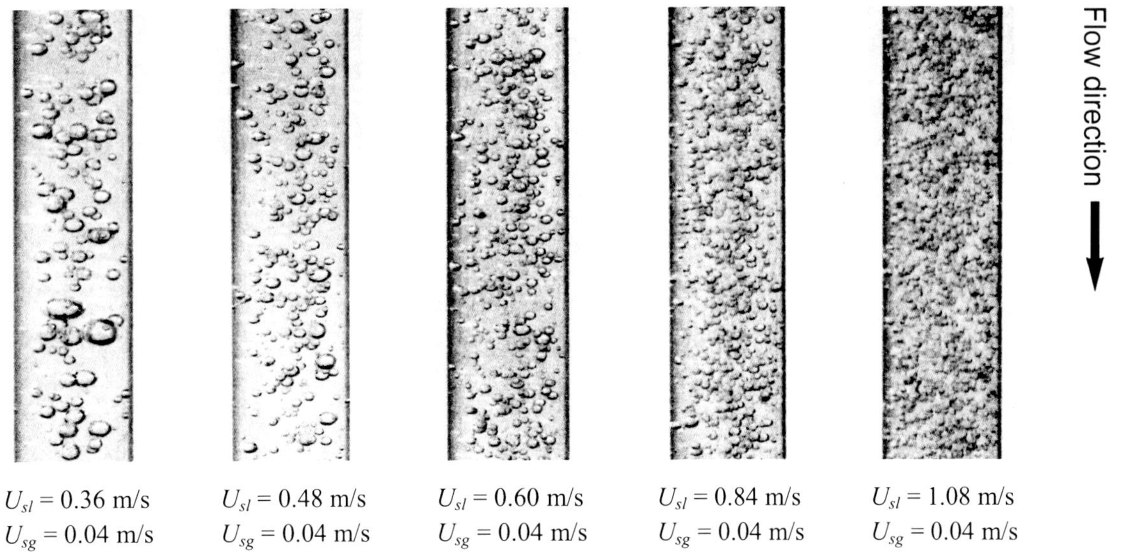

Figure 6: Downward vertical air-water bubble flow at U_{sg} = 0.04 m/s and varying U_{sl}.

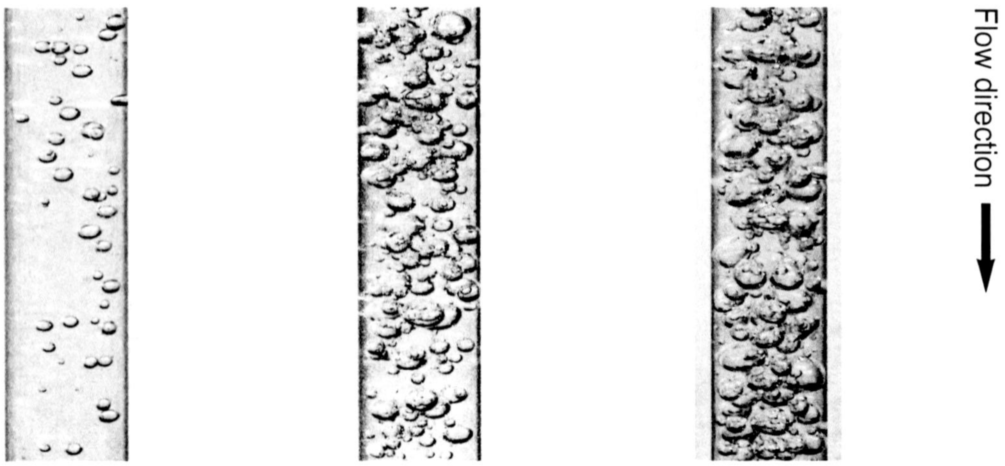

U_{sl} = 0.24 m/s U_{sg} = 0.008 m/s | U_{sl} = 0.24 m/s U_{sg} = 0.04 m/s | U_{sl} = 0.24 m/s U_{sg} = 0.08 m/s

Figure 7: Downward vertical air-water bubble flow at U_{sl} = 0.24 m/s and varying U_{sg}.

The experimental results of void fraction in upward and downward vertical and horizontal flows were measured from the test section for void fraction measurement and flow visualization, as illustrated in Fig. **2**. The variations of void fraction with superficial gas and liquid velocities for upward and downward vertical and horizontal flows were established from the experimental results (see Figs. **8-10**, respectively). As superficial liquid velocity increases, the increase in liquid holdup causes the void fraction versus superficial gas velocity curves to shift lower. At low superficial gas velocities ($U_{sg} < 2$ m/s) small increase in superficial gas velocity caused rapid increase in the void fraction. The rapid increase in the void fraction suggests difficulty in correlating void fraction accurately in this low superficial gas velocity region. For $U_{sg} > 2$ m/s, void fraction increases gradually with increasing superficial gas velocity.

The aforementioned characteristics of void fraction variation with superficial gas and liquid velocities are consistent for upward and downward vertical and horizontal flows, as shown in Figs. **8-10**, respectively. For upward vertical flow, the void fraction trends observed in the experimental results are similar to the observations of Jiang and Rezkallah [8], Mukherjee [9], Sujumnong [10], and Woods *et al.* [11]. In Figs. **8** and **9**, the groupings of various flow patterns on the variation of void fraction with superficial gas velocity curves are shown. For upward vertical flow, bubble and slug flows are confined to low-range superficial gas velocity with void fraction ranges of $0.05 < \alpha < 0.5$ and $0.2 < \alpha < 0.7$, respectively. Churn/froth flow is found in mid-range superficial gas velocity with void fraction range of $0.3 < \alpha < 0.8$. At a given superficial gas velocity, churn flow has higher void fraction than froth flow. Annular flow is in the high superficial gas velocity region with void fraction range of $0.7 < \alpha < 0.9$. For downward vertical flow, bubble and slug flows are confined to low-range superficial gas velocity with void fraction ranges of $0.02 < \alpha < 0.3$ and $0.2 < \alpha < 0.7$, respectively. Froth and falling film flows are found in mid-range superficial gas velocity with void fraction ranges of $0.4 < \alpha < 0.7$ and $0.7 < \alpha < 0.9$, respectively. Annular flow is in the high superficial gas velocity region with void fraction range of $0.7 < \alpha < 0.9$.

4. EXPERIMENTAL DATABASE

In an effort to make the evaluation of void fraction correlations unbiased, experimental data from various sources were compiled. Including the experimental data measured for present study, a total of 3385 experimental data points (1208 for upward vertical, 909 for downward vertical, and 1268 for horizontal) from 22 different sources were compiled. As summarized in Table **1**, the experimental data include void fraction measurements in pipe diameters ranging from 9.53 to 102 mm, and operating pressures ranging from 101 to 5990 kPa, with different gas-liquid combinations of upward and downward vertical and horizontal two-phase flows. Based on sources with reported experimental uncertainties, the experimental data have associated uncertainties of less than 10%.

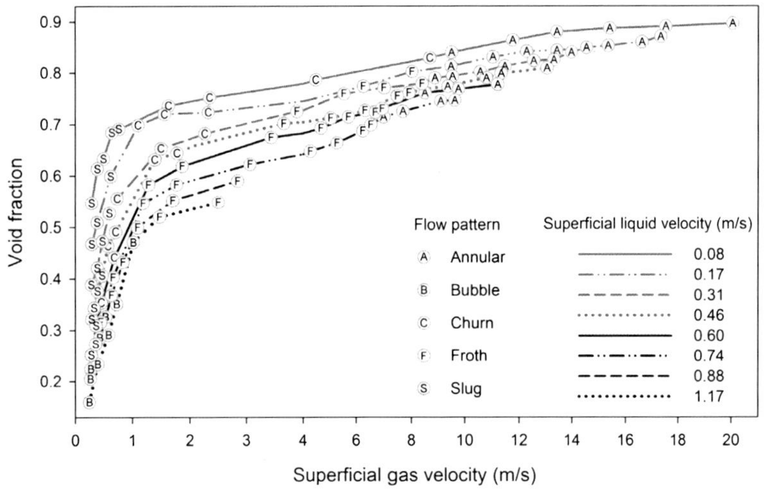

Figure 8: Variation of void fraction with superficial gas and liquid velocities for upward vertical flow.

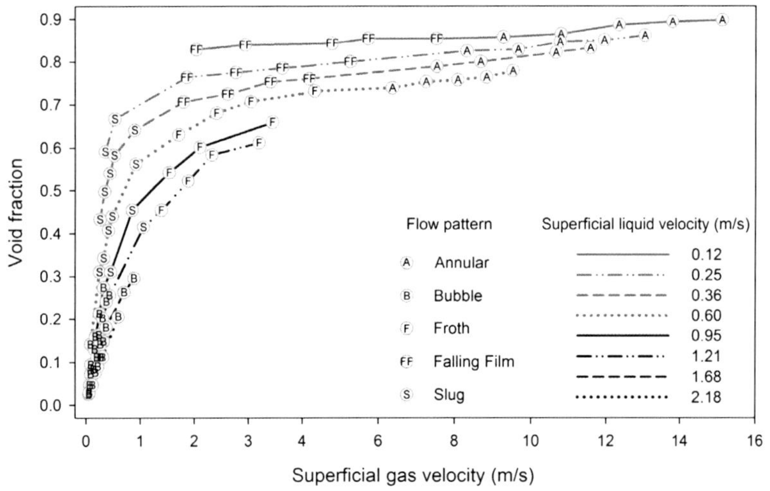

Figure 9: Variation of void fraction with superficial gas and liquid velocities for downward vertical flow.

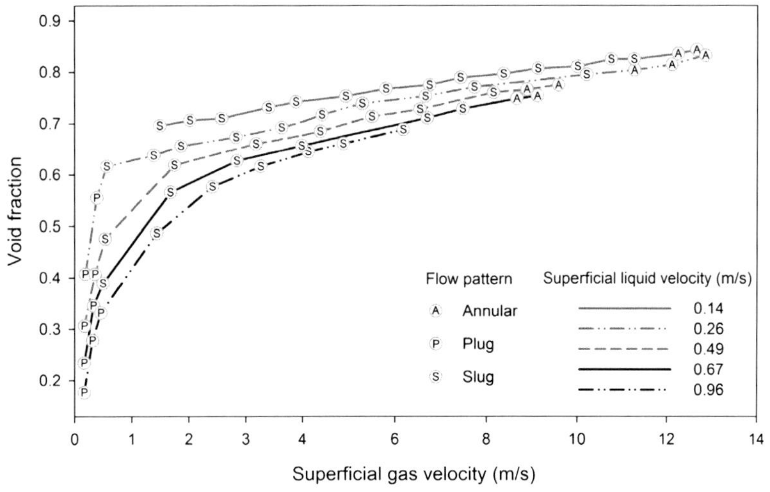

Figure 10: Variation of void fraction with superficial gas and liquid velocities for horizontal flow.

The compiled experimental data were measured at different experimental facilities with various measurement techniques: quick closing valve, gamma-ray absorption, capacitance sensor, manometer, neutron scattering, and conductance probe. The measurement technique of using manometer, also referred to as manometric method, to measure void fraction is only applicable for vertical flow and limited flow condition. In this measurement technique, pressure drop between pressure taps along the test section is measured. The measured pressure drop, where frictional and accelerational contributions are assumed negligible, is then used for determining the void fraction from $\Delta p = [\rho_g \alpha + \rho_l(1 - \alpha)] g H$. Thus, the void fraction determined from this method relies heavily on the assumption that frictional and accelerational contributions are negligible. The feasibility of this method is therefore primarily limited to vertical flow and flow condition in which the contribution of gravity significantly dominates over those of friction and acceleration. Miyahara *et al.* [12] indicated that void fraction results measured from the manometric method agree satisfactorily with results measured from conductance probe for void fraction below 0.5.

Experimental data measured for the present study were compared with experimental data from other sources at similar superficial velocities of gas and liquid for upward and downward vertical and horizontal flows. A total of 557 data points (210 for upward vertical, 252 for downward vertical, and 115 for horizontal) from various independent sources were used in this comparison. The experimental data from the various sources were measured at different experimental facilities for different pipe diameters (9.53 to 102 mm) and different gas-liquid flows.

For upward vertical flow, 210 experimental data points from other sources with similar superficial gas and liquid velocities were identified and compared with experimental data measured for the present study. The 210 experimental data points from eight different sources were measured for various pipe diameters (12.7 to 76 mm) and various operating pressure (101 to 3410 kPa). All 210 experimental data points are within ±10% agreement with the experimental data from present study. The result of this comparison is shown in Fig. **11**. The comparison of the experimental data from present study and other sources, for this flow orientation, has a Root-Mean-Square (RMS) error of 4.2%, and 77% of the 210 experimental data points are within ±5% agreement.

For downward vertical flow, 252 experimental data points from other sources with similar superficial gas and liquid velocities were identified and compared with experimental data measured for the present study. The 252 experimental data points from eight different sources were measured for various pipe diameters (9.53 to 90 mm) and various operating pressure (101 to 501 kPa). All 252 experimental data points are within ±10% agreement with the experimental data from present study (see Fig. **12**), and 67% of the 252 experimental data points are within ±5% agreement. The comparison of the experimental data from present study and other sources, for this flow orientation, has a Root-Mean-Square (RMS) error of 4.8%.

For horizontal flow, 115 experimental data points from other sources with similar superficial gas and liquid velocities were identified and compared with experimental data measured for the present study. The 115 experimental data points from six different sources were measured for various pipe diameters (19 to 102 mm) and various operating pressure (101 to 2850 kPa). All 115 experimental data points are within ±10% agreement with the experimental data from present study, and 72% of the 115 experimental data points are within ±7% agreement. The result of this comparison is shown in Fig. **13**. The comparison of the experimental data from present study and other sources, for this flow orientation, has a Root-Mean-Square (RMS) error of 6.2%.

The RMS errors found from the comparisons of experimental data from other sources and present study, for upward and downward vertical and horizontal flow, are between 4.2 to 6.2%. The RMS errors of these comparisons are comparable with the uncertainties associated with the experimental data of present study, which is within ±4.16%. The agreement between the experimental data from present study and other sources validated the goodness of the data compiled for this study.

Table 1: Experimental database used for comparison with results from present study and void fraction correlations.

Source	Pipe diameter (mm)	Fluids	No. of data points	Operating pressure (kPa)
Upward vertical flow (1208 data points)				
Present study [1]	12.7	Air-water	153	114–260
Beggs [13] [1]	25.4 & 38.1	Air-water	27	527–677
Chokshi [14] [2]	76	Air-water	103	1220–3410
Fernandes [15] [1]	50.7	Air-water	88	
Isbin *et al.* [16] [2]	22.1	Steam-water	22	101
Mukherjee [9] [3]	38.1	Air-kerosene	65	267–609
Oshinowo [17] [1]	25.4	Air-water	153	134–206
	25.4	Air-glycerin	172	141–199
Schmidt *et al.* [18] [2]	54.5	Nitrogen-water	20	
Nguyen [19] [1]	45.5	Air-water	224	107–125
Sujumnong [10] [1]	12.7	Air-glycerin	77	102–307
	12.7	Air-water	104	101–343
Downward vertical flow (909 data points)				
Present study [1]	12.7	Air-water	193	113–237
Beggs [13] [1]	25.4 & 38.1	Air-water	26	437–662
Hernandez *et al.* [20] [1]	50.8	Air-water	39	
Jiang and Rezkallah [8] [2]	9.53	Air-water	81	
Lorenzi and Sotgia [21] [4]	32, 44 & 90	Air-water	141	
Mukherjee [9] [3]	38.1	Air-kerosene	54	253–567
	38.1	Air-lube oil	48	282–456
Nguyen [19] [1]	45.5	Air-water	79	101–105
Oshinowo [17] [1]	25.4	Air-glycerin	78	145–203
	25.4	Air-water	112	134–205
Paras [22] [5]	19.1	Air-water	35	
Usui and Sato [23] [6]		Air-water	25	
Horizontal flow (1268 data points)				
Present study [1]	12.7	Air-water	184	110–242
Abdul-Majeed [24] [1]	50.8	Air-kerosene	88	197–919
Badie *et al.* [25] [2]	78	Air-oil*	30	
	78	Air-water	36	
Beggs [13] [1]	25.4 & 38.1	Air-water	58	358–680
Chen *et al.* [26] [3]	77.9	Air-kerosene	48	175–233
Eaton [27] [1]	54.5 & 102	Natural gas-water	238	2110–5990
Franca and Lahey [28] [1]	19	Air-water	88	101–116
Minami and Brill [29] [1]	77.9	Air-kerosene	57	301–667
	77.9	Air-water	54	320–589
Mukherjee [9] [3]	38.1	Air-kerosene	75	194–634
Nguyen [19] [1]	45.5	Air-water	270	101–115
Ottens [30]	51	Air-water	42	

[1] Quick closing valve, [2] Gamma-ray absorption, [3] Capacitance sensor, [4] Manometer, [5] Neutron scattering, [6] Conductance probe. * Shell Tellus 22 oil.

The results from the comparisons of experimental data from other sources and present study (see Figs. **11**, **12**, and **13**) also suggest, at least within the confines of these comparisons, that void fraction is not affected by pipe diameter (9.53 to 102 mm), operating pressure (101 to 3410 kPa), and type of gas-liquid flow. In addition, referring to the information listed in Table **1**, results shown in Figs. **11-13** indicate that experimental void fraction results measured *via* different techniques are generally in good agreement. This observation confirms with the findings of Jiang and Rezkallah [8] and Schmidt *et al.* [18], which had compared void fractions measured by gamma-ray absorption technique and quick closing valve technique.

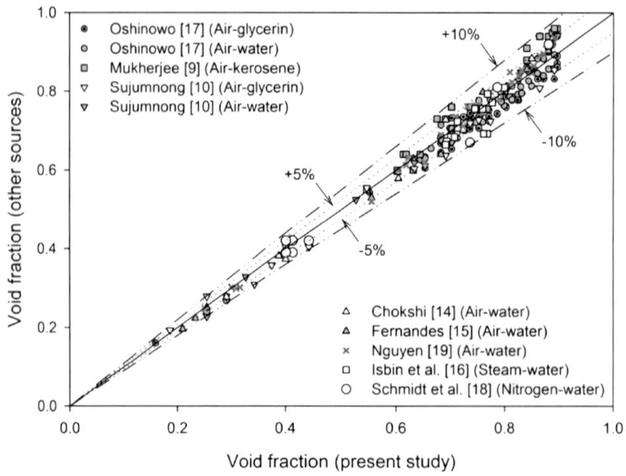

Figure 11: Comparison between experimental data from present study and other sources (see Table **1**) for upward vertical flow.

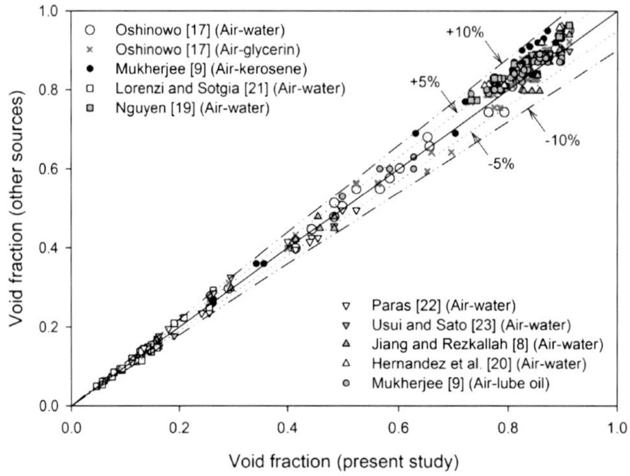

Figure 12: Comparison between experimental data from present study and other sources (see Table **1**) for downward vertical flow.

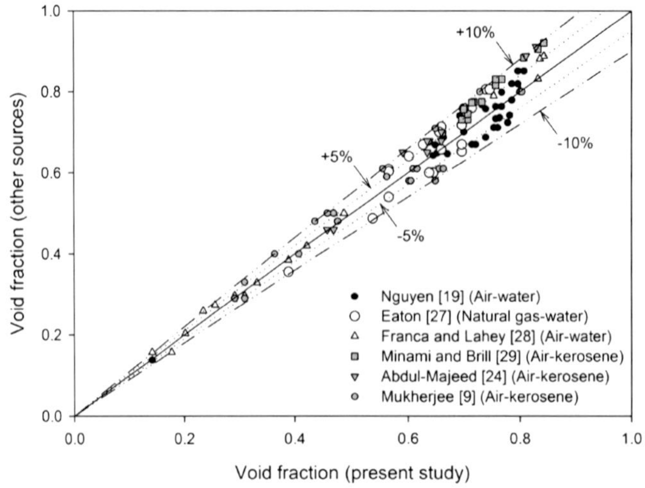

Figure 13: Comparison between experimental data from present study and other sources (see Table **1**) for horizontal flow.

5. EVALUATION OF VOID FRACTION CORRELATIONS WITH EXPERIMENTAL DATA

Fifty four void fraction correlations have been identified from the literature and considered in this analysis. These correlations were considered due to the reason that complete information on input parameters is available from the compiled experimental database; such that these correlations can be used to calculate void fractions and compared with measured results. Very few of the void fraction correlations identified for this study have been compared with experimental data that is as extensive as such that is summarized in Table **1**. Many of the correlations are "specialized" correlations and were originally proposed for specific conditions such as pipe orientation and flow pattern. Thus, their performances outside their specified conditions have not been scrutinized. In the interest of determining the "best" correlation for each pipe orientation (upward and downward vertical and horizontal), the correlations were subjected to two levels of assessment.

The first level is the evaluation of the correlations with experimental data on the basis of the overall performance, where the results predicted by the correlations for an entire database ($0 < \alpha < 1$) of each pipe orientation were compared with the measured results. The first level of assessment by itself is perfunctory, since it overlooks the strengths and weaknesses of the correlations in specific ranges of void fraction. The number of experimental data points is not uniformly distributed throughout the entire void fraction range of $0 < \alpha < 1$ (majority of the data points are in $0.75 < \alpha < 1$), hence relying solely on overall performances of the correlations could lead to biased interpretations. Subsequently, the second level is the analysis of the correlations in smaller void fraction ranges, by dividing the entire void fraction range into four categories: 0 to 0.25, 0.25 to 0.5, 0.5 to 0.75, and 0.75 to 1. Dividing the analysis into specific void fraction ranges would also reveal the most accurate correlations for each specific range; thus allowing the access to correlations with higher accuracies in specific void fraction range of interest.

Criteria for satisfactory performance were specified to aid the evaluation of the void fraction correlations with the experimental data (see Table **4**). The criteria summarized in Table **4** were not set arbitrarily, but rather based on how the correlations responded to the comparison with experimental data. For instance, in $0 < \alpha \leq 0.25$ the predictions by the correlations for upward vertical and horizontal flows have greater disparity than that of downward vertical flow. Hence, the criteria set for downward vertical flow are more restrictive than the criteria set for upward vertical and horizontal flows in $0 < \alpha \leq 0.25$ (see Table **4**). Similarly, the criteria set for $0.75 < \alpha < 1$ are more restrictive than the criteria set for $0 < \alpha \leq 0.25$ (see Table **4**), since the correlations performed with higher accuracies in $0.75 < \alpha < 1$ than in $0 < \alpha \leq 0.25$ for all three pipe orientations.

Out of the 54 void fraction correlations considered, 23 of them are mentioned in this article because they were found to compare satisfactorily with the compiled experimental results. Table **2** lists the 23 void fraction correlations to be discussed in this article and the reason in which these correlations were selected will be explained in this section. The sources for the other correlations that were analyzed and evaluated but not discussed here, because of unsatisfactory comparison with experimental results, are listed in Table **3**.

5.1. Performance of Correlations for the Entire Void Fraction Range

The 54 void fraction correlations considered in this analysis were compared with the data points in the experimental database summarized in Table **1** for upward and downward vertical and horizontal pipe orientations. Out of 54 correlations compared with the experimental data for $0 < \alpha < 1$, the correlations that compared satisfactorily with the experimental data and the comparison results are listed in Table **5**. The equations for the correlations listed in Table **5** are given in Table **2**.

For upward vertical flow, eight correlations were determined to be satisfactory (see Table **5**). For the number of data points predicted within the ±10% error band of the experimental data, the correlation of Rouhani and Axelsson [49] performed the best, with 856 (70.9%) data points predicted within the error band. A noticeable observation was made that the correlations of Rouhani and Axelsson [49] and Ishii [42] are the only correlations, among the 54 correlations considered in this study, that predicted at least 85, 75, and 65% of the data points within ±20, ±15, and ±10% error bands of the experimental data, respectively. Based on the evaluation thus far, the choice of which correlation to use lies with the user's interest of the accuracy in the correlation. If the interest of accuracy is between ±15 and ±20% error bands of the experimental data, then the correlation of Nicklin *et al.* [47] is preferable. If the interest is for higher accuracy with error bands of the experimental data between ±10 and ±15%, then the correlation of Rouhani and Axelsson [49] is preferable.

The similarity of the Nicklin *et al.* [47], Bonnecaze *et al.* [33], and Kokal and Stanislav [44] correlations is noteworthy. The correlations of Bonnecaze *et al.* [33] and Kokal and Stanislav [44] are slight variant of Nicklin *et al.* [47] correlation, hence all three correlations produced very similar results when compared with the experimental data (see Table **5**). The only variation between these three correlations is the relation for gas phase drift velocity (u_{gu}). Comparing the correlations of Nicklin *et al.* [47] and Bonnecaze *et al.* [33], there is an added term of $(1 - \rho_g / \rho_l)$ in the gas phase drift velocity (u_{gu}) relation for Bonnecaze *et al.* [33] correlation. In a similar manner, Kokal and Stanislav [44] correlation has an added term of $(1 - \rho_g / \rho_l)^{0.5}$ in the gas phase drift velocity (u_{gu}) relation of their correlation. Both additional terms in the gas phase drift velocity (u_{gu}) relations of Bonnecaze *et al.* [33] and Kokal and Stanislav [44] correlations do not impact the results significantly, since the density ratio (ρ_g / ρ_l) is very small for most gas-liquid combinations.

For downward vertical flow, seven correlations were determined to be satisfactory (see Table **5**). The correlation of Gomez *et al.* [37] predicted the most number of experimental data points within the ±20, ±15, and ±10% error bands as well as with the lowest RMS error. Following Gomez *et al.* [37], both correlations of Nicklin *et al.* [47] and Bonnecaze *et al.* [33] predicted 859 (94.5%) and 796 (87.6%) within ±20 and ±15%, respectively. In downward vertical flow, the gas phase drift velocity (u_{gu}) is in the downward direction, thus a negative sign applied to the u_{gu} for the correlations of [33, 44, 47, 49, 51], as indicated in Table **2**. Although in this first level of assessment, the Gomez *et al.* [37] correlation appears to be the best correlation for downward vertical flow, the second level of assessment is necessary to confirm its satisfactory performances in different void fraction ranges.

For horizontal flow, eight correlations were determined to be satisfactory (see Table **5**). The correlation of Woldesemayat and Ghajar [52] predicted the most number of experimental data points within the ±20, ±15, and ±10% error bands with RMS error of 16%. Following Woldesemayat and Ghajar [52], the Minami and Brill [29] correlation predicted 1173 (92.5%) and 1115 (87.9%) within ±20 and ±15%, respectively. Based on the evaluation thus far, the Woldesemayat and Ghajar [52] correlation is the best correlation, and further scrutiny in different void fraction ranges would evaluate its performances for these categories.

Table 2: List of void fraction correlations that performed satisfactorily and discussed in this study.

Source	Correlation	
Armand [31] – Massena [32]	$\alpha = (0.833 + 0.167x)\dfrac{U_{sg}}{U_{sg} + U_{sl}}$	(1)
Bonnecaze *et al.* [33]	$\alpha = \dfrac{U_{sg}}{C_0(U_{sg} + U_{sl}) + u_{gu}}$ where $C_0 = 1.2$	(2)
	$u_{gu} = 0.35\sqrt{gD}\,(1 - \rho_g / \rho_l)$	(2*a*)*
Chisholm [34]	$\alpha = \left[1 + \sqrt{1 - x(1 - \rho_l / \rho_g)}\,\dfrac{1-x}{x}\dfrac{\rho_g}{\rho_l}\right]^{-1}$	(3)
Dix [35]	$\alpha = \dfrac{U_{sg}}{C_0(U_{sg} + U_{sl}) + u_{gu}}$	(4)
	$C_0 = \dfrac{U_{sg}}{U_{sg} + U_{sl}}[1 + (U_{sl} / U_{sg})^b]$ where $b = (\rho_g / \rho_l)^{0.1}$	(4*a*)

Table 2: *cont….*

	$u_{gu} = 2.9\left(g\sigma\frac{\rho_l-\rho_g}{\rho_l^2}\right)^{0.25}$	(4*b*)
El-Boher *et al.* [36]	$\alpha = \left[1+0.27\left(\frac{x}{1-x}\frac{\rho_l}{\rho_g}\right)^{-0.69}(Fr_{sl})^{-0.177}\left(\frac{\mu_l}{\mu_g}\right)^{0.378}\left(\frac{Re_{sl}}{We_{sl}}\right)^{0.067}\right]^{-1}$	(5)
	where $Fr_{sl} = \frac{U_{sl}^2}{gD}$, $Re_{sl} = \frac{\rho_l U_{sl} D}{\mu_l}$, $We_{sl} = \frac{\rho_l U_{sl}^2 D}{\sigma}$	
Gomez *et al.* [37]	$\alpha = \frac{U_{sg}}{C_0(U_{sg}+U_{sl})+u_{gu}}$ where $C_0 = 1.15$	(6)
	$u_{gu} = 1.53\left(g\sigma\frac{\rho_l-\rho_g}{\rho_l^2}\right)^{0.25}(1-\alpha)^{0.5}\sin\theta$	(6*a*)
Greskovich and Cooper [38]	$\alpha = \frac{U_{sg}}{C_0(U_{sg}+U_{sl})+u_{gu}}$ where $C_0 = 1.0$	(7)
	$u_{gu} = 0.671\sqrt{gD}(\sin\theta)^{0.263}$	(7*a*)
Guzhov *et al.* [39]	$\alpha = 0.81\frac{U_{sg}}{U_{sg}+U_{sl}}\left(1-e^{-2.2\sqrt{Fr}}\right)$ where $Fr = \frac{(U_{sg}+U_{sl})^2}{gD}$	(8)
Hibiki and Ishii [40]	$\alpha = \frac{U_{sg}}{C_0(U_{sg}+U_{sl})+u_{gu}}$	(9)
	$C_0 = (1.2-0.2\sqrt{\rho_g/\rho_l})(1-e^{-18\alpha})$	(9*a*)
	$u_{gu} = \left(4g\sigma\frac{\rho_l-\rho_g}{\rho_l^2}\right)^{0.25}(1-\alpha)^{1.75}$	(9*b*)
Huq and Loth [41]	$\alpha = 1-\frac{2(1-x)^2}{1-2x+\sqrt{1+4x(1-x)(\rho_l/\rho_g-1)}}$	(10)
Ishii [42] [1]	$\alpha = \frac{U_{sg}}{C_0(U_{sg}+U_{sl})+u_{gu}}$	(11)
	$C_0 = \min(C_{01}, C_{02})$	(11*a*)
	$u_{gu} = \min(u_{gu1}, u_{gu2})$	(11*b*)
	See Ohkawa and Lahey [43] for C_{01}, C_{02}, u_{gu1}, and u_{gu2}	
Kokal and Stanislav [44]	$\alpha = \frac{U_{sg}}{C_0(U_{sg}+U_{sl})+u_{gu}}$ where $C_0 = 1.2$	(12)
	$u_{gu} = 0.345[gD(1-\rho_g/\rho_l)]^{0.5}$	(12*a*)*
Lockhart and Martinelli [45] [2]	$\alpha = \left[1+0.28\left(\frac{1-x}{x}\right)^{0.64}\left(\frac{\rho_g}{\rho_l}\right)^{0.36}\left(\frac{\mu_l}{\mu_g}\right)^{0.07}\right]^{-1}$	(13)

Table 2: *cont….*

Minami and Brill [29]	$\alpha = \exp\left[-\left(\frac{\ln Z + 9.21}{8.7115}\right)^{4.3374}\right]$	(14)
	Z is a combination of dimensionless numbers, see [29]	
Morooka *et al.* [46]	$\alpha = \frac{U_{sg}}{C_0(U_{sg} + U_{sl}) + u_{gu}}$ where $C_0 = 1.08$, $u_{gu} = 0.45$	(15)
Mukherjee [9]	$\alpha = 1 - \exp\left[(C_1 + C_2 \sin\theta + C_3 \sin^2\theta + C_4 N_{l\mu})\frac{N_{gu}^{C_5}}{N_{lu}^{C_6}}\right]$	(16)
	$N_{gu} = U_{sg}[\rho_l / (g\sigma_l)]^{1/4}$	(16*a*)
	$N_{lu} = U_{sl}[\rho_l / (g\sigma_l)]^{1/4}$	(16*b*)
	$N_{l\mu} = \mu_l(\rho_l \sigma_l^3)^{-1/4}$	(16*c*)
	$C_1, C_2, \ldots, C_6$ are constants given by Mukherjee [9]	
Nicklin *et al.* [47]	$\alpha = \frac{U_{sg}}{C_0(U_{sg} + U_{sl}) + u_{gu}}$ where $C_0 = 1.2$	(17)
	$u_{gu} = 0.35\sqrt{gD}$	(17*a*)*
Premoli *et al.* [48][3]	$\alpha = \left\{1 + \frac{\rho_g}{\rho_l}\left(\frac{1-x}{x}\right)\left[1 + K\left(\frac{Y}{1+CY} - CY\right)^{0.5}\right]\right\}^{-1}$	(18)
	$K = 1.578 Re^{-0.19}(\rho_l / \rho_g)^{0.22}$	(18*a*)
	$C = 0.0273 We Re^{-0.51}(\rho_l / \rho_g)^{-0.08}$	(18*b*)
	where $y = \left(\frac{1-x}{x}\frac{\rho_g}{\rho_l}\right)^{-1}$, $Re = \frac{GD}{\mu_l}$, $We = \frac{G^2 D}{\rho_l \sigma}$	
Rouhani and Axelsson [49]	$\alpha = \frac{U_{sg}}{C_0(U_{sg} + U_{sl}) + u_{gu}}$	(19)
	$C_0 = 1 + 0.2(1-x)(gD\rho_l^2 / G^2)^{0.25}$	(19*a*)[4]
	$C_0 = 1 + 0.2(1-x)$	(19*b*)[5]
	$u_{gu} = 1.18\left(g\sigma\frac{\rho_l - \rho_g}{\rho_l^2}\right)^{0.25}$	(19*c*)*
Smith [50]	$\alpha = \left\{1 + \frac{\rho_g}{\rho_l}\left(\frac{1-x}{x}\right)\left[0.4 + 0.6\left(\frac{\rho_l/\rho_g + 0.4(1/x - 1)}{1 + 0.4(1/x - 1)}\right)^{0.5}\right]\right\}^{-1}$	(20)
Sun *et al.* [51]	$\alpha = \frac{U_{sg}}{C_0(U_{sg} + U_{sl}) + u_{gu}}$	(21)
	$C_0 = [0.82 + 0.18(p_{sys} / p_{cr})]^{-1}$	(21*a*)

Table 2: *cont….*

	$$u_{gu} = 1.41\left(g\sigma\frac{\rho_l - \rho_g}{\rho_l^2}\right)^{0.25}$$	(21*b*)*
Woldesemayat and Ghajar [52]	$$\alpha = \frac{U_{sg}}{C_0(U_{sg} + U_{sl}) + u_{gu}}$$	(22)
	$$C_0 = \frac{U_{sg}}{U_{sg} + U_{sl}}\left[1 + \left(\frac{U_{sl}}{U_{sg}}\right)^{(\rho_g/\rho_l)^{0.1}}\right]$$	(22*a*)
	$$u_{gu} = 2.9(1.22 + 1.22\sin\theta)^{p_{atm}/p_{sys}}\left[\frac{gD\sigma(1+\cos\theta)(\rho_l - \rho_g)}{\rho_l^2}\right]^{0.25}$$ The leading constant of 2.9 in Eq. (22*b*) carries a unit of $m^{-0.25}$	(22*b*)
Yamazaki and Yamaguchi [53]	$$\frac{\alpha}{(1-\alpha)(1-K\alpha)} = \frac{\alpha_h}{1-\alpha_h} \text{ where } \alpha_h = \frac{U_{sg}}{U_{sg} + U_{sl}}$$	(23)
	$K = 2.0 - 0.4/\alpha_h$ for $\alpha_h \le 0.2$	(23*a*)
	$K = -0.25 + 1.25\alpha_h$ for $\alpha_h \ge 0.2$	(23*b*)

[1] As given by Ohkawa and Lahey [43], [2] As given by Butterworth [54], [3] As given by Woldesemayat and Ghajar [52], [4] As given by Rouhani [55], in this analysis, it is used for $0 < \alpha \le 0.25$, [5] As given by Woldesemayat and Ghajar [52], in this analysis it is used for $\alpha > 0.25$.
* For downward vertical flow, a negative sign is applied on these equations for u_{gu}.

Table 3. Sources for void fraction correlations that were evaluated but did not perform satisfactorily.

Source	Source	Source
Armand [31] – Chisholm [56] [1]	Bankoff [57] [1]	Baroczy [58] [1,2]
Beggs [13] [1]	Bestion [59] [1,3]	Czop *et al.* [60] [1]
Dimentiev *et al.* [61] [1,4]	Filimonov *et al.* [62]	Gardner [63] [1]
Hughmark [64] [1]	Inoue *et al.* [65] [1,3]	Jowitt *et al.* [66] [1,3]
Kowalczewski [1,5]	Kütüçüğlu [1,5]	Lahey and Moody [67] [6,7]
Madsen [68] [1]	Maier and Coddington [69] [1,3]	Mattar and Gregory [70] [1]
Moussali [1,5]	Neal and Bankoff [71] [1]	Ohkawa and Lahey [43] [7]
Orell and Rembrand [72] [7]	Shvarts *et al.* [73] [7]	Sonnenburg [74] [3,7]
Spedding and Chen [75] [1]	Spedding *et al.* [76] [1]	Sterman [77] [1]
Takeuchi *et al.* [78] [7]	Thom [79] [1,2]	Wilson *et al.* [80] [1]
Yeh and Hochreiter [81] [7]		

[1] Also given by Woldesemayat and Ghajar [52], [2] Also given by Butterworth [54], [3] Also given by Coddington and Macian [82], [4] Also given by Kataoka and Ishii [83], [5] Also given by Isbin and Biddle [84], [6] Also given by Ohkawa and Lahey [43], [7] Also given by Godbole [85].

5.2. Performance of Correlations for $0 < \alpha \le 0.25$

Due to lower values of void fraction, the percentage error in prediction with respect to measured data was expected to be more prominent than for the case with higher void fraction values. The high percentage error from the correlations in this region is in part a result of the rapid increase in void fraction at low superficial gas velocity (see Figs. **8-10**). The rapid increase in void fraction makes correlating void fraction accurately in this region difficult. Due to the relatively higher percentage error in prediction with respect to experimental data in this low range of $0 < \alpha \le 0.25$, larger error bands and RMS error were used for the criteria to determine a satisfactory correlation (see Table **4**).

For upward vertical flow, 199 experimental data points from the entire database were categorized in the void fraction range of $0 < \alpha \leq 0.25$. Six correlations that are listed in Table **5** performed satisfactorily in the $0 < \alpha \leq 0.25$ range (see Table **6**). Among the 54 correlations considered in this study, correlations of Greskovich and Cooper [38], Nicklin *et al.* [47], and Rouhani and Axelsson [49] are among the three best correlations in the $0 < \alpha \leq 0.25$ range. The results of the comparison of the three aforementioned correlations with experimental data are listed in Table **7**. Although the performance of Greskovich and Cooper [38] correlation is not listed in Table **5**, it is among the three best correlations in the void fraction range of $0 < \alpha \leq 0.25$ and it predicted the most number of experimental data points within ±10% error band. The equations for all three correlations are listed in Table **2**.

For downward vertical flow, 237 experimental data points from the entire database were categorized in the void fraction range of $0 < \alpha \leq 0.25$. Four correlations that are listed in Table **5** performed satisfactorily in the $0 < \alpha \leq 0.25$ range (see Table **6**). Among the 54 correlations considered in this study, correlations of Bonnecaze *et al.* [33], Gomez *et al.* [37], and Nicklin *et al.* [47] are among the three best correlations in the $0 < \alpha \leq 0.25$ range. The results of the comparison of the three aforementioned correlations with experimental data are listed in Table **8**. The equations for all three correlations are listed in Table **2**.

Table 4: Criteria for Determining Satisfactory Correlation.

Category	Criteria for satisfactory correlation in error band and RMS error
Upward vertical flow	
Entire range	at least 85 & 75% of data points predicted within ±20 & ±15%, RMS error ≤ 30%
$0 < \alpha \leq 0.25$	at least 80% of data points predicted within ±30%, RMS error ≤ 60%
$0.25 < \alpha \leq 0.5$	at least 80% of data points predicted within ±20%, RMS error ≤ 20%
$0.5 < \alpha \leq 0.75$	at least 80% of data points predicted within ±15%, RMS error ≤ 15%
$0.75 < \alpha < 1$	at least 80% of data points predicted within ±10%, RMS error ≤ 10%
Downward vertical flow	
Entire range	at least 90 & 80% of data points predicted within ±20 & ±15%, RMS error ≤ 15%
$0 < \alpha \leq 0.25$	at least 80% of data points predicted within ±20%, RMS error ≤ 15%
$0.25 < \alpha \leq 0.5$	at least 90 & 80% of data points predicted within ±20 & ±15%, RMS error ≤ 15%
$0.5 < \alpha \leq 0.75$	at least 90 & 80% of data points predicted within ±20 & ±15%, RMS error ≤ 15%
$0.75 < \alpha < 1$	at least 95 & 80% of data points predicted within ±15 & ±10%, RMS error ≤ 10%
Horizontal flow	
Entire range	at least 90 & 70% of data points predicted within ±20 & ±10%, RMS error ≤ 30%
$0 < \alpha \leq 0.25$	at least 60% of data points predicted within ±20%, RMS error ≤ 40%
$0.25 < \alpha \leq 0.5$	at least 70% of data points predicted within ±20%, RMS error ≤ 20%
$0.5 < \alpha \leq 0.75$	at least 90 & 80% of data points predicted within ±20 & ±15%, RMS error ≤ 15%
$0.75 < \alpha < 1$	at least 95 & 90% of data points predicted within ±15 & ±10%, RMS error ≤ 10%

For horizontal flow, there are only 40 experimental data points from the entire database were categorized in the void fraction range of $0 < \alpha \leq 0.25$. Only three correlations that are listed in Table **5** performed satisfactorily in the $0 < \alpha \leq 0.25$ range (see Table **6**). Among the 54 correlations considered in this study, correlations of Huq and Loth [41], Premoli *et al.* [48], and Rouhani and Axelsson [49] are among the three best correlations in the $0 < \alpha \leq 0.25$ range. Premoli *et al.* [48] correlation is among the three best correlations in the void fraction range of $0 < \alpha \leq 0.25$, but it is not among the correlations that performed satisfactorily for $0 < \alpha < 1$. The results of the comparison of the three aforementioned correlations with experimental data are listed in Table **9**. The equations for all three correlations are listed in Table **2**. The limited number of experimental data points in this void fraction range is noteworthy, and a more conclusive assessment can be obtained had there been more data points available for comparison.

5.3. Performance of Correlations for $0.25 < \alpha \leq 0.5$

For upward vertical flow, 190 experimental data points from the entire database were categorized in the void fraction range of $0.25 < \alpha \leq 0.5$. Five correlations that are listed in Table **5** performed satisfactorily in the $0.25 < \alpha \leq 0.5$ range (see Table **6**). Among the 54 correlations considered in this study, correlations of Bonnecaze *et al.* [33], Kokal and Stanislav [44], and Nicklin *et al.* [47] are among the three best correlations in the $0.25 < \alpha \leq 0.5$ range, and their performances for $0 < \alpha < 1$ are also listed in Table **5**. The equations for the three aforementioned correlations and the results of their comparison with experimental data are listed in Table **2** and Table **7**, respectively.

Table 5: Results of correlations that compared satisfactorily with the entire experimental database ($0 < \alpha < 1$).

	Percentage of data points predicted within			**RMS error**
Correlations	**±10%**	**±15%**	**±20%**	**(%)**
Upward vertical flow (1208 data points)				
Bonnecaze *et al.* [33]	62.4	84.4	91.7	23.9
Guzhov *et al.* [39]	55.0	77.6	88.7	24.6
Ishii [42]	66.7	80.5	87.3	28.0
Kokal and Stanislav [44]	62.3	84.4	91.6	24.0
Morooka *et al.* [46]	62.3	79.1	87.8	23.6
Nicklin *et al.* [47]	62.4	84.4	91.7	23.9
Rouhani and Axelsson [49]	70.9	86.8	91.1	25.2
Sun *et al.* [51]	58.4	78.1	91.1	23.9
Downward vertical flow (909 data points)				
Bonnecaze *et al.* [33]	68.0	87.6	94.5	10.3
Gomez *et al.* [37]	79.2	90.4	96.1	9.2
Kokal and Stanislav [44]	68.0	87.3	94.1	10.4
Nicklin *et al.* [47]	68.0	87.6	94.5	10.3
Rouhani and Axelsson [49]	76.3	86.8	91.0	11.2
Sun *et al.* [51]	65.6	84.0	91.7	11.2
Yamazaki and Yamaguchi [53]	70.5	82.2	89.0	13.7
Horizontal flow (1268 data points)				
Armand [31] – Massena [32]	78.7	86.4	90.5	13.6
Chisholm [34]	75.0	86.2	90.7	13.6
Huq and Loth [41]	71.8	83.2	89.9	13.3
Minami and Brill [29]	74.1	87.9	92.5	14.1
Mukherjee [9]	74.4	85.8	92.0	17.2
Rouhani and Axelsson [49]	78.2	87.2	91.6	12.8
Smith [50]	72.5	84.5	90.5	13.5
Woldesemayat and Ghajar [52]	82.5	89.1	93.0	16.0

$$\text{RMS error} = \sqrt{\frac{1}{N-1}\sum_{i=1}^{N}\left[\frac{(\alpha_{\text{calc}})_i - (\alpha_{\text{meas}})_i}{(\alpha_{\text{meas}})_i}\right]^2} \times 100\%$$

where N is the number of experimental data points

For downward vertical flow, 105 experimental data points from the entire database were categorized in the void fraction range of $0.25 < \alpha \leq 0.5$. All seven correlations that are listed in Table **5** performed satisfactorily in the $0.25 < \alpha \leq 0.5$ range (see Table **6**). Among the 54 correlations considered in this study, correlations of Bonnecaze *et al.* [33], Nicklin *et al.* [47], and Rouhani and Axelsson [49] are among the three best correlations in the $0.25 < \alpha \leq 0.5$ range, and their performances for $0 < \alpha < 1$ are also listed in Table **5**. The equations for the three aforementioned correlations and the results of their comparison with experimental data are listed in Table **2** and Table **8**, respectively.

For horizontal flow, 145 experimental data points from the entire database were categorized in the void fraction range of $0.25 < \alpha \leq 0.5$. Only three correlations that are listed in Table **5** performed satisfactorily in the $0.25 < \alpha \leq 0.5$ range (see Table **6**). Among the 54 correlations considered in this study, correlations of Minami and Brill [29], Mukherjee [9], and Woldesemayat and Ghajar [52] are among the three best correlations in the $0.25 < \alpha \leq 0.5$ range, and their performances for $0 < \alpha < 1$ are also listed in Table **5**. The equations for the three aforementioned correlations and the results of their comparison with experimental data are listed in Table **2** and Table **9**, respectively.

Table 6: Qualitative performance of satisfactory correlations (see Table **5**) in four void fraction ranges.

	Void fraction range			
Correlations	**0.00 to 0.25**	**0.25 to 0.50**	**0.50 to 0.75**	**0.75 to 1.00**
Upward vertical flow (1208 data points)				
Bonnecaze *et al.* [33]	S	S	S	NS
Guzhov *et al.* [39]	NS	NS	S	NS
Ishii [42]	NS	NS	S	S
Kokal and Stanislav [44]	S	S	S	NS
Morooka *et al.* [46]	S	NS	NS	S
Nicklin *et al.* [47]	S	S	S	NS
Rouhani and Axelsson [49]	S	S	S	S
Sun *et al.* [51]	S	S	S	NS
Downward vertical flow (909 data points)				
Bonnecaze *et al.* [33]	S	S	S	NS
Gomez *et al.* [37]	S	S	S	S
Kokal and Stanislav [44]	S	S	S	NS
Nicklin *et al.* [47]	S	S	S	NS
Rouhani and Axelsson [49]	NS	S	S	S
Sun *et al.* [51]	NS	S	S	NS
Yamazaki and Yamaguchi [53]	NS	S	S	S
Horizontal flow (1268 data points)				
Armand [31] – Massena [32]	S	NS	NS	S
Chisholm [34]	NS	NS	NS	S
Huq and Loth [41]	S	NS	NS	S
Minami and Brill [29]	NS	S	NS	NS
Mukherjee [9]	NS	S	NS	NS
Rouhani and Axelsson [49]	S	NS	S	S
Smith [50]	NS	NS	NS	S
Woldesemayat and Ghajar [52]	NS	S	S	S

NS = not satisfactory, S = satisfactory

5.4. Performance of Correlations for $0.5 < \alpha \leq 0.75$

For upward vertical flow, 351 experimental data points from the entire database were categorized in the void fraction range of $0.5 < \alpha \leq 0.75$. Seven correlations that are listed in Table **5** performed satisfactorily in the $0.5 < \alpha \leq 0.75$ range (see Table **6**). Out of the 54 correlations considered in this study, correlations of Bonnecaze *et al.* [33], Kokal and Stanislav [44], and Nicklin *et al.* [47] are among the three best correlations in the $0.5 < \alpha \leq 0.75$ range, and their performances for $0 < \alpha < 1$ are also listed in Table **5**. The equations for the three aforementioned correlations and the results of their comparison with experimental data are listed in Table **2** and Table **7**, respectively.

For downward vertical flow, 126 experimental data points from the entire database were categorized in the void fraction range of $0.5 < \alpha \leq 0.75$. All seven correlations that are listed in Table **5** performed satisfactorily in the $0.5 < \alpha \leq 0.75$ range (see Table **6**). Among the 54 correlations considered in this study, correlations of Nicklin *et al.* [47], Rouhani and Axelsson [49], and Woldesemayat and Ghajar [52] are among the three best correlations in the $0.5 < \alpha \leq 0.75$ range. Woldesemayat and Ghajar [52] correlation is the best correlations in the void fraction range of $0.5 < \alpha \leq 0.75$, but it is not among the correlations that performed satisfactorily for $0 < \alpha < 1$. The equations for the three aforementioned correlations and the results of their comparison with experimental data are listed in Table **2** and Table **8**, respectively.

Table 7: Results of three best correlations from comparison with experimental data in specific void fraction ranges for upward vertical flow.

Correlations	Percentage of data points predicted within ±10%	±15%	±20%	RMS error (%)
0 < α ≤ 0.25 (199 data points)				
Greskovich and Cooper [38]	45.2	57.8	67.8	55.1
Nicklin *et al.* [47]	37.2	54.3	70.9	53.4
Rouhani and Axelsson [49]	40.2	62.8	71.9	56.5
0.25 < α ≤ 0.5 (190 data points)				
Bonnecaze *et al.* [33]	53.2	80.5	85.3	16.9
Kokal and Stanislav [44]	53.2	81.6	84.7	16.9
Nicklin *et al.* [47]	53.2	80.5	85.3	16.8
0.5 < α ≤ 0.75 (351 data points)				
Bonnecaze *et al.* [33]	80.3	93.4	96.6	9.7
Kokal and Stanislav [44]	80.1	93.2	96.6	9.7
Nicklin *et al.* [47]	80.3	93.4	96.6	9.7
0.75 < α < 1 (468 data points)				
Armand [31] – Massena [32]	94.7	99.4	100	5.3
Dix [35]	93.8	98.7	99.8	5.2
Rouhani and Axelsson [49]	94.0	99.4	99.6	5.6

Table 8: Results of three best correlations from comparison with experimental data in specific void fraction ranges for downward vertical flow.

Correlations	Percentage of data points predicted within ±10%	±15%	±20%	RMS error (%)
0 < α ≤ 0.25 (237 data points)				
Bonnecaze *et al.* [33]	48.5	73.4	88.2	13.5
Gomez *et al.* [37]	62.9	79.3	93.7	11.4
Nicklin *et al.* [47]	48.5	73.4	88.2	13.5
0.25 < α ≤ 0.5 (105 data points)				
Bonnecaze *et al.* [33]	75.2	91.4	96.2	10.2
Nicklin *et al.* [47]	75.2	91.4	96.2	10.2
Rouhani and Axelsson [49]	77.1	91.4	94.3	9.9
0.5 < α ≤ 0.75 (126 data points)				
Nicklin *et al.* [47]	77.0	85.7	92.1	9.3
Rouhani and Axelsson [49]	75.4	86.5	95.2	9.7
Woldesemayat and Ghajar [52]	80.2	92.1	97.6	8.2
0.75 < α < 1 (441 data points)				
Gomez *et al.* [37]	89.3	97.7	99.3	7.0
Rouhani and Axelsson [49]	91.2	97.5	98.2	6.8
Woldesemayat and Ghajar [52]	91.4	95.2	97.3	7.4

For horizontal flow, 289 experimental data points from the entire database were categorized in the void fraction range of $0.5 < \alpha \leq 0.75$. Only two correlations that are listed in Table **5** performed satisfactorily in the $0.5 < \alpha \leq 0.75$ range (see Table **6**). Among the 54 correlations considered in this study, correlations of Rouhani and Axelsson [49], Sun *et al.* [51], and Woldesemayat and Ghajar [52] are among the three best correlations in the $0.5 < \alpha \leq 0.75$ range. Correlation of Sun *et al.* [51] is not among the correlations that performed satisfactorily for $0 < \alpha < 1$, but it is among the three best correlations in the void fraction range of $0.5 < \alpha \leq 0.75$. The equations for the three aforementioned correlations and the results of their comparison with experimental data are listed in Table **2** and Table **9**, respectively.

5.5. Performance of Correlations for $0.75 < \alpha < 1$

For upward vertical flow, 468 experimental data points from the entire database were categorized in the void fraction range of $0.75 < \alpha < 1$. Out of the eight correlations listed in Table **5**, only three correlations performed satisfactorily in the $0.75 < \alpha < 1$ range (see Table **6**). Out of all the correlations considered in this study, correlations of Armand [31] – Massena [32], Dix [35], and Rouhani and Axelsson [49] are among the three best correlations in the $0.75 < \alpha < 1$ range. Correlations of Armand [31] – Massena [32] and Dix [35] are among the three best correlations in the void fraction range of $0.75 < \alpha < 1$, but they are not among the correlations that performed satisfactorily for $0 < \alpha < 1$. The equations for the three aforementioned correlations and the results of their comparison with experimental data are listed in Table **2** and Table **7**, respectively.

For downward vertical flow, 441 experimental data points from the entire database were categorized in the void fraction range of $0.75 < \alpha < 1$. Three correlations that are listed in Table **5** performed satisfactorily in the $0.75 < \alpha < 1$ range (see Table **6**). Among the 54 correlations considered in this study, correlations of Gomez *et al.* [37], Rouhani and Axelsson [49], and Woldesemayat and Ghajar [52] are among the three best correlations in the $0.75 < \alpha < 1$ range. Woldesemayat and Ghajar [52] correlation is not among the correlations that performed satisfactorily for $0 < \alpha < 1$, but it is among the three best correlations in the void fraction range of $0.75 < \alpha < 1$. The equations for the three aforementioned correlations and the results of their comparison with experimental data are listed in Table **2** and Table **8**, respectively.

Table 9: Results of three best correlations from comparison with experimental data in specific void fraction ranges for horizontal flow.

	Percentage of data points predicted within			RMS error
Correlations	**±10%**	**±15%**	**±20%**	**(%)**
$0 < \alpha \leq 0.25$ (40 data points)				
Huq and Loth [41]	27.5	47.5	65.0	33.9
Premoli *et al.* [48]	35.0	50.0	62.5	37.4
Rouhani and Axelsson [49]	32.5	50.0	62.5	34.2
$0.25 < \alpha \leq 0.5$ (145 data points)				
Minami and Brill [29]	43.4	61.4	75.9	17.6
Mukherjee [9]	54.4	65.5	82.1	14.6
Woldesemayat and Ghajar [52]	47.6	67.6	80.7	16.2
$0.5 < \alpha \leq 0.75$ (289 data points)				
Rouhani and Axelsson [49]	70.6	84.8	91.3	11.7
Sun *et al.* [51]	69.2	87.2	90.3	12.3
Woldesemayat and Ghajar [52]	73.7	86.2	92.7	12.0
$0.75 < \alpha < 1$ (794 data points)				
Armand [31] – Massena [32]	97.0	99.9	100	4.9
Lockhart and Martinelli [45]	95.0	98.6	99.6	4.6
Premoli *et al.* [48]	97.2	98.9	99.1	4.1

For horizontal flow, 794 experimental data points from the entire database were categorized in the void fraction range of $0.75 < \alpha < 1$. Six correlations that are listed in Table **5** performed satisfactorily in the $0.75 < \alpha < 1$

range (see Table **6**). Among the 54 correlations considered in this study, correlations of Armand [31] – Massena [32], Lockhart and Martinelli [45], and Premoli *et al.* [48] are among the three best correlations in the $0.75 < \alpha < 1$ range. Correlations of Lockhart and Martinelli [45] and Premoli *et al.* [48] are among the three best correlations in the void fraction range of $0.75 < \alpha < 1$, but they are not among the correlations that performed satisfactorily for $0 < \alpha < 1$. The equations for the three aforementioned correlations and the results of their comparison with experimental data are listed in Table **2** and Table **9**, respectively.

5.6. Summary of Correlations Performances

The correlations listed in Table **5** were selected on the basis of overall performance, which overlooks the strengths and weaknesses in specific ranges of void fraction. For upward vertical flow, this overall comparison with the experimental data yielded 2 correlations with best prediction of experimental data, namely correlations of Nicklin *et al.* [47] and Rouhani and Axelsson [49]. Also, the overall comparisons showed that correlations of Gomez *et al.* [37] and Woldesemayat and Ghajar [52] have the best predictions for downward vertical and horizontal flows, respectively. The subsequent logical approach was to analyze the selected correlations in ranges, by dividing the entire void fraction range into four ranges: 0 to 0.25, 0.25 to 0.5, 0.5 to 0.75, and 0.75 to 1. The qualitative outcomes of the correlations listed in Table **5** and their performances, whether satisfactory or not satisfactory, were summarized in Table **6**.

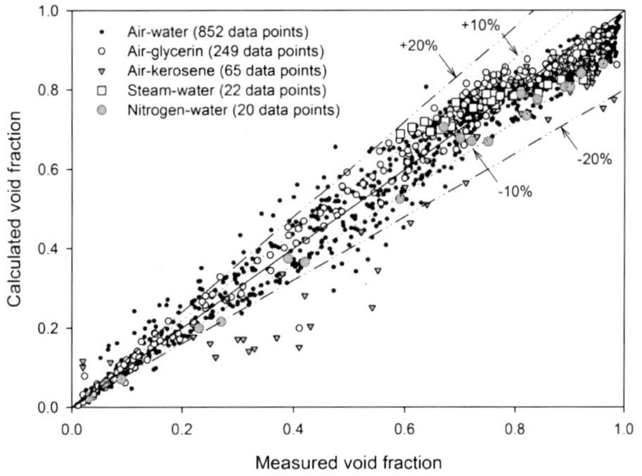

Figure 14: Comparison of the Rouhani and Axelsson [49] correlation with 1208 experimental data points for upward vertical flow.

For upward vertical flow, the comparison of the void fraction correlations with experimental data in each of the four specific ranges confirmed that the correlation of Rouhani and Axelsson [49] is the best correlation. This finding is based on the correlation's performance as one of the best for $0 < \alpha < 1$ (see Table **5**) and also the only correlation that is satisfactory for each of the four void fraction ranges (see Table 6). It is worthwhile to note that the Nicklin *et al.* [47] correlation performed best for predicting void fraction within $0.25 \leq \alpha \leq 0.75$. The correlation of Nicklin *et al.* [47] performed unsatisfactorily in $0.75 < \alpha < 1$, and this is because the correlation uses a constant distribution parameter ($C_0 = 1.2$). It is worth noting that the Nicklin *et al.* [47] correlation was originally developed for slug flow, and this analysis showed that it is the best performing correlation for $0.25 \leq \alpha \leq 0.75$, which is the void fraction range for slug and churn/froth flows (see Fig. **8**). Comparison of the correlation of Rouhani and Axelsson [49] with the experimental data is shown in Fig. **14**. The results in Fig. **14** show that the Rouhani and Axelsson [49] correlation predicted the experimental data generally well for different pipe diameters (12.7 to 76 mm), operating pressures (101 to 3413 kPa), and gas-liquid combinations.

For downward vertical flow, the comparison of the void fraction correlations with experimental data in each of the four specific ranges confirmed that the Gomez *et al.* [37] correlation is the best correlation. The correlation of Gomez *et al.* [37] has the least disparity when compared with experimental data for $0 < \alpha < 1$ (see Table **5**), and it is also the only correlation that performed satisfactorily in each four void fraction ranges (see Table **6**). The comparison of the Gomez *et al.* [37] correlation with the experimental data is shown in Fig. **15**. The results in Fig. **15** show that the Gomez *et al.* [37] correlation predicted the experimental data generally well for different pipe diameters (9.53 to 90 mm), operating pressures (101 to 501 kPa), and gas-liquid combinations.

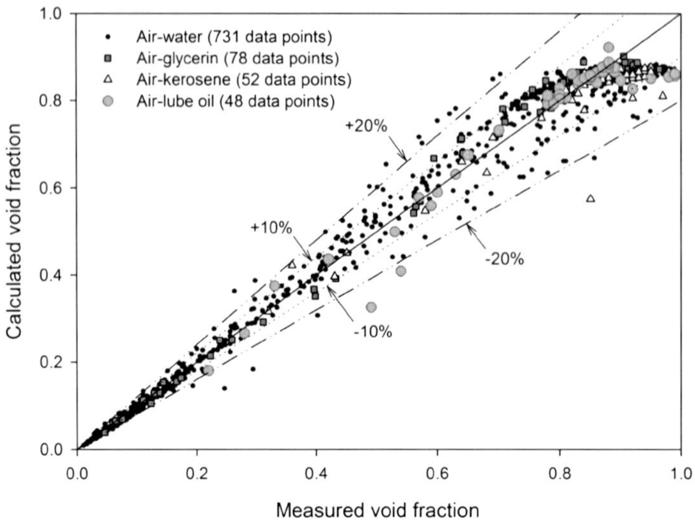

Figure 15: Comparison of the Gomez *et al.* [37] correlation with 909 experimental data points for downward vertical flow.

For horizontal flow, the correlation of Woldesemayat and Ghajar [52] has the best predictions for $0 < \alpha < 1$. The comparison of the void fraction correlations with experimental data in each of the four specific ranges showed that the Woldesemayat and Ghajar [52] correlation is the best correlation. It is the only correlation that performed satisfactorily for three different ranges of void fraction (see Table **6**), and it performed unsatisfactorily for $0 < \alpha \leq 0.25$. The number of experimental data points available for $0 < \alpha \leq 0.25$ is very limited, and a more conclusive assessment can be obtained if more number of data points are available for comparison. The comparison of the Woldesemayat and Ghajar [52] correlation with the experimental data is shown in Fig. **16**. The results in Fig. **16** show that the Woldesemayat and Ghajar [52] correlation performed satisfactorily for different pipe diameters (19 to 102 mm), operating pressures (101 to 2850 kPa), and gas-liquid combinations.

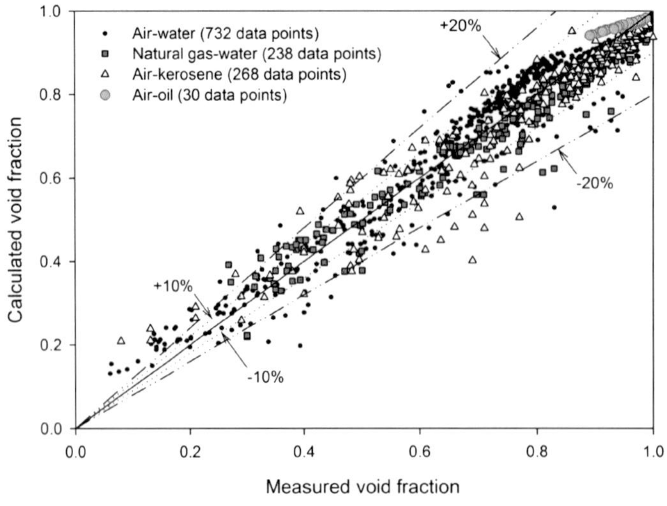

Figure 16: Comparison of the Woldesemayat and Ghajar [52] correlation with 1268 experimental data points for downward vertical flow.

6. CONCLUDING REMARKS

A very extensive comparison of most of the void fraction correlations available in the open literature was made against 3385 experimental data points for upward and downward vertical and horizontal two-phase flows. The best performing correlations considering the total number of data points predicted as well as their relative consistency in performance were highlighted. Furthermore the best correlations for four different void fraction ranges, namely 0 to 0.25, 0.25 to 0.5, 0.5 to 0.75, and 0.75 to 1, were highlighted as well as the best flow pattern specific correlations. Results of these categorical comparisons would allow the access to correlations with higher accuracies for specific void fraction range of interest.

In addition to identifying correlations with minimum disparity when compared to experimental data, the effort of this undertaking also resulted in the compilation of an extensive experimental void fraction database. The experimental database is not only useful for validating void fraction correlations, but also for extending the understanding of the physics for two-phase flow. For instance, experimental void fraction results can be used to aid in the development of the theory-based closure relations for the wall shear and interfacial shear stresses in terms of liquid holdup and liquid and gas velocities [86]. Another example of the usefulness of experimental void fraction results is in aiding the development of two-phase flow models; such as a recently developed unified annular flow model by Thome and Cioncolini [87], which integrated together models for predicting the entrained liquid fraction, the wall shear stress and the velocity profile in the annular liquid film. Experimental void fraction results are also useful in the analysis of two-phase pressure drop.

ACKNOWLEDGEMENTS

The authors thank Swanand Bhagwat, Pranav Godbole, and Nishant Mathure for their contributions in the experimental part for void fraction as well as the analysis of void fraction correlations in upward [85] and downward vertical and horizontal two-phase flows. Sincere thanks are offered to Emerson Micro Motion for generously donating one of the Coriolis flow meters. The authors are grateful to Mr. Martin Mabry for his assistance in procuring these meters. Special thanks is also extended to Dr. Neima Brauner and colleagues from Tel-Aviv University for sharing the experimental void fraction data of Badie *et al.* [25] and Ottens [30].

NOMENCLATURE

C_0	two-phase distribution parameter, dimensionless
D	pipe inside diameter, m
Fr	Froude number, dimensionless
g	gravitational acceleration, m s^{-2}
G	mass flux or mass velocity, kg m^{-2} s^{-1}
H	distance between pressure taps, m
p	pressure, Pa
Re	Reynolds number, dimensionless
u_{gu}	gas phase drift velocity, m s^{-1}
U	velocity, m s^{-1}
V	volume, m^3
We	Weber number, dimensionless
x	flow quality, dimensionless

Greek Symbols

α	void fraction, dimensionless
θ	inclination angle, rad.
μ	dynamic viscosity, kg m^{-1} s^{-1}
ρ	density, kg m^{-3}

σ surface tension, N m^{-1}

Subscripts

atm	atmosphere
cr	critical
g	gas
h	homogeneous
l	liquid
sg	superficial gas
sl	superficial liquid
sys	system
t	total

REFERENCES

[1] W. L. Cook, "An experimental apparatus for measurement of pressure drop, void fraction and non-boiling two-phase heat transfer and flow visualization in pipes for all inclinations," M.S. thesis, Oklahoma State University, Stillwater, Oklahoma, 2008.

[2] S. J. Kline and F. A. McClintock, "Describing uncertainties in single-sample experiments," *Mech. Eng.*, vol. 75, pp. 3-8, 1953.

[3] J. M. Mandhane, G. A. Gregory, and K. Aziz, "A flow pattern map for gas-liquid flow in horizontal pipes," *Int. J. Multiphase Flow,* vol. 1, pp. 537-553, 1974.

[4] D. Barnea, O. Shoham, Y. Taitel, and A. E. Dukler, "Flow pattern transition for gas-liquid flow in horizontal and inclined pipes. Comparison of experimental data with theory," *Int. J. Multiphase Flow,* vol. 6, pp. 217-225, 1980.

[5] Y. Taitel, D. Bornea, and A. E. Dukler, "Modelling flow pattern transitions for steady upward gas-liquid flow in vertical tubes," *AIChE J.*, vol. 26, pp. 345-354, 1980.

[6] J. Weisman and S. Y. Kang, "Flow pattern transitions in vertical and upwardly inclined lines," *Int. J. Multiphase Flow,* vol. 7, pp. 271-291, 1981.

[7] T. Oshinowo and M. E. Charles, "Vertical two-phase flow. Part i. Flow pattern correlations," *Can. J. Chem. Eng.*, vol. 52, pp. 25-35, 1974.

[8] Y. Jiang and K. S. Rezkallah, "A study on void fraction in vertical co-current upward and downward two-phase gas-liquid flow—I: Experimental results," *Chem. Eng. Comm.*, vol. 126, pp. 221-243, 1993.

[9] H. Mukherjee, "An experimental study of inclined two-phase flow," Ph.D. thesis, University of Tulsa, Tulsa, Oklahoma, 1979.

[10] M. Sujumnong, "Heat transfer, pressure drop and void fraction in two-phase, two-component flow in a vertical tube," Ph.D. thesis, University of Manitoba, Winnipeg, Manitoba, Canada, 1998.

[11] G. S. Woods, P. L. Spedding, J. K. Watterson, and R. S. Raghunathan, "Vertical two phase flow," *Dev. Chem. Eng. Mineral Proc.*, vol. 7, pp. 7-16, 1999.

[12] T. Miyahara, M. Tsuzaki, T. Saito, Y. Matsuba, and T. Takahashi, "Froth characteristics of shallow bubble columns with sieve plates," *Chem. Eng. J.*, vol. 40, pp. 21-29, 1989.

[13] H. D. Beggs, "An experimental study of two phase flow in inclined pipes," Ph.D. thesis, University of Tulsa, Tulsa, Oklahoma, 1972.

[14] R. Chokshi, "Prediction of pressure drop and liquid holdup in vertical two-phase flow through large diameter tubing," Ph.D. thesis, University of Tulsa, Tulsa, Oklahoma, 1994.

[15] R. C. Fernandes, "Experimental and theoretical studies of isothermal upward gas-liquid flows in vertical tubes," Ph.D. thesis, University of Houston, Houston, Texas, 1981.

[16] H. S. Isbin, N. C. Shear, and K. C. Eddy, "Void fractions in steam water two-phase flow," *AIChE J.*, vol. 3, pp. 136-142, 1957.

[17] O. Oshinowo, "Two-phase flow in a vertical tube coil," Ph.D. thesis, University of Toronto, Toronto, Canada, 1972.

[18] J. Schmidt, H. Giesbrecht, and C. W. M. van der Geld, "Phase and velocity distributions in vertically upward high-viscosity two-phase flow," *Int. J. Multiphase Flow,* vol. 34, pp. 363-374, 2008.

[19] V. T. Nguyen, "Two-phase, gas-liquid co-current flow: An investigation of holdup, pressure drop and flow pattern in a pipe at various inclinations," Ph.D. Thesis, University of Auckland, Auckland, New Zealand, 1975.

[20] A. Hernandez, L. Gonzalez, and P. Gonzalez, "Experimental research on downward two-phase flow," in *Proc. 2002 SPE Ann. Technical Conf. & Exh.*, San Antonio, TX, 2002, pp. 1503-1508.

[21] A. Lorenzi and G. Sotgia, "Downward two-phase flow: Experimental investigation," *Energia Nucleare,* vol. 23, pp. 396-401, 1976.

[22] G. J. Paras, "Characterization of downward two phase flow by neutron noise analysis," M.S. thesis, University of Washington, Seattle, Washington, 1982.

[23] K. Usui and K. Sato, "Vertically downward two-phase flow. (i). Void distribution and average void fraction," *J. Nucl. Sci. Tech.,* vol. 26, pp. 670-680, 1989.

[24] G. H. Abdul-Majeed, "Liquid holdup in horizontal two-phase gas-liquid flow," *J. Petroleum Sci. Eng.,* vol. 15, pp. 271-280, 1996.

[25] S. Badie, C. P. Hale, C. J. Lawrence, and G. F. Hewitt, "Pressure gradient and holdup in horizontal two-phase gas-liquid flows with low liquid loading," *Int. J. Multiphase Flow,* vol. 26, pp. 1525-1543, 2000.

[26] X. T. Chen, X. D. Cai, and J. P. Brill, "Gas-liquid stratified-wavy flow in horizontal pipelines," *J. Energy Resources Tech.*, vol. 119, pp. 209-216, 1997.

[27] B. A. Eaton, "The prediction of flow patterns, liquid holdup and pressure losses occurring during continuous two-phase flow in horizontal pipelines," Ph.D. thesis, University of Texas, Austin, Texas, 1966.

[28] F. Franca and R. T. Lahey Jr., "The use of drift-flux techniques for the analysis of horizontal two-phase flows," *Int. J. Multiphase Flow,* vol. 18, pp. 787-801, 1992.

[29] K. Minami and J. P. Brill, "Liquid holdup in wet-gas pipelines," *SPE Production Eng.,* vol. 2, pp. 36-44, 1987.

[30] M. Ottens, "Gas-liquid flow through pipes and t-junctions," Ph.D. thesis, University of Amsterdam, Amsterdam, Netherlands, 1998.

[31] A. A. Armand, "The resistance during the movement of a two-phase system in horizontal pipes," *Izvestiya Vsesoyuznogo Teplotekhnicheskogo Instituta,* vol. 1, pp. 16-23 (AERE-Lib/Trans 828), 1946.

[32] W. A. Massena, "Steam-water pressure drop and critical discharge flow—a digital computer program," Report HW-65706, Hanford Atomic Products Operation, Richland, WA, 1960.

[33] R. H. Bonnecaze, W. Erskine, and E. J. Greskovich, "Holdup and pressure drop for two-phase slug flow in inclined pipelines," *AIChE J.,* vol. 17, pp. 1109-1113, 1971.

[34] D. Chisholm, "Research note: Void fraction during two-phase flow," *J. Mech. Eng. Sci.,* vol. 15, pp. 235-236, 1973.

[35] G. E. Dix, "Vapor void fractions for forced convection with subcooled boiling at low flow rates," Ph.D. thesis, University of California, Berkeley, California, 1971.

[36] A. El-Boher, S. Lesin, Y. Unger, and H. Branover, "Experimental studies of liquid metal two-phase flows in vertical pipes," in *Proc. the 1st World Conf. Experimental Heat Transfer, Fluid Mechanics, and Thermodynamics*, Dubrovnik, Yugoslavia, 1988, pp. 312-319.

[37] L. E. Gomez, O. Shoham, Z. Schmidt, R. N. Chokshi, and T. Northug, "Unified mechanistic model for steady-state two-phase flow: Horizontal to vertical upward flow," *SPEs J.*, vol. 5, pp. 339-350, 2000.

[38] E. J. Greskovich and W. T. Cooper, "Correlation and prediction of gas-liquid holdups in inclined upflows," *AIChE J.*, vol. 21, pp. 1189-1192, 1975.

[39] A. L. Guzhov, V. A. Mamayev, and G. E. Odishariya, "A study of transportation in gas liquid systems," in *10th International Gas Union Conference*, Hamburg, Germany, 1967.

[40] T. Hibiki and M. Ishii, "Distribution parameter and drift velocity of drift-flux model in bubbly flow," *Int. J. Heat Mass Transfer,* vol. 45, pp. 707-721, 2002.

[41] R. Huq and J. L. Loth, "Analytical two-phase flow void prediction method," *J. Thermophysics Heat Transfer,* vol. 6, pp. 139-144, 1992.

[42] M. Ishii, "One-dimensional drift-flux model and constitutive equations for relative motion between phases in various two-phase flow regimes," Report ANL-77-47, Argonne National Lab., 1977.

[43] K. Ohkawa and R. T. Lahey, "The analysis of ccfl using drift-flux models," *Nucl. Eng. Des.,* vol. 61, pp. 245-255, 1980.

[44] S. L. Kokal and J. F. Stanislav, "An experimental study of two-phase flow in slightly inclined pipes—ii. Liquid holdup and pressure drop," *Chem. Eng. Sci.,* vol. 44, pp. 681-693, 1989.

[45] R. W. Lockhart and R. C. Martinelli, "Proposed correlation of data for isothermal two-phase two-component flow in pipes," *Chem. Eng. Progr.*, vol. 45, pp. 39-48, 1949.

[46] S. Morooka, T. Ishizuka, M. IIzuka, and K. Yoshimura, "Experimental study on void fraction in a simulated bwr fuel assembly (evaluation of cross-sectional averaged void fraction)," *Nucl. Eng. Des.,* vol. 114, pp. 91-98, 1989.

[47] D. J. Nicklin, J. O. Wilkes, and J. F. Davidson, "Two-phase flow in vertical tubes," *Chem. Eng. Res. Des.,* vol. 40, pp. 61-68, 1962.

[48] A. Premoli, D. Francesco, and A. Prima, "An empirical correlation for evaluating two-phase mixture density under adiabatic conditions," in *European Two-Phase Flow Group Meeting*, Milan, Italy, 1970.

[49] S. Z. Rouhani and E. Axelsson, "Calculation of void volume fraction in the subcooled and quality boiling regions," *Int. J. Heat Mass Transfer,* vol. 13, pp. 383-393, 1970.

[50] S. L. Smith, "Void fractions in two-phase flow: A correlation based upon an equal velocity head model," *Proc. Institution Mech. Engineers,* vol. 184, pp. 647-664, 1969.

[51] K. H. Sun, R. B. Duffey, and C. M. Peng, "The prediction of two-phase mixture level and hydrodynamically-controlled dryout under low flow conditions," *Int. J. Multiphase Flow,* vol. 7, pp. 521-543, 1981.

[52] M. A. Woldesemayat and A. J. Ghajar, "Comparison of void fraction correlations for different flow patterns in horizontal and upward inclined pipes," *Int. J. Multiphase Flow,* vol. 33, pp. 347-370, 2007.

[53] Y. Yamazaki and K. Yamaguchi, "Characteristics of cocurrent two-phase downflow in tubes: Flow pattern, void fraction and pressure drop," *J. Nucl. Sci. Tech.,* vol. 16, pp. 245-255, 1979.

[54] D. Butterworth, "A comparison of some void-fraction relationships for co-current gas-liquid flow," *Int. J. Multiphase Flow,* vol. 1, pp. 845-850, 1975.

[55] S. Z. Rouhani, "Modified correlations for void fraction and two-phase pressure drop," Report AE-RTV-841, AB Atomenergi, Stockholm, Sweden, March 1969.

[56] D. Chisholm, *Two-phase flow in pipelines and heat exchangers.* London: George Godwin in association with The Institution of Chemical Engineers, 1983.

[57] S. G. Bankoff, "A variable density single-fluid model for two-phase flow with particular reference to steam-water flow," *J. Heat Transfer,* vol. 82, pp. 265-272, 1960.

[58] C. J. Baroczy, "Correlation of liquid fraction in two-phase flow with application to liquid metals," Report NAA-SR-8171, Atomics International. Division of North American Aviation, Inc., Canoga Park, CA, 1963.

[59] D. Bestion, "The physical closure laws in the cathare code," *Nucl. Eng. Des.,* vol. 124, pp. 229-245, 1990.

[60] V. Czop, D. Barbier, and S. Dong, "Pressure drop, void fraction and shear stress measurements in an adiabatic two-phase flow in a coiled tube," *Nucl. Eng. Des.,* vol. 149, pp. 323-333, 1994.

[61] B. A. Dimentiev, R. S. Lepilin, and A. A. Loginov, "An investigation of hydrodynamic process of bubbling through a vapor liquid mixture of considerable height," *Nauch. Dokl. Vish. Shkol-Energetica,* vol. 2, 1959.

[62] A. I. Filimonov, M. M. Przhizhalovski, E. P. Dik, and J. N. Petrova, "The driving head in pipes with a free interface in the pressure range from 17 to 180 atm," *Teploenergetika,* vol. 4, pp. 22-26, 1957.

[63] G. C. Gardner, "Fractional vapour content of a liquid pool through which vapour is bubbled," *Int. J. Multiphase Flow,* vol. 6, pp. 399-410, 1980.

[64] G. A. Hughmark, "Holdup in gas-liquid flow," *Chem. Eng. Progr.,* vol. 58, pp. 62-65, 1962.

[65] A. Inoue, T. Kurosu, M. Yagi, S.-C. Morooka, A. Hoshide, T. Ishizuka, and K. Yoshimura, "In-bundle void measurement of a bwr fuel assembly by an x-ray ct scanner: Assessment of bwr design void correlation and development of new void correlation," in *Proc. the 2nd ASME-JSME Nucl. Eng.Joint Conf.*, San Francisco, CA, USA, 1993, pp. 39-45.

[66] D. Jowitt, C. A. Cooper, and K. G. Pearson, "The thetis 80% blocked cluster experiment, part 5: Level swell experiments," AEEW-R 1767, UKAEA Atomic Energy Establishment Winfrith, Safety and Engineering Science Division, Winfrith, UK, 1984.

[67] R. T. Lahey and F. J. Moody, "The thermal-hydraulics of a boiling water nuclear reactor," *Am. Nucl. Soc. Monograph,* 1977.

[68] N. Madsen, "A void-fraction correlation for vertical and horizontal bulk-boiling of water," *AIChE J.,* vol. 21, pp. 607-608, 1975.

[69] D. Maier and P. Coddington, "Review of wide range void correlations against an extensive data base of rod bundle void measurements," in *Proc. the 5th Int. Conf. Nucl. Eng. (ICONE 5)*, Nice, France, 1997.

[70] L. Mattar and G. A. Gregory, "Air-oil slug flow in an upward-inclined pipe—i: Slug velocity, holdup and pressure gradient," *J. Can. Petroleum Techn.,* vol. 13, pp. 69-76, 1974.

[71] L. G. Neal and S. G. Bankoff, "Local parameters in cocurrent mercury-nitrogen flow: Parts i and ii," *AIChE J.,* vol. 11, pp. 624-635, 1965.

[72] A. Orell and R. Rembrand, "A model for gas-liquid slug flow in a vertical tube," *Ind. & Eng. Chem. Fund.s,* vol. 25, pp. 196-206, 1986.

[73] A. L. Shvarts, G. M. Anosova, and G. S. Levin, "Generalization of experimental data for the void fraction with vapor-liquid flow in vertical and inclined tubes," *Therm. Eng.,* vol. 40, pp. 689-691, 1993.

[74] H. G. Sonnenburg, "Full-range drift-flux model based on the combination of drift-flux theory with envelope theory," in *Proc. 4th Int. Topical Meeting on Nuclear Reactor Thermalhydraulics (NURETH-4)*, Karlsruhe, Germany, 1989, pp. 1003-1009.

[75] P. L. Spedding and J. J. J. Chen, "Holdup in two phase flow," *Int. J. Multiphase Flow,* vol. 10, pp. 307-339., 1984.

[76] P. L. Spedding, D. R. Spence, and N. P. Hands, "Prediction of holdup in two-phase gas-liquid inclined flow," *Chem. Eng. J.,* vol. 45, pp. 55-74, 1990.

[77] L. S. Sterman, "The generalization of experimental data concerning the bubbling of vapor through liquid," *Soviet Physics: Technical Phys.,* vol. 1, pp. 0038-5662, 1956.

[78] K. Takeuchi, M. Y. Young, and L. E. Hochreiter, "Generalized drift flux correlation for vertical flow," *Nucl. Sci. Eng.,* vol. 112, pp. 170-180, 1992.

[79] J. R. S. Thom, "Prediction of pressure drop during forced circulation boiling of water," *Int. J. Heat Mass Transfer,* vol. 7, pp. 709-724, 1964.

[80] J. F. Wilson, R. J. Grenda, and J. F. Patterson, "Steam volume fraction in a bubbling two-phase mixture," *Trans. Am.Nucl. Soc.,* vol. 4, pp. 356-357, 1961.

[81] H.-C. Yeh and L. E. Hochreiter, "Mass effluence during flecht forced reflood experiments," *Nucl. Eng. Des.,* vol. 60, pp. 413-429, 1980.

[82] P. Coddington and R. Macian, "A study of the performance of void fraction correlations used in the context of drift-flux two-phase flow models," *Nucl. Eng. Des.,* vol. 215, pp. 199-216, 2002.

[83] I. Kataoka and M. Ishii, "Drift flux model for large diameter pipe and new correlation for pool void fraction," *Int. J. Heat Mass Transfer,* vol. 30, pp. 1927-1939, 1987.

[84] H. S. Isbin and D. Biddle, "Void-fraction relationships for upward flow of saturated, steam-water mixtures," *Int. J. Multiphase Flow,* vol. 5, pp. 293-299, 1979.

[85] P. V. Godbole, "Study of flow patterns and void fraction in vertical upward two-phase flow," M.S. thesis, Oklahoma State University, Stillwater, Oklahoma, 2009.

[86] A. Ullmann and N. Brauner, "Closure relations for two-fluid models for two-phase stratified smooth" *Int. J. Multiphase Flow,* vol. 32, pp. 82-105, 2006. and stratified wavy flows.

[87] J. R. Thome and A. Cioncolini, "Void fraction prediction in annular two-phase flow using an algebraic turbulence model," *Microgravity Sci. Tech.,* vol. 22, pp. 425-431, 2010.

INDEX

A

B

C

Lixin Cheng and Dieter Mewes (Eds)

G

H

I

J

L

M

N

O

P

R

S

W